FINANCIAL
AND ACTUARIAL
STATISTICS

AN INTRODUCTION
SECOND EDITION

DALE S. BOROWIAK

University of Akron
Ohio, USA

ARNOLD F. SHAPIRO

Pennsylvania State University
USA

CRC Press
Taylor & Francis Group
Boca Raton London New York

CRC Press is an imprint of the
Taylor & Francis Group, an **informa** business
A CHAPMAN & HALL BOOK

CRC Press
Taylor & Francis Group
6000 Broken Sound Parkway NW, Suite 300
Boca Raton, FL 33487-2742

First issued in paperback 2020

Version Date: 20130923

ISBN 13: 978-0-367-57626-4 (pbk)
ISBN 13: 978-1-4200-8580-8 (hbk)

Library of Congress Cataloging-in-Publication Data

Borowiak, Dale S., 1952-
  Financial and actuarial statistics : an introduction / Dale S. Borowiak, Arnold F. Shapiro. -- Second Edition.
    pages cm
  Includes bibliographical references and index.
  ISBN 978-1-4200-8580-8 (hardback)
  1. Finance--Statistical methods. 2. Insurance--Statistical methods. 3. Finance--Mathematical models. 4. Insurance--Mathematical models. I. Shapiro, Arnold. II. Title.

HG176.5.B67 2013
332.01'5195--dc23

2013037012

Visit the Taylor & Francis Web site at
http://www.taylorandfrancis.com

and the CRC Press Web site at
http://www.crcpress.com

Contents

Preface..ix

1 Statistical Concepts 1
  1.1   Probability 1
  1.2   Random Variables 7
      1.2.1   Discrete Random Variables 8
      1.2.2   Continuous Random Variables 10
      1.2.3   Mixed Random Variables 13
  1.3   Expectations... 14
  1.4   Moment Generating Function 20
  1.5   Survival Functions 22
  1.6   Nonnegative Random Variables 25
      1.6.1   Pareto Distribution 25
      1.6.2   Lognormal Distribution 26
      1.6.3   Weibull Distribution 26
      1.6.4   Gompertz Distribution... 27
      1.6.5   Makeham Distribution 28
  1.7   Conditional Distributions 29
  1.8   Joint Distributions 31
  Problems.. 36
  Excel Problems 38
  Solutions... 38

2 Statistical Techniques... 41
  2.1   Sampling Distributions and Estimation 41
      2.1.1   Point Estimation... 42
      2.1.2   Confidence Intervals.. 44
      2.1.3   Percentiles and Prediction Intervals 45
      2.1.4   Confidence and Prediction Sets 46
  2.2   Sums of Independent Variables 49
  2.3   Order Statistics and Empirical Prediction Intervals 54
  2.4   Approximating Aggregate Distributions 57
      2.4.1   Central Limit Theorem 57
      2.4.2   Haldane Type A Approximation 61
      2.4.3   Saddlepoint Approximation.. 62
  2.5   Compound Aggregate Variables... 65
      2.5.1   Expectations of Compound Aggregate Variables............. 65
      2.5.2   Limiting Distributions for Compound Aggregate
              Variables 66
  2.6   Regression Modeling... 70

        2.6.1   Least Squares Estimation...71
        2.6.2   Regression Model-Based Inference 74
    2.7 Autoregressive Systems 75
    2.8 Model Diagnostics 78
        2.8.1   Probability Plotting 79
        2.8.2   Generalized Least Squares Diagnostic 83
        2.8.3   Interval Data Diagnostic..84
    Problems...87
    Excel Problems 88
    Solutions...90

3 **Financial Computational Models**..93
    3.1 Fixed Financial Rate Models 94
        3.1.1   Financial Rate-Based Calculations 94
        3.1.2   General Period Discrete Rate Models 99
        3.1.3   Continuous-Rate Models 100
    3.2 Fixed-Rate Annuities 101
        3.2.1   Discrete Annuity Models 101
        3.2.2   Continuous Annuity Models 104
    3.3 Stochastic Rate Models...106
        3.3.1   Discrete Stochastic Rate Model................................106
        3.3.2   Continuous Stochastic Rate Models........................112
        3.3.3   Discrete Stochastic Annuity Models........................114
        3.3.4   Continuous Stochastic Annuity Models..................116
    Problems...117
    Excel Problems 119
    Solutions...120

4 **Deterministic Status Models** 123
    4.1 Basic Loss Model..123
        4.1.1   Deterministic Loss Models..124
        4.1.2   Stochastic Rate Models..126
    4.2 Stochastic Loss Criterion..128
        4.2.1   Risk Criteria 129
        4.2.2   Percentile Criteria 130
    4.3 Single-Risk Models 131
        4.3.1   Insurance Pricing..131
        4.3.2   Investment Pricing..135
        4.3.3   Options Pricing 136
        4.3.4   Option Pricing Diagnostics 139
    4.4 Collective Aggregate Models 140
        4.4.1   Fixed Number of Variables..141
        4.4.2   Stochastic Number of Variables................................143
        4.4.3   Aggregate Stop-Loss Reinsurance and Dividends 145
    4.5 Stochastic Surplus Model..148

     4.5.1   Discrete Surplus Model 148

     4.5.2   Continuous Surplus Model 152

   Problems 155

   Excel Problems 158

   Solutions 159

5 Future Lifetime Random Variables and Life Tables 163

   5.1   Continuous Future Lifetime 164

   5.2   Discrete Future Lifetime 167

   5.3   Force of Mortality 169

   5.4   Fractional Ages 175

   5.5   Select Future Lifetimes 177

   5.6   Survivorship Groups 179

   5.7   Life Models and Life Tables 182

   5.8   Life Table Confidence Sets and Prediction Intervals 185

   5.9   Life Models and Life Table Parameters 187

     5.9.1   Population Parameters 188

     5.9.2   Aggregate Parameters 191

     5.9.3   Fractional Age Adjustments 193

   5.10 Select and Ultimate Life Tables 194

   Problems 198

   Excel Problems 200

   Solutions 200

6 Stochastic Status Models 203

   6.1   Stochastic Present Value Functions 204

   6.2   Risk Evaluations 205

     6.2.1   Continuous-Risk Calculations 205

     6.2.2   Discrete Risk Calculations 206

     6.2.3   Mixed Risk Calculations 207

   6.3   Percentile Evaluations 208

   6.4   Life Insurance 210

     6.4.1   Types of Unit Benefit Life Insurance 212

   6.5   Life Annuities 215

     6.5.1   Types of Unit Payment Life Annuities 217

     6.5.2   Apportionable Annuities 220

   6.6   Relating Risk Calculations 223

     6.6.1   Relations among Insurance Expectations 223

     6.6.2   Relations among Insurance and Annuity Expectations ... 225

     6.6.3   Relations among Annuity Expectations 226

   6.7   Actuarial Life Tables 227

   6.8   Loss Models and Insurance Premiums 230

     6.8.1   Unit Benefit Premium Notation 232

     6.8.2   Variance of the Loss Function 235

   6.9   Reserves 237

6.9.1   Unit Benefit Reserves Notations 240
6.9.2   Relations among Reserve Calculations 241
6.9.3   Survivorship Group Approach to Reserve Calculations 243
6.10   General Time Period Models... 244
     6.10.1 General Period Expectation.. 245
     6.10.2 Relations among General Period Expectations 246
6.11   Expense Models and Computations 249
Problems.. 252
Excel Problems 254
Solutions... 254

7   **Advanced Stochastic Status Models**.. 257
7.1   Multiple Future Lifetimes... 257
     7.1.1   Joint Life Status 258
     7.1.2   Last Survivor Status 260
     7.1.3   General Contingent Status.................................. 263
7.2   Multiple-Decrement Models.. 264
     7.2.1   Continuous Multiple Decrements 264
     7.2.2   Forces of Decrement 266
     7.2.3   Discrete Multiple Decrements 268
     7.2.4   Single-Decrement Probabilities 269
     7.2.5   Uniformly Distributed Single-Decrement Rates............. 271
     7.2.6   Single-Decrement Probability Bounds 273
     7.2.7   Multiple-Decrement Life Tables.......................... 275
     7.2.8   Single-Decrement Life Tables.............................. 278
     7.2.9   Multiple-Decrement Computations 279
7.3   Pension Plans.. 280
     7.3.1   Multiple-Decrement Benefits............................... 281
     7.3.2   Pension Contributions... 285
     7.3.3   Future Salary-Based Benefits and Contributions............. 287
     7.3.4   Yearly Based Retirement Benefits....................... 288
Problems.. 290
Excel Problems 291
Solutions... 292

8   **Markov Chain Methods**.. 295
8.1   Introduction to Markov Chains.. 296
8.2   Nonhomogeneous Stochastic Status Chains.............................. 297
     8.2.1   Single-Decrement Chains 298
     8.2.2   Actuarial Chains.. 299
     8.2.3   Multiple-Decrement Chains................................. 300
     8.2.4   Multirisk Strata Chains.. 303
8.3   Homogeneous Stochastic Status Chains 307

  8.3.1 Expected Curtate Future Lifetime.................................309
  8.3.2 Actuarial Chains...310
 8.4 Survivorship Chains...312
  8.4.1 Single-Decrement Models...313
  8.4.2 Multiple-Decrement Models 314
  8.4.3 Multirisk Strata Models 315
 Problems...316
 Excel Problems 317
 Solutions...320

9 Scenario and Simulation Testing 323
 9.1 Scenario Testing 323
  9.1.1 Deterministic Status Scenarios 324
  9.1.2 Stochastic Status Scenarios..325
  9.1.3 Stochastic Rate Scenarios...328
 9.2 Simulation Techniques 330
  9.2.1 Bootstrap Sampling 331
  9.2.2 Simulation Sampling 332
  9.2.3 Simulation Probabilities...335
  9.2.4 Simulation Prediction Intervals......................................337
 9.3 Investment Pricing Applications 340
 9.4 Stochastic Surplus Application 343
 9.5 Future Directions in Simulation Analysis..................................344
 Problems...346
 Excel Problems 348
 Solutions...350

10 Further Statistical Considerations 353
 10.1 Mortality Adjustment Models...354
  10.1.1 Linear Mortality Acceleration Models............................355
  10.1.2 Mean Mortality Acceleration Models 357
  10.1.3 Survival-Based Mortality Acceleration Models 360
 10.2 Mortality Trend Modeling..361
 10.3 Actuarial Statistics 364
  10.3.1 Normality-Based Prediction Intervals............................365
  10.3.2 Prediction Set-Based Prediction Intervals......................366
  10.3.3 Simulation-Based Prediction Intervals...........................368
 10.4 Data Set Simplifications 370
 Problems...371
 Excel Problems 371
 Solutions...373

**Appendix A: Excel Statistical Functions, Basic Mathematical
Functions, and Add-Ins** 375

Appendix B: Acronyms and Principal Sections 377

References 379

Symbol Index 385

Index 389

Preface

Financial and actuarial modeling is an ever-changing field with an increased reliance on statistical techniques. This is seen in the changing of competency exams, especially at the upper levels, where topics include more statistical concepts and techniques. In the years since the first edition was published statistical techniques such as reliability measurement, simulation, regression, and Markov chain modeling have become more prominent. This influx in statistics has put an increased pressure on students to secure both strong mathematical and statistical backgrounds and the knowledge of statistical techniques in order to have successful careers.

As in the first edition, this text approaches financial and actuarial modeling from a statistical point of view. The goal of this text is twofold. The first is to provide students and practitioners a source for required mathematical and statistical background. The second is to advance the application and theory of statistics in financial and actuarial modeling.

This text presents a unified approach to both financial and actuarial modeling through the utilization of general status structures. Future time-dependent financial actions are defined in terms of a status structure that may be either deterministic or stochastic. Deterministic status structures lead to classical interest and annuity models, investment pricing models, and aggregate claim models. Stochastic status structures are used to develop financial and actuarial models, such as surplus models, life insurance, and life annuity models.

This edition is updated with the addition of nomenclature and notations standard to the actuarial field. This is essential to the interchange of concepts and applications between actuarial, financial, and statistical practitioners. Throughout this edition exercise problems have been added along with solutions listing detailed equation links. After each chapter a series of application problems listed as "Excel Problems," along with solutions listing useful library functions, are newly included. Specific changes in this edition, listed by chapter, are now discussed.

Chapter 1 from the first edition is now split into two new chapters. Chapter 1 gives basic statistical theory and applications. Additional examples to help prepare students for the initial actuarial exams are also given along with a new section on nonnegative variables, namely, the Pareto, lognormal, and Weibull. Chapter 2 consists of statistical models and techniques including a new section on model diagnostics. Probability plotting, least squares, and interval data diagnostics are explored. In Chapter 4 the discussions of option pricing and stochastic surplus models are expanded. New discussions include option pricing diagnostics and upper and lower bounds on the probability of ruin for standard surplus models. Further, ruin computations for aggregate sums are demonstrated. Discussions of advanced actuarial models, specifically multiple future lifetime and multiple decrement models, are collected in a new Chapter 7. Pension system modeling rounds out this chapter as a natural extension of a multiple decrement system.

This edition includes a new chapter introducing Markov chains and demonstrating actuarial applications. In Chapter 8 both homogeneous and nonhomogeneous chains are presented for single-decrement and multiple-decrement models used to compute survival and decrement probabilities based on life table data. Actuarial chains are introduced that lead to computing techniques for standard present value expectations. The concept of multirisk strata modeling using Markov chains is introduced with actuarial computing techniques. Group survivorship chains and applications are presented and used to model population decrement characteristics by year for single and multiple decrements as well as multirisk strata models.

In Chapter 9 the discussion of scenario testing is reorganized by deterministic status and stochastic status designations. Discussions of simulation techniques have been expanded. Simulation prediction intervals based on nonparametric techniques have been added. Applications of investment pricing and stochastic surplus models have been expanded. Further, the concept of actuarial statistics for a collection of stochastic status models is introduced. For the aggregate sum of present values prediction intervals are developed using asymptotic theory and simulation techniques.

The major differences between this edition and the second edition are

- Problems dealing with standard probability and statistics theory have been added to the text and exercises. Solutions to exercise problems with detailed equation links are given. For example, the distribution for aggregate sums using the moment generating function is demonstrated for standard statistical distributions.

- Discussions of nonnegative random variables, Pareto, lognormal, Weibull (in Section 1.6), and left truncated normal have been added. These are utilized in actuarial and financial applications. Diagnostic procedures such as probability plotting (Section 2.8.1) and generalized least squares (Section 2.8.2) are presented and demonstrated on these models.

- The maximum likelihood approach to parameter estimation is discussed along with asymptotic applications (Section 2.5.2). Confidence sets and prediction intervals are developed for maximum estimators (Section 2.1.1). Applications include prediction intervals for actuarial variables based on life table data (Section 5.3).

- Nonparametric prediction intervals are discussed in Section 2.3.

- Option pricing diagnostics have been added.

- Discussion of surplus models and ruin computations are expanded. A lower bound on the probability of ruin (Section 3.5.1) and the continuous surplus model (Section 3.5.2) are now discussed. Applications of ruin computations for aggregate sums are discussed in Section 3.5.
- Variance of the loss function associated with standard actuarial models is discussed.
- In Chapter 7 a discussion of discrete Markov chains and actuarial applications is presented. Both homogeneous (Section 7.3) and non-homogeneous (Section 7.2) chains are presented for single-decrement and multiple-decrement models used to compute survival and decrement probabilities based on life table data. Actuarial chains are introduced that lead to computing techniques for standard actuarial present value expectations. The concept of multirisk strata modeling using Markov chains is introduced with actuarial computing techniques. Group survivorship chains and applications are presented and used to model population decrement characteristics by year for single and multiple decrements as well as multirisk strata models in Section 7.4.
- The discussion of scenario testing is reorganized by deterministic status and stochastic status designations.
- The discussion of simulation techniques has been expanded. Simulation prediction intervals based on nonparametric techniques have been added in Section 8.2.4. Applications of investment pricing (Section 8.4) and surplus models (Section 8.3) have been expanded.
- The concept of actuarial statistics for a collection of stochastic status models is introduced. For the aggregate sum of present values prediction intervals are developed using asymptotic theory (Section 9.3.2) and simulation techniques (Section 9.3.3).
- Excel exercises have been included in the exercise section of each section. These are short exercises that demonstrate the computations discussed in this text and give the student exposure to Excel functions and statistical computations.
- Discussions of both the Gompertz and Makeham distributions are added.

The authors thank the people at Taylor & Francis. In particular, we acknowledge the efforts of David Grubbs, who showed interest in this work and demonstrated great patience. Further, we thank Amber Donley for her work as project coordinator.

1

Statistical Concepts

The modeling of financial and actuarial systems starts with the mathematical and statistical concepts of actions and associated variables. There are two types of actions in financial and actuarial statistical modeling, referred to as nonstochastic or deterministic and stochastic. Stochastic actions possess an associated probability structure and are described by statistical random variables. Nonstochastic actions are deterministic in nature without a probability attachment. Interest and annuity calculations based on fixed time periods are examples of nonstochastic actions. Examples of stochastic actions and associated random variables are the prices of stocks at some future date, the age of death of an insured life, and the time of occurrence and severity of an accident.

This chapter presents the basic statistical concepts, basic probability and statistical tools, and computations that are utilized in the analysis of stochastic variables. For the most part, the concepts and techniques presented in this chapter are based on the frequentist approach to statistics and are limited to those that are required later in the analysis of financial and actuarial models. A goal of this chapter is to present statistical basic theories and concepts applied in a unifying approach to both financial and actuarial modeling.

We start this chapter with a brief introduction to probability in Section 1.1 and then proceed to the various statistical topics. Standard statistical concepts such as discrete, continuous, and mixed random variables and statistical distributions are discussed in Sections 1.2.1, 1.2.2, and 1.2.3. Expectations of random variables are introduced in Section 1.3, and moment generating functions and their applications are explored in Section 1.4. The specific random variables useful in actuarial and economic sciences and their distributions, namely, Pareto, lognormal, and Weibull, are discussed in Sections 1.6.1, 1.6.2, and 1.6.3, respectively. The chapter ends with an introduction to conditional distributions in Section 1.7 and joint distributions of more than one random variable in Section 1.8.

1.1 Probability

This section presents a brief introduction to some basic ideas and concepts in probability. Many probability texts give a broader background, but a review is useful since the basis of statistical inference is contained in probability theory. The results discussed either are used directly in the latter part of this book or give insight useful for later topics. Some of these topics may be review for the reader, and we refer to Larson (1995) and Ross (2002) for further background in basic probability.

For a random process let the set of all possible outcomes comprise the sample space, denoted Ω. Subsets of the sample space, consisting of some or all of the possible outcomes, are called events. Primarily, we are interested in assessing the likelihood of events occurring. Basic set operations are defined on the events associated with a sample space. For events A and B the union of A and B, $A \cup B$, is comprised of all outcomes in A, B, or common to both A and B. The intersection of two events A and B is the set of all outcomes common to both A and B and is denoted $A \cap B$. The complement of event A is the event that A does not occur and is A^c.

In general, we wish to quantify the likelihood of particular events taking place. This is accomplished by defining a stochastic or probability structure over the set of events, and for any event A, the probability of A, measuring the likelihood of occurrence, is denoted $P(A)$. Taking an empirical approach, if the random process is observed repeatedly, then as the number of trials or samples increases, the proportion of time A occurs within the trials approaches the probability of A or $P(A)$. In classical statistics this is referred to as the weak law of large numbers. This concept is the basis of modern simulation techniques and is explored in Chapter 9.

There are certain mathematical properties that every probability function, more formally referred to as a probability measure, follow. A probability measure, P, is a real-valued set function where the domain is the collection of relevant events where:

1. $P(A) \geq 0$ for all events A.
2. $P(\Omega) = 1$.
3. Let $A_1, A_2, \ldots$ be a collection of disjoint events, i.e., $A_i \cap A_j = \emptyset$ for $i \neq j$.

Then

$$P\left(\bigcup_{i=1}^{\infty} A_i\right) = \sum_{i=1}^{\infty} P(A_i) \tag{1.1}$$

Conditions 1–3 are called the axioms of probability, and 3 is referred to as the countably additive property. The application of (1.1) is demonstrated in the following example.

Example 1.1

A life insurance company has different types of policies where life insurance and auto insurance are denoted by LI and AI, respectively, while all other policies are denoted by O. A review of their accounts reveals that

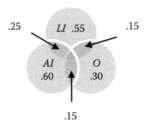

FIGURE 1.1
Venn diagram.

55% have *LI*, 60% have *AI*, and 30% have other types. Further, 25% have both *LI* and *AI*, 15% have *LI* and *O*, and 15% have *AI* and *O*. These are described in Figure 1.1 in terms of a Venn diagram.

To find the percentage of policies that have all three types, namely, *LI*, *AI*, and *O*, we construct disjoint sets and apply (1.1) and the principle of inclusion-exclusion (see Rohatgi, 1976, p. 27). Here

$$1 = P(LI \cup AI \cup O) = P(LI) + P(AI) + P(O) - P(LI \cap AI)$$

$$- P(LI \cap O) - P(O \cap AI) + P(O \cap AI \cap LI)$$

and so

$$P(AI \cap LI \cap O) = 1 - .55 - .60 - .30 + .25 + .15 + .15 = .10$$

Thus, 10% of the policies have all three types.

In applications probability measures are constructed in two ways. The first is based on assumed functional structures derived from physical laws and is mathematically constructed. The second, more statistical in nature, relies on observed or empirical data. Both methods are utilized in financial and actuarial modeling, and an introductory example is now given.

Example 1.2

A survey of $n = 25$ people in a particular age group, or strata, is taken. Let K denote the number of whole future years an individual holds a particular stock. Thus, K is an integer future lifetime and can take on values 0, 1, ... From the survey data a table of frequencies (Table 1.1), given by $f(k)$, for chosen values of k is constructed. The relative frequency concept is used to estimate probabilities when the choices corresponding to individual outcomes are equally likely. Thus, $P(K = k) = f(k)/n$. For example, the probability a person sells the stock in less than 1 year is the proportion $P(K = 0) = 2/25 = .08$. The probability a stock is held for 4 or more years is $P(K \geq 4) = 6/25 = .24$.

TABLE 1.1

Survey of Future Holding Lifetimes of a Stock

$K = k$	0	1	2	3	4	5 or more
$f(k)$	2	4	5	8	4	2

Simple concepts, such as integer years presented in Example 1.2, introduce basic statistical ideas and notations used in the development of financial and actuarial models. Another is the concept of conditioning on relevant information leading to conditional probabilities and is central to financial and actuarial calculations. For two events, A and B, the conditional probability of A given the fact B has occurred is defined by

$$P(A|B) = P(A \cap B)/P(B) \tag{1.2}$$

provided $P(B)$ is not zero. Thus, from (1.2)

$$P(A \cap B) = P(B)\,P(A|B) \tag{1.3}$$

Two illustrative examples applying conditional probabilities in the context of actuarial and financial modeling are now presented.

Example 1.3

An auto insurance company classifies drivers in terms of risk categories A, B, and C. The proportion of policies associated with A is 25%, while B and C comprise 55 and 20% of the policies, respectively. Over a 6-month time period the accident rates for categories A, B, and C are 10, 5, and 1%, respectively. To find the proportion of policyholders that have accidents over a 6-month time period we apply (1.3) and (1.1).

$$P(\text{Accident}) = P(\text{Accident} \cap A) + P(\text{Accident} \cap B) + P(\text{Accident} \cap C)$$
$$= P(A)\,P(\text{Accident } |A) + P(B)\,P(\text{Accident } |B)$$
$$+ P(C)\,P(\text{Accident } |C)$$
$$= .25(.1) + .55(.05) + .20(.01) = .0545$$

If a policyholder has an accident in the period, the probability he or she is in risk category B is computed using (1.2) as

$$P(B\ |\text{Accident}) = P(B)\,P(\text{Accident } |B)/P(\text{Accident}) = .50458$$

Example 1.4

Consider the conditions of the stock sales measurements of Example 1.2 where K denotes the number of whole years a stock is held. Given an individual holds a stock for the first year, the conditional probability of selling the stock in subsequent years is found using (1.2). For $K \geq 1$,

$$P(K = k | K \geq 1) = P(K = k)/P(K \geq 1) \qquad (1.4)$$

For example, the conditional probability of retaining possession of the stock for at least 4 additional years is $P(K \geq 5 | K \geq 1) = (2/25)/(23/25) = 2/23 = .087$.

The conditional probability concept can be utilized to compute joint probabilities corresponding to multiple events by extending (1.3). For a collection of events $A_1, A_2, \ldots, A_n$ the probability of all A_i, $i = 1, 2, \ldots, n$, occurring is

$$P(A_1 \cap A_2 \cap \ldots \cap A_n) = P(A_1)P(A_2 | A_1) \ldots P(A_n | A_1 \cap \ldots \cap A_{n-1})$$

Further, the idea of independence plays a central role in many applications. A collection of events $A_1, A_2, \ldots, A_n$ are completely independent or just independent if

$$P\left(\bigcap_{i=1}^{n} A_i\right) = \prod_{i=1}^{n} P(A_i) \qquad (1.5)$$

In practice many formulas used in the analysis of financial and actuarial actions are based on the ideas of conditioning and independence. A clear understanding of these concepts aids in the mastery of future statistical, financial, and actuarial topics.

General properties and formulas of probability systems follow from the axioms of probability. Two such properties frequently used in the application and development of statistical models are now given. First, letting the complement of event A be A^c, from the axioms of probability 1 and 3,

$$P(A) = 1 - P(A^c) \qquad (1.6)$$

Second, for two events A and B the probability of their union can be written as

$$P(A \cup B) = P(A) + P(B) - P(A \cap B) \qquad (1.7)$$

It is sometimes useful to use graphs of the sample space and the respective events, referred to as Venn diagrams, to view these probability rules. Figure 1.2 shows the Venn diagrams corresponding to rules (1.6) and (1.7). The reader is left to verify rules (1.6) and (1.7) using (1.1) and utilizing disjoint sets. These formulas have many applications, and we follow with two examples introducing two important actuarial multiple life structures.

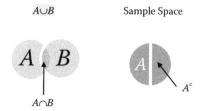

FIGURE 1.2
Venn diagram for rules (1.6) and (1.7).

Example 1.5

In general nomenclature, we let (x) denote a life aged x. Parties (x) and (y) enter into a financial contract that pays a benefit predicated on their survival for an additional k years. Let the events be $A = \{(x)$ lives past age $x + k\}$ and $B = \{(y)$ lives past age $y + k\}$. We consider two different types of contract conditions where the events A and B are considered independent:

1. Joint life conditions requires both people to survive an additional k years. The probability of paying the benefit, using (1.5), is $P(A \cap B) = P(A)P(B)$.
2. Last survivorship conditions requires at least one person to survive an additional k years. The event the benefit is paid with probability (1.7) is $P(A \cup B) = P(A) + P(B) - P(A \cap B)$.

In particular, let the frequencies presented in Table 1.1 hold where two integer-valued future stock whole-year lifetimes are given by K_1 and K_2. Thus, for any individual stock the probability of holding the stock for at least 3 years is $P(K(x) \geq 3) = 14/25 = .56$. From 1 the probability of holding both an additional 3 years is

$$P(K_1(x) \geq 3) \cap (K_2(x) \geq 3) = P(K(x) \geq 3)^2 = .56^2 = .3136$$

From (1.7) the probability at least one of the two is held for an additional 3 years is

$$P((K_1(x) \geq 3) \cup (K_2(x) \geq 3)) = 2P(K(x) \geq 3) - P(K(x) \geq 3)^2 = 2(.56) - .56^2 = .8064$$

These basic probabilistic concepts easily extend to more than two future lifetime variables.

Example 1.6

An insurance company issues insurance policies to a group of individuals. Over a short period, such as a year, the probability of a claim for any policy is .1. The probability of no claim in the first 3 years is found assuming independence and applying (1.5)

$$P(\text{No claims in 3 years}) = .9^3 = .729$$

Also, using (1.6), the probability of at least one claim in 3 years is

$$P(\text{At least one claim in 3 years}) = 1 - P(\text{No claims in 3 years}) = .271$$

In the balance of this chapter we turn our attention to statistical topics useful to the financial and actuarial fields.

1.2 Random Variables

In financial and actuarial modeling there are two types of variables, stochastic and deterministic. Deterministic variables lack any stochastic structure. Random variables are variables that possess some stochastic structure. Random variables include the future lifetime of an individual with a particular health status, the value of a stock after 1 year, and the amount of a health insurance claim. In general notation, random variables are denoted by uppercase letters, such as X or T, and fixed constants take the form of lowercase letters, like x and t. There are three types of random variables characterized by the structure of their domains. These include the typical discrete and continuous random variables, and the combinations of discrete and continuous variables, referred to as mixed random variables. For a general discussion of random variables and corresponding properties we refer to Hogg and Tanis (2010, Chapters 3 and 4) and Rohatgi (1976, Chapter 2).

In financial and actuarial modeling the time until a financial action occurs may be associated with a stochastic event. In actuarial science nomenclature a status model defines conditions describing future financial actions. An action is initiated when the conditions of the status change. This general structure of a status and economic actions is used to unite financial and actuarial modeling in a common framework, and we refer to Bowers et al. (1997, p. 257) for a more detailed description. For example, with a life insurance policy the initial status condition is the act of the person surviving. After the death of the person the status condition changes and an insurance benefit is paid. Similarly, in finance an investor may retain a particular stock, thereby ownership defining the initial status, until the price of the stock reaches a particular level. Upon reaching the target price the ownership of the stock changes, thereby signifying a change in status. In general the specific conditions that dictate one or more financial actions are referred to as a status and the lifetime of a status is a random variable, which we denote by T.

1.2.1 Discrete Random Variables

A discrete random variable, denoted X, takes on a countable number of values or outcomes, and associated with each outcome is a corresponding probability. The collection of these probabilities comprises the classical probability mass function (pmf) denoted $f(x)$

$$f(x) = P(X = x) \qquad (1.8)$$

for possible outcome values x. We refer to (1.8) as a probability mass function or just pmf. The support of $f(x)$, denoted by S, is the domain set on which $f(x)$ is positive. From the association between the random variable and the probability axioms 1–3 we see that $f(x) \geq 0$ for all x in S and the sum of (1.8) over all elements in S is 1.

In many settings the analysis of a financial or actuarial model depends on the integer-valued year a status changes. For example, an insurance policy may pay a fixed benefit at the end of the year of death. The variable K is the year of payment as measured from the date the policy was issued so that $K = 1, 2, \ldots$. We follow with examples in the context of life insurance that demonstrate these concepts and introduce standard probability measures and their corresponding pmfs.

Example 1.7

In the case of the death of an insured life within 5 years of the issue of the policy, a fixed amount or benefit, b, is paid. The benefit is paid at the end of the year of death. If the policyholder survives 5 years, amount b is immediately paid. Let K denote the time a payment is made, so that $K = 1, 2, \ldots, 5$ and the support is $S = \{1, 2, 3, 4, 5\}$. Let the probability of death in a year be q and the probability of no death be p, so that $0 \leq p \leq 1$ and $q = 1 - p$. The probability structure is contained in the pmf of K, which, for demonstrational purposes and not representative of human lifetimes, takes the geometric random variable form, given by

$$P(K = k) = f(k) = \begin{cases} q\,p^{k-1}, & k = 1, 2, 3, 4 \\ q\,p^4 + p^5, & k = 5 \end{cases}$$

The pmf can be used to assess the expected cost and statistical aspects of the policy. The graph of the pmf is given in Figure 1.3 and is typical of a discrete random variable where the probabilities are represented as spikes at the support points of the pmf.

Example 1.8

Over a short time period a collection of m insurance policies is considered. For policy i, $i = 1, 2, \ldots, m$, let the random variable $X_i = 1$ if a claim is made and 0 in the event of no claim. Also, for each i let $P(X_i = 1) = q$ and

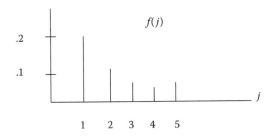

FIGURE 1.3
Discrete pmf.

$P(X_i = 0) = p = 1 - q$. These are Bernoulli random variables $X_1, X_2, ..., X_m$ and are assumed to be independent. The binomial random variable is

$$N = \sum_{i=1}^{m} X_i \qquad (1.9)$$

and designates the number of claims out of the m policies. Here N is discrete on support $S = \{0, 1, ..., m\}$ with parameters m and q. The pmf gives the probability that $N = n$ and is

$$f(n) = \frac{m!}{n!(m - n)!} \, q^n (1 - q)^{m-n} \qquad (1.10)$$

for $n = 0, 1, ..., m$. Thus, N is a binomial random variable with parameters n and q. Its statistical aspects are left to later discussions.

Example 1.9

Let N denote the number of insurance claims over a specific time period, where N takes on a Poisson distribution. Here the pmf of N is based on support $S = \{0, 1, ...\}$ and takes the form

$$f(n) = \exp(-\lambda) \, \lambda^n / n! \qquad (1.11)$$

for parameter $\lambda > 0$. Hence, the probability of no claims is $P(N = 0) = f(0) = \exp(-\lambda)$.

The Poisson probability structure can be derived from a set of conditions, referred to as the Poisson postulates, that imply that the Poisson pmf is appropriate to model discrete random processes. There are many classical examples of modeling random structures with a Poisson random variable, a detailed description of which is given by Helms (1997, p. 271). A typical problem involving the Poisson pmf might equate individual probabilities. For example, suppose the event $\{N = 3\}$ is four times as likely as $\{N = 2\}$. So $P(N = 3) = 4\,P(N = 2)$ and (1.11) implies $2\,\lambda^2 = \lambda^3/6$ and $\lambda = 12$.

1.2.2 Continuous Random Variables

For a continuous random variable, X, the stochastic structure differs from the discrete random variable where the domain consists of one or more intervals. The cumulative distribution function (cdf) associated with X is defined as probability the random variable attains at most a fixed quantity and is given by

$$F(x) = P(X \leq x) \tag{1.12}$$

for constant x. We remark that the cdf is defined over the entire real line. In the continuous random variable case the probability density function (pdf) corresponding to X is a nonnegative function, $f(x)$, where the probability of an interval corresponds to the area under $f(x)$. Hence, we have $f(x) \geq 0$ and the total area under $f(x)$ is 1. The support of $f(x)$, denoted S, designates the set where $f(x)$ is positive. If $f(x)$ is differentiable over the interval $[a, b]$ contained in S,

$$P(a \leq X \leq b) = \int_a^b f(s)\,ds = F(b) - F(a) \tag{1.13}$$

In Figure 1.4 probability (1.13) is represented as the area under the curve $f(x)$ between a and b. Thus, the cdf $F(x)$ is the antiderivative of the pdf $f(x)$ over support S. Standard continuous statistical models are introduced in the next set of examples.

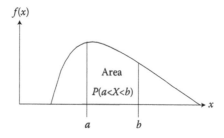

FIGURE 1.4
Continuous pdf.

Example 1.10

The continuous random variable X is uniform on support $S = [a, b]$, $a < b$, denoted by $X \sim U[a, b]$, when the pdf takes the form

$$f(x) = \begin{cases} \dfrac{1}{b-a} & \text{for } a \leq x \leq b \\ 0 & \text{otherwise} \end{cases} \tag{1.14}$$

The cdf, from definition (1.12), is defined by

$$F(x) = \begin{cases} 0 & \text{for} & x < a \\ \dfrac{x-a}{b-a} & \text{for} & a \leq x \leq b \\ 1 & \text{for} & b \leq x \end{cases} \tag{1.15}$$

We remark that mathematically the cdf is bounded by 1, nondecreasing, and right continuous. Figure 1.5 is a graph of both the pdf and cdf given by (1.14) and (1.15) when $b = 3$ and $a = 1$. The graphs given in Figure 1.5 are typical for continuous-type distributions where probabilities of events correspond to areas under $f(x)$ and cumulative probabilities are given by the cdf. The uniform distribution is often utilized in modeling probabilities when little or no information about the stochastic structure of a process is known.

Example 1.11

Let the lifetime associated with an insured event be T, where T follows an exponential distribution. The pdf is given by

$$f(t) = (1/\theta)\exp(-t/\theta) \tag{1.16}$$

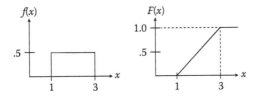

FIGURE 1.5
Continuous distributions.

for parameter constant $\theta > 0$, and the support is given by $S = \{t : t \geq 0\}$. The probability that T exceeds a constant c, called the reliability or survival to c, is

$$P(T > c) = 1 - F(c) = \int_{t=c}^{\infty} \frac{1}{\theta} e^{-t/\theta} dt = e^{-c/\theta} \qquad (1.17)$$

for $c > 0$. The exponential distribution has many applications (see Walpole et al., 1998, p. 166) and is frequently used in survival and reliability modeling.

Example 1.12

Let the future time of a economic action, T, approximately follow a normal distribution with parameters defined as the mean μ and standard deviation $\sigma > 0$, denoted $T \sim N(\mu, \sigma^2)$. The normal distribution is not often used to model future times and is used here for exposition purposes, and the parameters are such that $P(T < 0) = 0$ and the pdf associated with T takes the form

$$f(t) = \frac{\exp(- (t - \mu)^2/(2\sigma^2))}{(2\pi)^{1/2}\sigma} \qquad (1.18)$$

where the support is $S = (0, \infty)$. The pdf (1.18) is symmetric about the mean μ, and to compute probabilities the transformation to the standard normal random variable is required. The standard normal random variable, denoted Z, is a normal random variable that takes mean 0 and variance 1. The pdf associated with the standard normal random variable Z is denoted by $\phi(z)$. The Z random variable associated with $T = t$ is given by the transformation $Z = (T - \mu)/\sigma$. The cdf for T is

$$P(T \leq t) = P\left(Z \leq \frac{t-\mu}{\sigma}\right) = \Phi\left(\frac{t-\mu}{\sigma}\right) \qquad (1.19)$$

for any real-valued t where Φ is the cdf of the standard normal random variable. The evaluation of Φ in (1.19) is achieved using numerical approximation methods and is given in tabular form or is found using computer packages such as Excel (see Problem 1.17). For example, let the lifetime associated with a status condition, T, be a normal random variable with parameters $\mu = 65$ and $\sigma = 10$. The probability that the condition holds beyond age 80 is, using (1.19),

$$P(T > 80) = 1 - P(T \leq 80) = 1 - \Phi((80 - 65)/10) = 1 - \Phi(1.5) = 1 - .93319 = .06681$$

Further, the probability the conditions of the status change between ages 70 and 90 is found as

$$P(70 < T < 90) = P(T < 90) - P(T < 70) = \Phi(2.5) - \Phi(.5) = .3023$$

The continuous nature of the above random variable implies that the probability of the conditions changing at an exact time is negligible.

1.2.3 Mixed Random Variables

Mixed random variables are a combination of both discrete and continuous random variables. If X is a mixed random variable, the support is partitioned into two disjoint parts: a discrete part and a continuous part. The mixed probability function (mpf) is the combination of the respective pmf and pdf. Applications of mixed-type random variables occur in actuarial modeling, with multiple-decrement modeling (see Section 7.2) being a prime example. Many authors approach mixed random variable problems from a statistical conditioning perspective, while we present a straightforward approach. The following simple example demonstrates the versatility of this variable.

Example 1.13

An insurance policy pays claims between \$100 and \$500. The amount of the claim, X, is defined as a mixed random variable. The discrete part defines the probability of $X = 0$ as .5 and of $X = \$100$ or \$500 as .2 with discrete support $S_d = \{0, 100, 500\}$. The continuous part is defined by a constant (or uniform) pdf over the interval (\$100, \$500) with value .00025, so the continuous support is $S_c = (0, 500)$. Hence, the mpf is defined as

$$f(x) = \begin{cases} 0 & \text{if} & X < \$0, 0 < X < \$100, X > \$500 \\ .5 & \text{if} & X = \$0 \\ .2 & \text{if} & X = \$100, X = \$500 \\ .00025 & \text{if} & \$100 < X < \$500 \end{cases}$$

where the support is $S = S_d \cup S_c$. Probabilities are computed using the procedures for discrete and continuous random variables. For example, the requirement that the total probability associated with X is 1 is verified as

$$1 = P(X = 0) + P(X = 100) + \int_{100}^{500} .00025 \, dx + P(X = 500)$$

Also, the probability that the claim is at most \$250 is the combination of discrete and continuous-type calculations:

$$P(X \le \$250) = P(X = 0) + P(X = 100) + \int_{100}^{250} .00025 \, ds = .5 + .2 + .0375 = .7375$$

This example, although simple, demonstrates the possible types of mixed discrete and continuous random variables that can be constructed.

There is a variation of the mixed-type random variable that utilizes both discrete and continuous random variables in defining the mpf. This plays a part in insurance modeling, and an example of this type of random variable structure follows in Example 1.14.

Example 1.14

A 1-year insurance policy pays a benefit B, in the event of an accident claim. The probability of a claim in the first year is q. Given there is a claim, let B be a continuous random variable with pdf $f(B)$. The claim variable can be written as $X = I\, B$, where the indicator function $I = 1$ if there is a claim and 0 if there is no claim. The mpf of X, approached from a conditioning point of view, as introduced in (1.2), is

$$f(x) = \begin{cases} p = 1 - q & \text{if} & X = 0 \\ q\ f(B) & \text{if} & X = B > 0 \end{cases} \tag{1.20}$$

The probability the claim is greater than $c > 0$ is $P(X > c) = q\, P(B > c)$. This setting for single insurance policies has many practical applications. One is the extension of stochastic models of the form (1.20) to include a collection or portfolio of many policies. These are referred to as collective aggregate models, and discussed in Section 4.4. Further, while not addressed here, adjustments must be made to account for the effect of interest.

As we have seen in some of the examples, the pmf, pdf, or mpf along with the cdf, $F(x)$ may be a function of one or more parameters. In practice the unknown parameters are estimated from empirical data. Probabilistic and statistical aspects of such estimation must be accounted for in financial and actuarial models.

1.3 Expectations

The propensities of a random variable or a function of a random variable to take on particular outcomes are often important to financial and actuarial modeling. The expectation is one method used to predict and assess outcomes of a random variable. In general, the expected value of function $g(x)$, if it exists, is denoted $E\{g(x)\}$. The possible types of random variables, discrete, continuous, and mixed random variables, produce different formulas for expected values. First, if X is discrete with support S_d and pmf $f(x)$,

$$E\{g(X)\} = \sum_{S_d} g(x) f(x) \tag{1.21}$$

Second, if X is continuous and the pdf $f(x)$ has support S_c,

$$E\{g(X)\} = \int_{S_c} g(x)\, f(x)\, dx \tag{1.22}$$

In the last case, if X is a mixed random variable the expected value is a combination of (1.21) and (1.22). If the support is $S = S_d \cup S_c$ and the mpf is $f(x)$, then

$$E\{g(X)\} = \sum_{S_d} g(x)\, f(x) + \int_{S_c} g(x)\, f(x)\, dx \tag{1.23}$$

In many situations, the central core of financial and actuarial analysis is the expectation of properly chosen random variables.

There are a few standard expectations that play an important role in analyzing data. Employing the identity function, $g(x) = x$, yields the expected value or mean of X given by

$$\mu = E\{X\} \tag{1.24}$$

The mean of X is a weighted average, with respect to the probability structure, over the support, and is a measure of the center of the pmf, pdf, or mpf. If $g(x) = X^r$, for positive integer r, then the expected value, $E\{X^r\}$, is referred to as the rth moment or a moment of order r of X. It is a mathematical property that if moments of order r exist, then moments of order s exist for $s \leq r$. Central moments of order r, for positive integer r, are defined by $E\{(X - \mu)^r\}$. The variance of X is a central moment with $r = 2$ and is denoted by $\text{Var}\{X\} = \sigma^2$, and after simplification the variance becomes

$$\sigma^2 = E\{(X - \mu)^2\} = E\{X^2\} - \mu^2 \tag{1.25}$$

The existence of the second moment implies existence of the variance. The standard deviation of X is $\sigma = \sigma^{1/2}$. The variance and standard deviation of a random variable measure the dispersion or variability associated with the random variable and the associated pmf, pdf, or mpf. The discrete case computation is demonstrated in the next example.

Example 1.15

Let N have a Poisson distribution, introduced in Example 1.9, with parameter λ with pmf given in (1.11). The mean of N is found using a Taylor series (see Problem 1.4):

$$\mu = \sum_{n=0}^{\infty} nf(n) = \exp(-\lambda) \sum_{n=1}^{\infty} n \frac{\lambda^n}{n!} = \exp(-\lambda)\, \lambda \exp(\lambda) = \lambda$$

In a similar manner, $E\{N^2\} = \lambda(\lambda + 1)$ so that from (1.25), $\sigma^2 = \lambda(\lambda + 1) - \lambda^2 = \lambda$. Hence, for the Poisson random variable the mean and the variance are equal and completely determine the distribution.

As mentioned earlier, the mean, μ, measures the "center" and the standard deviation, σ, measures the variability or dispersion associated with the distribution of random variable X. Other useful moment measurements are the skewness and the kurtosis, denoted by Sk and Ku and defined as

$$Sk = \frac{E\{(X - \mu)^3\}}{\sigma^3} \quad \text{and} \quad Ku = \frac{E\{(X - \mu)^4\}}{\sigma^4} \tag{1.26}$$

These moments are classically used to characterize distributions in terms of shape. For an applied discussion in the usage of (1.26) see McBean and Rovers (1998). Examples of moment computations in the continuous and mixed random variable cases are now given.

Example 1.16

Let X be uniform on (a, b) where the pdf is given by (1.14) in Example 1.10. The mean or expected value of X is

$$\mu = \int_a^b \frac{x}{b-a}\,dx = \left.\frac{x^2}{2(b-a)}\right|_a^b = \frac{b+a}{2} \tag{1.27}$$

Further, the second moment is

$$E\{X^2\} = \int_a^b \frac{x^2}{b-a}\,dx = \left.\frac{x^3}{3(b-a)}\right|_a^b = \frac{b^2 + ab + a^2}{3}$$

Using (1.25), the variance of X simplifies to

$$\sigma^2 = \frac{(b-a)^2}{12} \tag{1.28}$$

The special case of the uniform distribution over the unit interval where $a = 0$ and $b = 1$ has many applications and from (1.27) and (1.28) produces mean $\mu = 1/2$ and variance $\sigma^2 = 1/12$.

Example 1.17

Let continuous random variable X have an exponential pdf given in Example 1.11 where the pdf takes the form (1.16). To find the mean of X we use integration by parts:

$$\mu = \int_0^\infty \frac{1}{\theta} s \exp(-s/\theta)\,ds = \theta$$

Using integration by parts twice we find

$$E\{X^2\} = \int\limits_0^\infty \frac{1}{\theta} s^2 \exp(-s/\theta)\, ds = 2\theta^2$$

Hence, from (1.25) the variance is $\sigma^2 = 2\theta^2 - (\theta)^2 = \theta^2$. In fact, for positive integer r, the general moment formula is given by $E\{X^r\} = r\theta^r$. Applying the general moment formula, the skewness and kurtosis, defined by (1.26), can be computed.

Example 1.18

In this example we consider the mixed-variable case of Example 1.13 where the claim random variable X possesses supports for both the discrete and continuous parts defined by $S_d = \{0, 100, 500\}$ and $S_c = (100, 500)$, respectively. Here $P(X = 0) = .5$, $P(X = 100) = P(X = 500) = .2$, and X is uniform over $(100, 500)$. The mean takes the form

$$\mu = \sum_{S_d} x\; f(x) + \int\limits_{S_c} x\; f(x)\, dx = 0(.5) + .2(100) + .2(500) + \int\limits_{100}^{500} x(.00025)\, dx = \$150$$

In application, over many policies the average claim amount is the mean computed as \$150.00.

Two additional formulas are used in the computation of the expected value of a function of X when $X \geq 0$. Let X have pmf or pdf $f(x)$ with support S and cdf $F(x)$, and let $G(x)$ be monotone where $G(x) \geq 0$. If X is continuous we assume $G(x)$ is differentiable with $\frac{d}{dx}G(x) = g(x)$ and that $E\{G(X)\}$ exists; then using integration by parts, the expectation takes the form

$$E\{G(x)\} = G(0) + \int\limits_S g(x)[1 - F(x)]\, dx \tag{1.29}$$

In the case X is discrete with corresponding support on the nonnegative integers, then if the expectation exists, the expected value of $G(X)$ is

$$E\{G(X)\} = G(0) + \sum_S [1 - F(x)]\, \Delta(G(x)) \tag{1.30}$$

where $\Delta(G(x)) = G(x + 1) - G(x)$. The derivations of these expectation formulas are left to the reader, and we now present a series of examples.

Example 1.19

Let the number of claims over a period of time be N so that the support is $S = \{0, 1, ...\}$. The probability of no claim in the time period is denoted by p and the pmf corresponding to N is assumed to take the form of the discrete geometric distribution introduced in Example 1.7. The general pmf is given by

$$f(n) = \begin{cases} p\,q^n & \text{for} \quad n = 0, 1, ... \\ 0 & \text{otherwise} \end{cases}$$

for constants p, $0 < p < 1$, and $q = 1 - p$. To compute the cdf for positive integer n we apply the geometric summation formula

$$\sum_{j=s}^{n} a^j = \frac{a^s - a^{n+1}}{1 - a} \tag{1.31}$$

In fact, the limit of (1.31) as n approaches infinity exists provided $|a| < 1$ and is

$$\sum_{j=s}^{\infty} a^j = \frac{a^s}{1 - a} \tag{1.32}$$

In Problem 1.3 the reader is asked to verify (1.31). For nonnegative integer n, the cdf is a step function given by

$$F(n) = p \sum_{j=0}^{n} q^j = \frac{p(q^0 - q^{n+1})}{1 - q} = 1 - q^{n+1}$$

and so $1 - F(n) = q^{n+1}$. These formulas can be used to compute conditional probabilities. For example, if there is at most four claims over the time period, the probability of at least one claim is

$$\frac{P(1 \le N \le 4)}{P(N \le 4)} = \frac{F(4) - F(0)}{F(4)} = \frac{q - q^5}{1 - q^5}$$

To find the mean of N we employ (1.30), with $G(n) = n$, $g(n) = 1$, and $\Delta(G(n)) = 1$,

$$\mu = \sum_{n=1}^{\infty} q^{n+1} = \frac{q}{1 - q} = \frac{q}{p} \tag{1.33}$$

We note that as the probability of a claim, q, increases, the mean, (1.33), increases. The second moment can be computed by using (1.30) with $G(x) = x^2$.

Example 1.20

The number of insurance claims over distinct time periods is denoted by N and has the geometric pmf of the form $P(N = n) = p\,q^n$ for $S = \{0, 1, \ldots\}$. Over two distinct time periods the aggregate sum of the number of claims is $N = N_1 + N_2$, where N_i is the number of claims in the ith period for $i = 1$ and 2. If N_1 and N_2 are independent, then the pdf for N can be found as

$$P(N = n) = \sum_{j=0}^{n} P(N_1 = j)\,P(N_2 = n - j) = \sum_{j=0}^{n} pq^j\,pq^{n-j} = (n+1)\,p^2 q^n$$

Example 1.21

A security is purchased for \$5,000 in the hopes of increased value over time. The sale of the security occurs at future time T with corresponding mpf

$$f(t) = \begin{cases} a\,t^2 & \text{for} & 0 \le t < 5 \\ .4 & \text{for} & t = 5 \\ 0 & & \text{otherwise} \end{cases}$$

First, we find the constant a that makes $f(t)$ an mpf. Hence, we require

$$1 = .4 + \int_0^5 a\,t^2\,dt = .4 + a\,\frac{5^3}{3}$$

which implies $a = .6(3)/5^3 = .0144$. We now compute the mean and variance of T. The first moment is

$$\mu = E\{T\} = 5(.4) + (.0144) \int_0^5 t^3\,dt = 4.25$$

The second moment is

$$E\{T^2\} = 5^2(.4) + (.0144) \int_0^5 t^4\,dt = 19$$

and hence, the variance of T using (1.25) is

$$\text{Var}\{T\} = E\{T^2\} - \mu^2 = .9375$$

In applications, the value of the investment is a function of both the length of time held and the return rate on the investment. In Chapter 3 financial computations and concepts concerning the return and interest rates on investments are discussed.

Example 1.22

The insurance policy setting of Example 1.14 is revisited where the claim variable X has mpf (1.20). Over a short time period let the probability of a claim be q and the benefit paid be B. As presented, the claim variable $X = I B$ takes the mpf given by (1.20). The expected value of X is computed as

$$E\{X\} = 0\, p + q \int_S B\, f(B)\, dB = q\, E(B) \qquad (1.34)$$

where S is the support of B. The second moment is

$$E\{X^2\} = q \int_S B^2\, f(B)\, dB \qquad (1.35)$$

Hence, utilizing (1.34) and (1.35), the variance of X can be written as the combination of moments given by

$$\text{Var}\{X\} = q\, \text{Var}\{B\} + q(1-q)\, (E\{B\})^2 \qquad (1.36)$$

In particular, if B is an exponential random variable with mean 1,000 and $q = .1$, from Example 1.17 and (1.34) we compute $E\{X\} = .1(1,000) = 100$. Further, using (1.17) the probability the claim variable is more than 1,200 is computed as

$$P(X > 1,200) = q\, P(B > 1,200) = q[1 - P(B \le 1,200)] = .1\, \exp(-1,200/1,000) = .0301$$

Hence, in any time period the probability of a claim is small at 3%.

1.4 Moment Generating Function

A special widely used expectation in both the theoretical and applied settings is the moment generating function (mgf). If $g(X) = \exp(tx)$, then the mgf is found by computing the expectation given by (1.21), (1.22), or (1.23) in the discrete, continuous, or mixed random variable cases, respectively. For the mgf to exist it must exist or converge for values of t in the neighborhood of zero (see Rohatgi, 1976, p. 95). The mgf, when X is discrete, is defined by

$$M_x(t) = \sum_S \exp(tx)\, f(x) \qquad (1.37)$$

and in the case where X is continuous the mgf is

$$M_x(t) = \int_S \exp(tx)\, f(x)\, dx \qquad (1.38)$$

The mixed random variable case is a combination of (1.37) and (1.38), and all the formulas concerning the mgf assume the expectations exist. The mgf has many uses, such as moment computation and distribution derivation, that will be described later in this chapter. We now follow with a discussion of moment generating functions through a set of typical random variable examples.

Example 1.23

From Example 1.8 the number of claims N is a binomial random variable with parameters m and p and pmf taking form (1.10). The mgf is computed using the binomial theorem as

$$M_N(t) = \sum_{n=1}^{m} \frac{m!}{n!(m-n)}\, \exp(tn)\, p^n (1-p)^{m-n} = [p\exp(t) + (1-p)]^m \quad (1.39)$$

It is clear that the mgf evaluated at zero takes the value of 1. One of the primary uses of the mgf, as described next, is to compute the moments associated with N.

In either the discrete or continuous case the mgf can be used to find the moments of X. For positive integer r the rth moment, if it exists, can be computed using

$$E\{X^r\} = \frac{d^r}{dt^r}\, M(t)\Big|_{t=0} = M^{(r)}(0) \qquad (1.40)$$

In the above we assume the derivative exists in support of the random variable. The application of (1.40) is now demonstrated.

Example 1.24

Let X be a normal random variable, introduced in Example 1.12, with mean μ, standard deviation σ, and pdf (1.18). The mgf can be shown (see Hogg et al., 2005, p. 139) to be

$$M(t) = \exp(t\mu + t^2\sigma^2/2) \qquad (1.41)$$

and is defined on the entire real line. From (1.40) with $r = 1$ the mean is calculated as

$$M^{(1)}(0) = \exp(\mu t + t^2\mu^2/2)(\mu + t\sigma^2)\big|_{t=0} = \mu$$

To compute the second moment using (1.40) for $r = 2$ we find

$$M^{(2)}(0) = \exp(\mu t + t^2\sigma^2/2)[(\mu + t\sigma^2)^2 + \sigma^2]\big|_{t=0} = \mu^2 + \sigma^2$$

Hence, $\mathrm{Var}(X) = E\{X^2\} - \mu^2 = \sigma^2$. Formally, this demonstrates the fact that the parameters μ and σ^2 in the normal pdf correspond to the mean and variance.

The mgf, when it exists, is unique and can be employed to find the distribution of a random variable. If a random variable under examination yields an mgf that matches an mgf of a known pdf, then the pdf also matches. This is a commonly used technique when examining the distribution of sums of independent random variables. Furthermore, the continuity theorem states that if the limit of an mgf converges point-wise to a proper mgf, then the corresponding distributions converge. For more in-depth discussions of the uses of the mgf see either Rohatgi (1976, Section 4.6) or Hogg et al. (2005, Section 4.7).

An alternative to the mgf is the characteristic function defined by $E\{\exp(itX)\}$. This complex-valued function, similar to the mgf, determines existing moments and is unique in that it completely determines the distribution function of the associated random variable. In fact, an inversion formula exists that allows derivation of the associated distribution, assuming it exists, based on the characteristic function. The characteristic function has an advantage over the mgf in that it always exists, but in some cases tools and concepts from complex analysis are required. For a discussion of the characteristic function and application we refer to Laha and Rohatgi (1979).

1.5 Survival Functions

For random variable X, the survival or reliability function defines the probability that X exceeds a fixed value. For X the survival function associated with constant the x is

$$S(x) = P(X > x) \tag{1.42}$$

Since X is the age of death, then (1.42) gives the probability that the lifetime is greater than the constant x. Using the cdf $F(x) = P(X \le x)$ we note the relationship with the cdf, $S(x) = 1 - F(x)$. When X is continuous the relation-ship between $S(x)$ and the pdf $f(x)$ is

$$f(x) = -\frac{d}{dx} S(x) = \frac{d}{dx} F(x) \equiv F^{(1)}(x) \qquad (1.43)$$

The survival function is unique and is determined by the distribution of the random variable. The balance of this section presents examples of survival functions and moment computations utilizing (1.42).

Example 1.25

Let the lifetime of (x) be a continuous random variable, X, with support $S = (0, 100)$, where the associated survival function is given by

$$S(x) = \begin{cases} \left(1 - \dfrac{x}{100}\right)^2 & \text{for } 0 < x < 100 \\ 0 & \text{otherwise} \end{cases}$$

Using (1.43), the corresponding pdf is

$$f(x) = -\frac{d}{dx}\left(1 - \frac{x}{100}\right)^2 = \frac{1}{50}\left(1 - \frac{x}{100}\right)$$

for $0 < x < 100$. Many measurements can be made that characterize the distribution associated with X. One classical measurement that charac-terizes the center of the distribution, alternate to the mean, is the median. The median of X is the constant $x_{1/2}$, where $S(x_{1/2}) = 1/2$. In this case the median is computed to be $x_{1/2} = 100(1 - 1/2^{1/2}) = 29.29$.

If the support of the pmf or pdf is nonnegative, the survival function can be used to compute moments of random variables. Letting $G(X) = X$ in (1.29) or (1.30), alternative formulas for the mean of X can be constructed. The mean or expected value of X can be found by

$$E\{X\} = \sum_S S(x) \quad \text{or} \quad E\{X\} = \int_S S(x)\, dx \qquad (1.44)$$

where X is discrete or continuous, respectively, and the summation and the integral are over the support of $f(x)$. Other moments can also be computed using these formulas. For example, the second moment is either

$$E\{X^2\} = \sum_S (2x + 1)\, S(x) \quad \text{or} \quad E\{X^2\} = 2\int_S x\, S(x)\, dx \qquad (1.45)$$

depending on X being either discrete or continuous. These formulas, (1.44) and (1.45), in some cases ease computations and their derivations are left to the reader.

Example 1.26

Let the lifetime random variable of a status X have an exponential distribution with pdf given by (1.16). The survival function is $S(x) = \exp(-x/\theta)$, and using (1.44) the expected value of X is

$$E\{X\} = \int_0^\infty \exp(-u/\theta)\, du = \theta$$

Further, using (1.45), we find $E\{X^2\} = 2\,\theta^2$, so that $\mathrm{Var}\{X\} = \theta^2$. In this case the alternate computation is easier than directly applying the definition approach of formulas (1.21) and (1.22).

Example 1.27

Let $X \sim U(0, 100)$ so that the pdf, following Example 1.10, is $f(x) = 1/100$ for $0 \le x \le 100$. For $0 \le x \le 100$ the survival function is

$$S(x) = \int_X^{100} \frac{1}{100}\, dt = \frac{100 - x}{100} = 1 - \frac{x}{100}$$

for $0 < x$, 100. We now compute the mean and variance of X. From (1.44) the mean is

$$E\{X\} = \int_0^{100} \left(1 - \frac{x}{100}\right) dx = \left(x - \frac{x^2}{200}\right)\Big|_0^{100} = 50$$

From (1.45),

$$E\{X^2\} = 2 \int_0^{100} x\left(1 - \frac{x}{100}\right) dx = 2\left(\frac{x^2}{2} - \frac{x^3}{300}\right)\Big|_0^{100} = \frac{100^2}{3}$$

and $\mathrm{Var}\{X\} = 100^2/3 - 100^2/4 = 100^2/12 = 833.33$. The mean and variance are not only descriptive measurements of the distribution of a random variable but also can be used in statistical inference. Normality-based prediction intervals for random variables are based on the moments (see (2.16)).

1.6 Nonnegative Random Variables

Many of the random variables used in financial and actuarial modeling are by their nature nonnegative. Lifetimes of people and insurance claim amounts are nonnegative quantities and in stochastic applications are modeled using nonnegative random variables where their support is $S = \{x \geq 0\}$ or appropriate subsets. In many financial and actuarial applications nonnegative random variables are heavy tailed random variables where their corresponding distributions have heavier tails than the exponential distribution. For these random variables general moments and the mgf may not exist. Three such distributions, namely, the Pareto, lognormal, and Weibull, are discussed. More general distributions are discussed in Klugman et al. (2008, Appendices A and B).

1.6.1 Pareto Distribution

In practice the Pareto distribution has been used to model individual wealth (see Page and Kelley, 1971) and survival distributions (see Goovaerts and De Pril, 1980). A recent paper on applying the Pareto distribution to property evaluations is Guiahi (2007). The Pareto distribution, based on parameters $\alpha > 0$ and $\beta > 0$, is skewed to the right, and the pdf and cdf are

$$f(x) = \alpha\,\beta^\alpha\,x^{-(1+\alpha)} \quad \text{and} \quad F(x) = 1 - (\beta/x)^\alpha \tag{1.46}$$

for $x > \beta$. The mean and variance are

$$E\{X\} = \frac{\alpha\beta}{\alpha - 1} \quad \text{for} \quad \alpha > 1 \tag{1.47}$$

and

$$\mathrm{Var}\{X\} = \frac{\alpha\beta}{(\alpha - 1)^2(\alpha - 2)} \quad \text{for} \quad \alpha > 2 \tag{1.48}$$

For $0 < p < 1$ the $p100$th percentile, denoted x_p, is defined by $F(x_p) = p$, and utilizing (1.46) takes the form

$$x_p = \beta\,(1 - p)^{-1/\alpha} \tag{1.49}$$

The Pareto distribution was originally used to model the allocation of wealth where the shape of the pdf (1.46) dictates that a small area, close to zero, contains a high proportion of outcomes.

1.6.2 Lognormal Distribution

The lognormal distribution has been used to model many nonnegative stochastic variables, including the modeling of lifetimes of manufactured parts in engineering and the modeling of stochastic interest rates in finance. The lognormal distribution was applied to economic problems by Aichison and Brown (1957) and property evaluations in Guiahi (2007). Recent actuarial extensions of the lognormal distribution include Dufresne (2009). The distribution of the lognormal random variable X has parameters α and $\beta^2 > 0$ and is defined by the relationship $Y = \ln(X) \sim N(\alpha, \beta^2)$. The lognormal pdf is skewed to the right (although not as much as the Pareto) and takes the form

$$f(x) = (2\pi\,\beta^2)^{-1/2}\,x^{-1}\exp(-(\ln(x) - \alpha)/(2\beta^2)) \qquad (1.50)$$

where $S = \{x > 0\}$. The cdf is a function of the standard normal cdf $\Phi(z)$ and is

$$F(x) \;=\; \Phi\!\left(\frac{\ln(x) - \alpha}{\beta}\right) \qquad (1.51)$$

for $x > 0$. The mean and variance are found by applying the techniques of Section 1.3 and are

$$E\{X\} = \exp(\alpha + \beta^2/2) \quad \text{and} \quad Var\{X\} = [E\{X\}]^2\,(\exp(\beta^2) - 1) \qquad (1.52)$$

For $0 < p < 1$ the $p100$th percentile x_p is found inverting (1.51) and is

$$x_p = \exp(\alpha + z_p\,\beta) \qquad (1.53)$$

where z_p is the $p100$th percentile associated with the standard normal random variable. The shape of the pdf (1.50) can be heavy right-tailed for properly chosen parametric values and is used to model insurance claim amounts.

1.6.3 Weibull Distribution

The Weibull distribution has been widely used to model failure times in engineering and reliability due to its flexible nature. This is due to the two nonnegative parameters α and β, where α is the shape parameter and β is the scale parameter. The pdf of the Weibull random variable, for $\alpha > 0$ and $\beta > 0$, takes the form

$$f(x) = \alpha\,\beta^{-\alpha}\,x^{\alpha-1}\exp(-(x/\beta)^\alpha) \qquad (1.54)$$

with $S = \{x > 0\}$. Here $\alpha < 1$ implies a decreasing hazard function that is indicative of infant mortality and $\alpha > 1$ results in an increasing hazard function indicating wear-out mortality. In actuarial or life science modeling the hazard function is referred to as the force of mortality and is explored in depth in Section 5.3. The cdf is

$$F(x) = 1 - \exp(-(x/\beta)^{\alpha}) \tag{1.55}$$

The mean and variance are functions of the gamma function Γ and, as noted by Christensen (1984, p. 175), are

$$E\{X\} = \beta \, \Gamma\left(1 + \frac{1}{\alpha}\right) \tag{1.56}$$

and

$$\mathrm{Var}\{X\} = \beta^2\left[\Gamma\left(1 + \frac{2}{\alpha}\right) - \Gamma\left(1 + \frac{1}{\alpha}\right)^2\right] \tag{1.57}$$

Further, x_p, the $p100$th percentile, for $0 < p < 1$, is found inverting (1.55) and is

$$x_p = \beta \, [-\ln(1 - p)]^{1/\alpha} \tag{1.58}$$

The Weibull distribution is an extension of the exponential distribution introduced in Example 1.11.

1.6.4 Gompertz Distribution

A distribution often used to model adult mortality is the Gompertz distribution. This distribution results from the Gompertz mortality law discussed in Example 5.7. This distribution has been used in numerous actuarial science applications, and for modern treatments we refer to London (1997, p. 18), Willemse and Koppelaar (2000, pp. 168–179), and Dickson et al. (2009, p. 24). Associated with this distribution are two parameters, $\alpha > 0$ and $\beta > 0$, and one form of the pdf is

$$f(x) = \alpha e^{\beta x} \exp\left(-\frac{\alpha}{\beta}(e^{\beta x} - 1)\right) \tag{1.59}$$

for $x > 0$. For $x > 0$ the Gompertz cdf and the survival function (1.42) are

$$F(x) = 1 - \exp\left(-\frac{\alpha}{\beta}(e^{\beta x} - 1)\right) \quad \text{and} \quad S(x) = \exp\left(-\frac{\alpha}{\beta}(e^{\beta x} - 1)\right) \tag{1.60}$$

TABLE 1.2

Gompertz Distribution Percentiles

p	.01	.10	.25	.50	.75	.90	.95	.99
Percentile x_p	25.63	50.79	61.68	71.53	79.22	84.85	87.78	92.55

The mean and the variance are not easily computed. For $0 < p < 1$ the p100th percentile, denoted x_p, is defined by $F(x_p) = p$, and utilizing (1.46) takes the form

$$x_p = \ln(1 - (\beta/\alpha) \ln(1 - p))/\beta \tag{1.61}$$

In the example that follows the mortality structure of this distribution is demonstrated.

Example 1.28

For an adult distribution the individual lifetimes are modeled by the Gompertz distribution with $\alpha = .0001$ and $\beta = .09$. Percentiles for various survival probabilities are computed using (1.61) and are listed in Table 1.2. A graph of the percentiles is given in Figure 1.6. We note the s-shaped graph of the cdf curve typical for population mortality.

The Gompertz distribution is characterized by an exponentially increasing mortality rate as measured by the force of mortality, introduced in Section 5.3, with age. In particular, the force of mortality associated with the Gompertz distribution, and the associated mortality structure, is explored in Example 5.7 with the change of parameter $\alpha = B$ and $c = \exp(\beta)$.

1.6.5 Makeham Distribution

The Makeham distribution is an extension of the Gompertz distribution presented in the previous section. This distribution, based on the Makeham mortality law discussed in Example 5.7, has been used to model the mortality

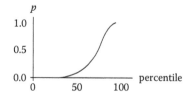

FIGURE 1.6

Percentile of the Gompertz distribution.

structure in actuarial science, and for classical and modern discussions, we refer to Makeham (1860), London (1997, p. 19), and Dickson et al. (2009, p. 35). The parameters are $\alpha > 0$, $\lambda > 0$, and $\beta > 0$, and for $x > 0$ the pdf is

$$f(x) = (\lambda + \alpha \exp(\beta x)) \exp\left(-\lambda x - \frac{\alpha}{\beta}(e^{\beta x} - 1)\right) \qquad (1.62)$$

The associated cdf and survival function for $x > 0$ are

$$F(x) = 1 - \exp\left(-\lambda x - \frac{\alpha}{\beta}(e^{\beta x} - 1)\right) \quad \text{and} \quad S(x) = \exp\left(-\lambda x - \frac{\alpha}{\beta}(e^{\beta x} - 1)\right) \qquad (1.63)$$

The formulas in (1.63) are extensions of the Gompertz cdf and survival function given in (1.60) through the addition of the λx component. The survival function associated with the Makeham distribution is the product of an exponential survival function and a Gompertz survival function and reduces to the Gompertz when $\lambda = 0$.

As age increases the instantaneous mortality rate or force of mortality, introduced in Section 5.3, of the Makeham distribution is a linear combination of the exponential distribution and the Gompertz distribution. This is demonstrated in Problem 7.4 in the context of multiple-decrement mortality modeling.

1.7 Conditional Distributions

Random variables associated with stochastic future actions commonly depend on separate variables. For example, for an insurance policy that pays a benefit at the time of death, variables, such as the age at the inception of a policy, are important. For this reason, the concept of conditional distributions for random variables is central in the study of financial and actuarial science. Nelson (2005) gives a review of conditioning with various distributions.

Let random variable X have pmf or pdf $f(x)$ with support S. Analogous to the formulas for the conditional probability given in (1.2), the conditional pdf of X, given $X > c$, is

$$f(x \mid X > c) = \frac{f(x)}{P(X > c)} \qquad (1.64)$$

where x is in the intersection S and $\{x > c\}$. The cdf and moments of the conditional distribution of X, given $X > c$, can be computed using (1.64) and the standard computational formulas. We follow with illustrative examples.

Example 1.29

Let X have an exponential distribution as discussed in Example 1.11 where the pdf is (1.16) and the survival function is (1.17). The pdf of X, given $X \geq c$, from (1.64), is

$$f(x \mid X > c) = \frac{f(x)}{S(c)} = \frac{1}{\theta} \exp(-(x-c)/\theta) \qquad (1.65)$$

From (1.65), if we let the future lifetime beyond c be $T = X - c$, then the pdf of T also is exponential with the identical parameter θ. This property is referred to as a lack of memory, since conditioning on $\{X > c\}$ produces a pdf that is a function of only the future lifetime $X - c$ and is independent of the values of c.

Example 1.30

Let X be a discrete geometric random variable with pmf given in Example 1.19, with pmf $f(x) = pq^x$ for $x = 0, 1, \ldots$. The survival function can be derived using the general summation formula (1.32) as

$$S(x) = q^{x+1} \qquad (1.66)$$

for $x = 0, 1, \ldots$. For a fixed positive integer c, the truncated distribution (1.64) becomes

$$f(x \mid X > c) = q^{x-c-1} p \quad \text{for} \quad x = c+1, c+2, \ldots \qquad (1.67)$$

As with the exponential distribution in the previous example, the conditional distribution takes the form of the initial distribution. The geometric random variable exhibits a lack of memory property in the discrete random variable setting where pmf (1.67) depends on $x - c$ regardless of the magnitude or value of c.

In many financial and actuarial models the future time at which conditions change is referred to as a lifetime variable where conditioning on relevant events affects the distribution. For example, the future lifetime associated with stochastic conditions is a conditional random variable where the conditioning is based on age. For $x > c$ the conditional survival function is

$$P(X > x \mid X > c) = \frac{S(x)}{S(c)} \qquad (1.68)$$

provided $x > c$ and c is in the support of the distribution of X. The conditioning concept and related formula are central to many financial and actuarial calculations presented in later chapters.

Example 1.31

In applications, the normal distribution with parameters μ and $\sigma^2 > 0$ may be altered by conditioning on the set of positive outcomes $\{x > 0\}$. The pdf of the left truncated normal random variable ($LTNRV$) is

$$f(x) = (2\pi\sigma^2)^{-1/2} \exp(-(x-\mu)^2/(2\sigma^2))/\Phi(\delta) \qquad (1.69)$$

for $S = \{x \geq 0\}$ and $\delta = \mu/\sigma$. Defining $\Delta(\delta) = \varphi(\delta)/\Phi(\delta)$, the mean and variance are found to be

$$\mu(\delta) = \mu + \sigma\,\Delta(\delta) \tag{1.70}$$

and

$$\sigma(\delta) = \sigma^2(1 - \delta\,\Delta(\delta) - \Delta(\delta)^2) \tag{1.71}$$

The mgf takes the form

$$M_x(t) = \exp(t\mu + t^2\sigma^2/2)\,\frac{\Phi(\delta + t\sigma)}{\Phi(\delta)} \tag{1.72}$$

Parameter δ measures the magnitude of the truncation. As δ approaches ∞ the *LTNRV* approaches the normal distribution. Derivation of (1.70), (1.71), and (1.72) along with applications are given by Borowiak and Das (2007).

1.8 Joint Distributions

In modeling of real data, such as that found in financial and actuarial fields, there is often more than one random variable required. The situation where we have two random variables, X and Y, is now considered, but the discussion can be extended to the general multiple random variable setting. Generally, these variables can be of any type, discrete, continuous, or mixed, and the joint pmf, pdf, or mpf is denoted $f(x, y)$. The basic concepts and formulas relating to one random variable are now extended to the multivariable case.

The initial concept in two-variable stochastic modeling is that the probability assessment that is (X, Y) falls in a defined set A. There are the following three possibilities depending on the type of random variables involved:

1. Discrete X and Y: $P(A) = \sum_A \sum f(x, y)$

2. Continuous X and Y: $P(A) = \iint_A f(x, y)\,dx\,dy$

3. Discrete X and continuous Y: $P(A) = \sum_A \int f(x, y)\,dy$

Here 1–3 define the probability structure of the joint random variable. Statistical concepts, such as marginal distributions, dependence, and independence, in the case of two random variables, are now discussed.

The marginal distributions are the distributions associated with individual random variables. Similar to the concepts and formulas for marginal probabilities there are three possible cases. The marginal pmfs or pdfs denoted $g(x)$ and $h(y)$ are given by:

4. Discrete X and Y: $g(x) = \sum_y f(x,y)$ and $h(y) = \sum_x f(x,y)$

5. Continuous X and Y: $g(x) = \int_y f(x,y)\,dy$ and $h(y) = \int_x f(x,y)\,dx$

6. Discrete X and continuous Y: $g(x) = \int_y f(x,y)\,dy$ and $h(y) = \sum_x f(x,y)$

Relationships between the variables are often important in statistical modeling and are associated with conditional distributions.

The conditional distributions and independence conditions follow the same pattern as those introduced in the probability measure given in (1.2). The conditional pmf or pdf of X given $Y = y$ explores the distribution of X, while Y is held fixed at value y and is

$$f(x|Y = y) = \frac{f(x,y)}{h(y)} \tag{1.73}$$

Further, X and Y are called independent if either

$$f(x, y) = g(x)\,h(y) \tag{1.74}$$

or

$$f(x|Y = y) = g(x) \tag{1.75}$$

over the support associated with the random variables. This definition of independence is analogous to the independence of sets definition given in (1.5). The above definitions, formulas, and concepts hold for all types of random variables. Two examples are now presented.

Example 1.32

Let X and Y have support $S = \{(x, y), \text{ where } x > 0, y > 0\}$ and joint pdf given by

$$f(x, y) = \frac{1}{\theta_1\theta_2}\exp(-x/\theta_1 - y/\theta_2) \quad \text{for} \quad x > 0 \text{ and } y > 0$$

for positive θ_1 and θ_2. Using condition (v) and related formulas we have

$$g(x) = \frac{1}{\theta_1}\exp(-x/\theta_1) \quad \text{and} \quad h(y) = \frac{1}{\theta_2}\exp(-y/\theta_2)$$

From criteria (1.74) it follows that X and Y are independent and the marginal distributions are both exponential in type.

Example 1.33

Conditions of insurance contracts are often defined separately for different groups of individuals. We consider an insurance policy where there are two risk categories or strata, $J = 1$ or 2, for proposed claims. The amount of a claim, X, in thousands of dollars, takes on values between 0 and 5 where the corresponding pdfs defined on their supports are for $J = 1$,

$$f_1(x) = 1/5 \quad \text{for } 0 \le x \le 5$$

and for $J = 2$,

$$f_2(x) = \begin{cases} .16x & \text{for} & 0 \le x \le 2.5 \\ -.16x + .8 & \text{for} & 2.5 < x < 5 \end{cases}$$

Assume the frequencies of the risk categories are defined by $P(J = 1) = .6$ and $P(J = 2) = .4$. The resulting joint mpf is a function of the two random variables J and X, and similar to the insurance modeling examples of Examples 1.14 and 1.21, the mpf is defined by

$$f(j,x) = \begin{cases} .6 f_1(x) & \text{for} & j = 1 & \text{and} & 0 \le x \le 5 \\ .4 f_2(x) & \text{for} & j = 2 & \text{and} & 0 \le x \le 5 \end{cases} \qquad (1.76)$$

The marginal pmf is computed as

$$g(j) = \int_0^5 f(j,x)\,dx = \begin{cases} .6 & \text{for} & j = 1 \\ .4 & \text{for} & j = 2 \end{cases}$$

and the marginal pdf is

$$h(x) = \sum_{j=1}^2 f(j,x) = \begin{cases} .12 + .064x & \text{for} & 0 \le x \le 2.5 \\ .44 - .064x & \text{for} & 2.5 < x \le 5 \end{cases} \qquad (1.77)$$

Probabilities of the basic events can be found using these distributions. The overall probability a claim is at most 3, from (1.77),

$$P(X \leq 3) = \int_{0}^{2.5} (.12 + .064x)\, dx + \int_{2.5}^{3} (.44 - .064x)\, dx = .632$$

Common statistical measurements, such as expectations, medians, and variances, can be computed using standard rules.

These basic statistical concepts and formulas extend to the setting of multiple random variables. In the case of two random variables, certain standard moments are useful to compute. The expected value of the function $g(x, y)$, if it exists, is denoted $E\{g(x, y)\}$. If X and Y are discrete with corresponding joint pmf $f(x, y)$ and support S, the expectation is

$$E\{g(x,y)\} = \sum_{S} \sum g(x,y)\, f(x,y) \tag{1.78}$$

In the continuous case, applying the joint pdf $f(x, y)$,

$$E\{g(x,y)\} = \int \int_{S} g(x,y)\, f(x,y)\, dx\, dy \tag{1.79}$$

while in the mixed-variable case the expectation takes the form

$$E\{g(x,y)\} = \sum_{S} \int g(x,y) f(x,y)\, dx \tag{1.80}$$

where $f(x, y)$ is the joint mpf. In these computations, we assume all the moments exist. Applying (1.78), (1.79), and (1.80), means and variances can be computed, but the relationship between variables must be considered in forming the proper summation and integral limits.

A measure of the linear relationship between X and Y is the covariance between X and Y denoted by $\text{Cov}\{X, Y\}$ or σ_{xy}. If the means of X and Y are given by μ_x and μ_y, respectively, then their covariance is defined as

$$\text{Cov}\{X, Y\} = E\{(X - \mu_x)(Y - \mu_y)\} = E\{XY\} - \mu_x \mu_y \tag{1.81}$$

If X and Y are independent, then $E\{XY\} = \mu_x \mu_y$ and $\sigma_{xy} = 0$. A scaled or normed measure of variable association based on (1.81) is the correlation coefficient denoted by ρ. This measure or parameter is used in correlation modeling, and the associated sample estimate is used as a diagnostic tool in regression modeling (see Section 2.6).

Many financial and actuarial modeling applications involve the sum of two or more random variables. Let two random variables X and Y have means μ_x and μ_y, variances σ_x^2 and σ_y^2, and covariance σ_{xy}. Since the expectation acts as a linear operator for the sum $S = X + Y$ (see (1.78), (1.79), and (1.80)), the expectation is

$$E\{S\} = E\{X\} + E(Y) = \mu_x + \mu_y \tag{1.82}$$

Further, applying the definition of variance, (1.25), squaring, and then taking the expectation, the variance of S can be written as

$$\text{Var}\{S\} = \sigma_x^2 + \sigma_y^2 + 2\sigma_{xy} \tag{1.83}$$

If the random variables are independent, then $\sigma_{xy} = 0$ and (1.83) indicates that the variance of the sum of variables is the sum of the separate variance terms. This concept extends to the general multiple random variable case in the same manner.

Expectations of functions of multiple random variables can be approached using the concept of conditioning. When two variables are involved, expectations are simplified by conditioning on one of the random variables inside the probability structure of the other. Let X and Y be random variables with joint pmf, pdf, or mpf $f(x, y) = h(y) f(x|y)$. We compute the mean and variance of functions of the form $g(x, y) = v(y)w(x, y)$. Here, for any type of random variable, extending Mood et al. (1974, p. 158),

$$E\{g(x, y)\} = E\{v(y) \, E\{w(x, y)|y\}\} \tag{1.84}$$

Also, the variance of $g(x, y)$, applying Mood et al. (1974, p. 159), is

$$\text{Var}\{g(x, y)\} = E\{v(y)^2 \, \text{Var}\{w(x, y)|y\}\} + \text{Var}\{v(y) \, E\{w(x, y)|y\}\} \tag{1.85}$$

Further, if $w(x, y) = w(x)$ and X and Y are independent, then

$$E\{g(x, y)\} = E\{v(y)\} \, E\{w(x)\} \tag{1.86}$$

and

$$\text{Var}\{g(x, y)\} = E\{v(y)^2\} \, \text{Var}\{w(x)\} + \text{Var}\{v(y)\}[E\{w(x)\}] \tag{1.87}$$

The general statistics theory required to derive the conditioning arguments is given in general theoretical statistical texts. The derivation of these formulas is left to the reader. Note that (1.86) and (1.87) can be used to derive (1.34) and (1.36).

We remark that the concepts and ideas presented in this section for the two variable settings can easily be extended to more than two random variables. For a review of joint distributions and their manipulations, see Hogg et al. (2005, Chapter 2), Hogg and Tanis (2010, Chapter 5), and Rohatgi (1996, Chapter 4).

Problems

1.1. An insurance company sells either auto or house insurance, but not both, to customers. Agent A has 74% auto policies and 26% house policies, while agent B has 60% auto policies and 40% house policies. Agent A or B is selected at random and then a policy from that agent is chosen at random. Find the probability (a) an auto policy is selected, and (b) the policy came from agent B if an auto policy was chosen.

1.2. In Problem 1.1 the insurance company now sells both auto and house insurance to the same customers. Agents A and B have the same number of policies. For A 60% of the policies contain auto insurance and 46% contain house insurance. Similarly, agent B has 80% auto and 35% house insurance policies. If a policy is selected at random, what is the probability it contains both auto and house insurance?

1.3. For real number a and positive integers $m \le n$ show that for the partial sum and associated series.

1.4. For continuously differentiable function $f(x)$ the general Taylor series is $f(x) = f(a) + f^{(1)}(a)(x-a)/1! + f^{(2)}(a)(x-a)^2/2! + \dots$.

   (a) Find the Taylor series approximation for $\exp(x)$ with $a = 0$.

   (b) For the Poisson pdf given in (1.11) show $f(0) + f(1) + \dots$ converges to 1.

   (c) For the geometric distribution given in Example 1.19 find the mgf and use it to compute the mean and variance.

1.5. For random variable X find the mean and variance, μ and σ^2, for pmf or pdf (a) $f(x) = 6x(1-x)$ for $S = \{x \mid 0 \le x \le 1\}$, and (b) $f(x) = 1/5$ for $S = \{1, 2, \dots, 5\}$.

1.6. In Example 1.10 let $X \sim U[0, 1]$. Compute (a) μ and σ^2, (b) $P(X > 1/4)$, and (c) $P(X > 1/2 \mid X > 1/4)$.

1.7.

   a. Compute the mean and variance of an exponential random variable with pdf (1.16).

   b. For the Gompertz distribution verify the relation between the pdf and cdf of (1.59) and (1.60).

   c. For the Makeham distribution verify the relation between the pdf and cdf of (1.62) and (1.63).

1.8.

   a. Write *Sk* and *Ku* as defined by (1.26) in terms of central moments.

     b. Use (1.29) and (1.30) to show the alternate forms of the first and second moments given in (1.44) and (1.45).

1.9. Consider the Poisson random variable, introduced in Example 1.8, with pmf given by (1.11).

     a. Find the mgf $M_N(t)$.

     b. Applying (1.40) use (a) to compute μ and σ^2.

1.10. Let X be the amount of an insurance claim where $P(X=0) = .4$, $P(X = \$1{,}000) = .2$ and claim amounts are uniform between \$200 and \$1,000.

     a. Give the mpf $f(x)$.

     b. What is the probability the claim is more than \$500?

     c. Find $E\{X\}$ and $\mathrm{Var}\{X\}$.

1.11. Let X have survival function $S(x) = 1 - x^2/2{,}500$ for $S = (0, 50)$. Find (a) the pdf $f(x)$, (b) μ and σ^2, and (c) the 75th percentile for X, denoted $x_{.75}$.

1.12. Let X be discrete uniform with pmf $f(x) = 1/25$ for $S = \{0, 1, \ldots, 24\}$. Find (a) $F(x)$, (b) $S(x)$, and (c) $E\{X\}$ and $\mathrm{Var}\{X\}$.

1.13. For an insurance policy the claim random variable over a short period is outlined in Example 1.14 and takes $q = .1$. If there is a claim the amount of the claim is uniform over \$0 to \$500. Over the time period let the claim variable be X.

     a. Find the expected value of X as in (1.34).

     b. Use (1.36) to compute the variance of X.

     c. Compute the 95th percentile for the claim variable X denoted $x_{.95}$.

1.14. Let random variables X and Y have joint pdf $f(x, y) = a$ for $S = \{0 < x < y < 1\}$.

     a. Find constant a.

     b. Find the marginal pdfs $g(x)$ and $h(y)$.

     c. Compute the covariance defined by (1.81).

     d. Are X and Y independent? Explain.

1.15. Let X and Y have joint mpf $f(x, y) = (\lambda^x/(\theta x!))\exp(-\lambda - y/\theta)$, where $S = \{y \geq o \text{ and } x = 0, 1, \ldots\}$.

     a. Find the marginal pmf $g(x)$ and pdf $h(y)$.

     b. Are X and Y independent? Explain.

1.16. Let X and Y be two random variables and consider a function of the form $g(x, y) = v(y)w(x, y)$.

     a. Conditioning on $Y = y$, show (1.84) and (1.85).

     b. Further, if X and Y are independent and $w(x, y) = w(x)$, show that (1.86) and (1.87) hold.

Excel Problems

1.17. Let $X \sim N(100, 100)$ as given in Example 1.12. Compute (a) $P(X > 115)$, (b) $P(X \le 83)$, and (c) $P(85 < X < 105)$. Excel: NORMDIST$(x,100,10,1)$ = $\Phi((x{-}100)/10)$.

1.18. Let $N \sim P(\lambda)$ as defined by (1.11). If the mean is 4, compute the probability N (a) is at least 2 and (b) at most 3. Excel: POISSON$(n,\lambda,1)$ = $P(N \le n)$.

Solutions

1.1. (a) $(1/2)(.74) + (1/2)(.6) = .67$, (b) $(1/2)(.6)/.67 = .4477$.

1.2. Using (1.7) $P(\text{auto and house}|A) = .6 + .46 - 1 = .06$ and B $P(\text{auto and house}|B) = .8 + .35 - 1 = .15$. Since A and B have the same number of policies, $P(\text{auto and house}) = .5(.06 + .15) = .105$

1.3. Take $S_{m,n} - a\,S_{m,n} = (1 - a)\,S_{m,n}$ and then divide by $(1 - a)$.

1.4. a. $\exp(x) = 1 + x + x^2/2! + x^3/3! + \dots.$

     c. $MN(t) = p\,\Sigma\,(q\,\exp(t))^n = p(1 - q\,\exp(t))^{-1}$ and using (1.40) $E\{N\} = qp^{-1}$, $E\{N^2\} = 2q^2/p^2 + q/p$, and (1.25) gives $\text{Var}\{N\} = q\,p^{-2}$.

1.5. a. $E\{X\} = 1/2$ and $E\{X^2\} = 3/10$ and apply (1.25).

     b. $E\{X\} = 3$ and $\text{Var}\{X\} = 2$.

1.6. a. $\mu = \frac{1}{2}$ and $\sigma^2 = 1/12$.

     b. $\frac{3}{4}$.

     c. Following (1.50), 2/3.

1.7. Using integration by parts $E\{T\} = \theta$, $E\{T^2\} = 2\theta^2$, so $\text{Var}\{T\} = \theta^2$.

1.8. a. We use $E\{(X - \mu)^3\} = E\{X^3\} - 3\mu\,E\{X^2\} + 2\,\mu^3$ and $E\{(X - \mu)^4\} = E\{X^4\} - 4\mu\,E\{X^3\} + 6\,\mu^2\,E\{X^2\} - 3\,\mu^4$.

1.9. Using 1.2,

     a. $MN(t) = e^{-\lambda}\,\Sigma\,(e^t\lambda)^n/n! = \exp(\lambda(e^t - 1))$.

     b. $E\{N\} = MN^{(1)}(0) = \lambda$, $MN^{(2)}(0) = \lambda^2 + \lambda$ so $\text{Var}\{N\} = \lambda$.

1.10. a. Following Example 1.13, $f(x) = .4$ if $x = 0$, $= c$ for $200 < x < 1{,}000$, and $= .2$ if $x = 1{,}000$, where $c = .0005$.

     b. $.0005(500) + .2 = .45$.

     c. Following Example 1.18, $E\{X\} = 440$, $E\{X^2\} = 365{,}333.3$, and $\text{Var}\{X\} = 171{,}733.3$.

1.11. a. $f(x) = -\,S^{(1)}(x) = x/1{,}250$.

     b. $E\{X\} = 33.33$, $E\{X^2\} = 1{,}250$, and $\text{Var}\{X\} = 138.889$

     c. $x_{.75} = (.75(2{,}500))^{1/2} = 43.301$.

1.13. a. $.1(250) = 25.$

   b. $\text{Var}\{X\} = 7{,}708.333.$

   c. $05 = .1(500 - x_{.95})/500 = 250.$

1.14. a. $a = 2.$

   b. $g(x) = 2(1 - x)$ for $0 < x < 1$, $h(y) = 2y$ for $0 < y < 1.$

   c. $E\{XY\} = 1/4$, $E\{X\} = 1/3$, $E\{Y\} = 2/3$, so $\text{Cov}\{X,Y\} = 1/4 - 2/9 = 1/36.$

   d. No, since $\text{Cov}(X,Y)$ is not zero.

1.15. a. $g(x) = e^{-\lambda}\,\lambda^x/x!$, $h(y) = \theta^{-1}\exp(-y/\theta).$

   b. Yes, since $f(x, y) = g(x)\,h(y).$

1.17. a. 0669.

   b. .0446.

   c. .6247.

1.18. a. .9084.

   b. .4335.

2

Statistical Techniques

There are two conceptual views in statistical modeling, referred to, respectively, as frequentist and Bayesian. In both approaches stochastic actions are modeled utilizing one or more statistical distributions that are functions of unknowns called parameters. In frequentist statistics the parameters are considered to be fixed constants and information about the parameters, such as point estimates and bounds, is entirely a function of observed data. Throughout this text the frequentist approach is utilized where information contained in sample data, such as financial or actuarial records, is used to construct well-known statistics. In Bayesian statistics the parameters themselves are modeled as random variables that are associated with specified prior probability distributions, and probability estimation and statistical inference take on a more mathematical flavor. For a review of Bayesian methods see Bickel and Doksum (2001, Section 1.2) and Hogg et al. (2005, Section 8.1).

The first chapter gave an introduction to the probabilistic and statistical theory utilized throughout the text. In this chapter we discuss statistical techniques required in many financial and actuarial procedures. This chapter starts with a discussion of sampling distributions associated with observed statistics in Section 2.1. Estimation is introduced in Section 2.1.1, leading to confidence intervals in Section 2.1.2. Notations and basic theory of percentiles, prediction intervals, and confidence sets are given in Sections 2.1.3 and 2.1.4. The discussion shifts to analysis of sums of random variables, referred to as an aggregate variable. The distribution of the aggregate variable and approximation techniques for a fixed number of variables are explored in Section 2.4. In particular, the central limit theorem and saddlepoint approximations are discussed in Sections 2.4.1 and 2.4.3. Introductions to regression modeling and autoregressive systems are presented in Sections 2.6 and 2.7. The chapter ends with a discussion of modeling diagnostics in Section 2.8. Theory and applications associated with probability plotting, least squares, and interval data diagnostics are given in Sections 2.8.1, 2.8.2, and 2.8.3.

2.1 Sampling Distributions and Estimation

The computed values of financial or actuarial statistics depend on the observed samples. The probability distribution associated with a statistic may depend on unknown parameters and is referred to as the sampling distribution of the statistic. The topic of statistical estimation for unknown parameters, as well as useful functions of parameters, is very broad, and we focus on applications of some of the areas in financial and actuarial modeling. In this section, the typical topics of point and interval estimation along with percentile measurements are presented. For a basic review of the principles of estimation we refer to Rohatgi (1976, Chapter 8), Hogg and Tanis (2010, Chapter 6), and Hogg et al. (2005, Chapter 7).

2.1.1 Point Estimation

In frequentist statistical theory parameters are considered fixed constants that characterize the distribution of a random variable. Common parameters explored include the mean, μ, the variance, σ^2, and the standard deviation, σ. In financial and actuarial modeling applications, other relevant quantities such as percentile points required for model analysis are investigated. Generally, statistics are defined as functions based on sample data and are used, in part, to estimate unknown parameters. A point estimator is a single-valued statistic that is used as an estimate of a parameter. Some common statistics are now discussed.

A random sample is a collection of variables, $X_1, \ldots, X_n$, where each random variable is independent and comes from the same distribution and the sample size is fixed. This is referred to as independent and identically distributed (iid) random variables with probability mass function (pmf) (pmf), probability density function (pdf), or mixed probability function (mpf) $f(x, \theta)$ based on parameter $\theta = (\theta_1, \theta_2, \ldots, \theta_m)$, which we denote by $X_i \sim$ iid $f(x, \theta)$ for $i = 1, 2, \ldots, n$. The likelihood function and using independence is

$$L(\theta) = \prod_{i=1}^{n} f(x_i, \theta) \tag{2.1}$$

The maximum likelihood estimator (mle), when it exists, is a parameter estimate denoted by $\hat{\theta}$ that maximizes (2.1). In many applications the mle is found by maximizing the log-likelihood function

$$l(\theta) = \ln(L(\theta)) = \sum_{i=l}^{n} \ln(f(x_i, \theta)) \tag{2.2}$$

Under general conditions, mles have optimal properties that include desirable asymptotic qualities as the sample size increases. We follow with two classical cases of parameter estimation.

Example 2.1

Let random variables X_i ~ iid normally distributed random variables for $i = 1, 2, ..., n$, with mean μ and standard deviation σ and pdf (1.18) with parameter $\theta = (\mu, \sigma^2)$. The maximization of (2.2) with respect to θ results in the following mle estimators for μ and σ^2:

$$\bar{X} = \frac{1}{n}\sum_{i=1}^{n} X_i \quad \text{and} \quad \hat{\sigma}^2 = \frac{(n-1)}{n} S^2 \quad \text{for} \quad S^2 = \frac{1}{n-1}\sum_{i=1}^{n}(X_i - \bar{X})^2 \quad (2.3)$$

where S^2 is referred to as the sample variance. The sample standard deviation is $S = (S^2)^{1/2}$. Using basic expectation properties and the iid assumption, we find that

$$E\{\bar{X}\} = \mu, \quad \text{Var}\{\bar{X}\} = \frac{\sigma^2}{n} \quad \text{and} \quad E\{S^2\} = \sigma^2 \quad (2.4)$$

For example, let the future lifetimes of a series of people, past an initial age, be 11, 15, 21, 25, 27, 30, 32, 36, 39, and 42. This is a sample of size $n = 10$, and the sums and sums of squares are $\sum X = 278$ and $\sum X^2 = 8,646$, respectively. From (2.3) the sample mean and variance are

$$\bar{X} = 27.8 \quad \text{and} \quad S^2 = 101.96 \quad (2.5)$$

The sample standard deviation is $s = 101.9^{1/2} = 10.095$. Many statistical techniques and inferences, such as confidence intervals, prediction intervals, and hypothesis tests, utilize these basic statistics.

Example 2.2

Consider the binomial random variable introduced in Example 1.8 where, following standard notation, p and q are interchanged and the pmf follows (1.10). There are two parameters, the sample size n and the proportion p. The mle estimator is found by setting the derivative of (2.2), with respect to p, to zero and solving, which results in the sample proportion

$$\hat{p} = \frac{1}{n}\sum_{i=1}^{n} X_i \quad (2.6)$$

Using the iid property we find the mean and variance of (2.6):

$$E\{\hat{p}\} = p \quad \text{and} \quad \text{Var}\{\hat{p}\} = \frac{p(1-p)}{n} \quad (2.7)$$

As the number of samples increase, (2.6) converges in probability to the theoretical probability p. This is the weak law of large numbers and is the intuitive concept for the long-run relative frequency approach to probability that is central to frequentist probability and statistics.

There are many properties associated with the point estimators, such as consistency, sufficiency, and unbiasedness. A statistic, W, is unbiased for parameter θ if

$$E\{W\} = \theta \tag{2.8}$$

Based on (2.4) and (2.7), $\bar{X}$, S^2 and $\hat{p}$ are unbiased parameters for μ, σ^2, and p. The unbiased property implies that the mean of the sampling distribution associated with the statistic matches the parameter to be estimated. To choose between unbiased estimators for a given parameter we select the estimator with the smallest variance. These estimators are referred to as minimum variance unbiased, or best, estimators.

Another useful property of point estimators is the idea of consistency. A statistic is consistent for a parameter if, as the sample size increases, the statistic converges in probability to that parameter (see Hogg et al., 2005, p. 206). In this situation, statistical theory supports the substitution of a statistic for the unknown parameter. This is used in the construction of limiting distributions for useful statistics. Examples of consistent estimators are the sample mean, variance, standard deviation, and sample proportions. For a review of point estimators and related properties see Hogg et al. (2005, Chapters 6 and 7) and Bickel and Doksum (2001, Chapter 2).

2.1.2 Confidence Intervals

Point estimates for parameters lack information about their variability and reliability. To address this, confidence sets leading to interval estimates for unknown parameters are constructed. The theoretical construction of these intervals is based on constructing a statistic, referred to as a pivot, whose distribution is independent of the unknown parameter. For a theoretical discussion of the construction of confidence intervals see Rohatgi (1976, Sections 11.2 and 11.3).

In this section, a single parameter θ is estimated with a confidence interval (*CI*) of the form $a \leq \theta \leq b$. The probability the interval contains the unknown parameter is called the confidence coefficient or confidence level. Hence, a confidence interval for θ, with confidence coefficient $1 - \alpha$, satisfies

$$1 - \alpha = P(a \leq \theta \leq b) \tag{2.9}$$

We remark that the parameter is a fixed value, while the interval is dependent on the selected sample observations and is therefore a random set. The classical case of normally distributed data is now considered.

Example 2.3

A random sample of size n is taken from a normal distribution where the mean, μ, is to be estimated and the standard deviation, σ, is known. The confidence interval is constructed using the pivot statistic

$$Z = n^{1/2} \frac{\bar{X} - \mu}{\sigma} \qquad (2.10)$$

which is distributed as a standard normal random variable. The distribution of the pivot statistic is independent of any unknown measurements. For a fixed level or confidence coefficient $1 - \alpha$, the $(1 - \alpha/2)100$th percentile of (2.10), denoted $z_{1-\alpha/2}$, is found using Excel, and thus

$$1 - \alpha = P(-z_{1-\alpha/2} \le Z \le z_{1-\alpha/2}) \qquad (2.11)$$

Using (2.10) in (2.11) and solving for μ yields the confidence interval

$$\bar{x} - z_{1-\alpha/2} \frac{\sigma}{n^{1/2}} \le \mu \le \bar{x} + z_{1-\alpha/2} \frac{\sigma}{n^{1/2}} \qquad (2.12)$$

In the case where the standard deviation σ is unknown and the sample size n is small, the t random variable is applied to form a confidence interval for μ. In that case, the $(1 - \alpha/2)100$th percentile value, based on a t distribution with $n - 1$ degrees of freedom, is used in place of the standard normal percentile in (2.12).

2.1.3 Percentiles and Prediction Intervals

In financial and actuarial applications attention often is focused on outcomes of random variables and not directly on their parameters. Percentiles for a random variable give a relative measure for realized outcomes. For a random variable or statistic, X, and fixed α, $0 \le \alpha \le 1$, the $(1 - \alpha)100$th percentile, denoted by $x_{1-\alpha}$, is defined by the relation

$$1 - \alpha = P(X \le x_{1-\alpha}) \qquad (2.13)$$

The 50th percentile, $x_{.5}$, is the median of a continuous distribution. An interval that covers a fixed proportion of the distribution associated with a random variable X is referred to as a prediction interval (*PI*). For α, $0 \le \alpha \le 1$, from (2.13) an equal tail area *PI* is

$$x_{\alpha/2} \le X \le x_{1-\alpha/2} \quad \text{since} \quad 1 - \alpha = P(x_{\alpha/2} \le X \le x_{1-\alpha/2}) \qquad (2.14)$$

We remark that one-sided prediction intervals for population measurements can also be constructed. This statistical interval construction is closely related to the idea of tolerance intervals for statistics and distributions where coverage probabilities are defined and analyzed (see Hogg et al., 2005, p. 252).

Example 2.4

In this example, let the lifetime variable associated with status conditions, T, be a normal random variable with mean μ and standard deviation σ. The $p100$th percentile follows from the normal cumulative distribution function (cdf) (1.19) and is

$$t_p = \mu + z_p\,\sigma \tag{2.15}$$

where z_p is the $p100$th percentile for the standard normal random variable. To construct a *PI* for T of the form (2.14), the percentiles $t_{\alpha/2}$ and $t_{1-\alpha/2}$ are applied using (2.15), resulting in

$$t_{\alpha/2} = \mu - \sigma\,z_{\alpha/2} \le T \le t_{1-\alpha/2} = \mu + \sigma\,z_{1-\alpha/2} \tag{2.16}$$

where $\alpha = \Phi(z_\alpha)$. For example, we find a 95% *PI* for future lifetime T where $\mu = 50$ and $\sigma = 10$. Using Excel (see Problem 2.10) we find $z_{.975} = 1.96$ so that $t_{.025} = 50 - 1.96(10) = 30.4$ and $t_{.975} = 50 + 1.96(10) = 69.6$. Thus, 95% of the lifetimes for the status condition fall between 30.4 and 69.6. In applications only lower or upper bounds on the lifetime may be of interest.

2.1.4 Confidence and Prediction Sets

In general theory, random sets are functions of random variables whose distribution depends on one or more parameters. In this section, two types of random sets and their applications are presented. In the first, sample data are used to construct a set of statistically reasonable parameter values referred to as a confidence set. Relevant diagnostic measurements based on an assumed distributional structure are optimized over the confidence set. In the second, the parameter vector is assumed to be known and statistically reasonable values of relevant statistics are found that are referred to as a prediction set. Optimizing values of future statistical measurements over the prediction set results in the construction of prediction intervals. In practice, statistical software, such as the Solver program in Excel (see Problem 2.11) is used to obtain optimum values. The general framework of random set construction is now explored.

Consider an iid collection of random variables X_i for $i = 1, 2, \ldots, n$ whose pmf, pdf, or mpf is a function of the parameter vector $\theta = (\theta_1, \theta_2, \ldots, \theta_m)$. The random sample is represented in the data vector $\mathbf{X} = (X_1, X_2, \ldots, X_n)$. In the construction of confidence and prediction sets, a vector consisting of statistics $\mathbf{S}(\mathbf{X})_i$ for $1 \le i \le r$, denoted $\mathbf{S}(\mathbf{X}) = (S(\mathbf{X})_1, S(\mathbf{X})_2, \ldots, S(\mathbf{X})_r)$, is utilized. Analogous to the construction of confidence intervals in Example 2.3, a statistic $W(\mathbf{S}(\mathbf{X}),\theta)$, where the associated cdf is independent of both $\mathbf{S}(\mathbf{X})$ and θ, is utilized as the pivot. For $0 < p < 1$, a $(1-p)100$th percentile for $W(\mathbf{S}(\mathbf{X}),\theta)$ is denoted w_{1-p} where

$$1 - p = P(W(\mathbf{S}(\mathbf{X}),\theta) \le w_{1-p}) \tag{2.17}$$

This construction is used in various statistical applications. The asymptotic theory of maximum likelihood estimators yields a general setting for the construction of confidence sets. An outline of this construction is now presented.

Let the maximum likelihood estimator for θ be $S(X) = \hat{\theta}$. As demonstrated by Hogg et al. (2005, pp. 325–327), under general conditions the asymptotic distribution of the mle is multivariate normal with asymptotic mean θ and asymptotic dispersion matrix denoted by Σ. Applying results involving the distribution of quadratic forms in normal variables for large sample size, we form an approximate chi square random variable with m degrees of freedom:

$$W(\hat{\theta}, \theta) = (\hat{\theta} - \theta)\Sigma^{-1}(\hat{\theta} - \theta)^t \sim \chi^2_m \qquad (2.18)$$

Background material on the distribution of quadratic forms is given by Graybill (1976) and Searle (1971). The confidence and prediction sets utilize (2.18) as a statistical pivot where the $(1 - \alpha)100$th percentile is denoted $w_{1-\alpha} = \chi^2_{1-\alpha,m}$. Pivot (2.18) is now used to form desired confidence and prediction sets.

In the first technique, we treat the observed statistics as fixed at $S(X) = s(x)$ in (2.17), resulting in the general $(1 - \alpha)100\%$ confidence set for θ:

$$CS_{1-\alpha} = \{\theta : W(s(X), \theta) \leq w_{1-\alpha}\} \qquad (2.19)$$

In many applications the confidence set (2.19) leads to a confidence ellipsoid in the unknown parameters. Applications involve forming interval estimates for functions of the underlying distribution, such as confidence intervals for reliabilities and percentiles. In the second approach, the parameter θ is assumed to be fixed in (2.17). A general $(1 - \alpha)100\%$ prediction set, PS, for $S(X)$ is

$$PS_{1-\alpha} = \{S(X) : W(S(X), \theta) \leq w_{1-\alpha}\} \qquad (2.20)$$

Applications of (2.20) arise in financial and actuarial studies where future or present values of economic actions under assumed distributions are to be estimated. These are demonstrated in the following examples.

Example 2.5

Consider a general random sample of size n from a normal distribution with parameter $\theta = (\mu, \sigma)$. Maximizing (2.2) the mle estimators using the notations in (2.3) are

$$\hat{\theta} = (\hat{\mu}, \hat{\sigma}) \quad \text{where} \quad \hat{\mu} = \bar{X} \quad \text{and} \quad \hat{\sigma} = \left(\frac{n-1}{n}\right)^{1/2} S \qquad (2.21)$$

Following Hogg et al. (2005, p. 345) the asymptotic dispersion matrix is found to be

$$\Sigma = \begin{pmatrix} \dfrac{\sigma^2}{n} & 0 \\ 0 & \dfrac{\sigma^2}{2n} \end{pmatrix} \tag{2.22}$$

In pivot (2.18), we take $S(\hat{\mathbf{X}}) = (\bar{X}, \sigma)$, resulting in the random variable and $(1 - \alpha)100$th percentile:

$$W(S(X), \theta) = \frac{n(\bar{X} - \mu)^2}{\sigma^2} + \frac{2n(\hat{\sigma} - \sigma)^2}{\sigma^2} \quad \text{and} \quad w_{1-\alpha} = \chi^2_{1-\alpha, 2} \tag{2.22}$$

This leads to the confidence inequality

$$\frac{n(\bar{X} - \mu)^2}{\sigma^2} + \frac{2n(\hat{\sigma} - \sigma)^2}{\sigma^2} \le \chi^2_{1-\alpha, 2} \tag{2.24}$$

For observed parameter estimators (2.21), a $(1 - \alpha)100\%$ confidence set consisting of the parameter θ satisfying (2.24) is (2.19). For illustration we form a *PI* for the percentiles defined by (2.15). For $p100$th percentile x_p the $(1 - \alpha)100\%$ *PI* is defined by

$$LB(x_p) \le x_p = \mu + z_p\,\sigma \le UB(x_p) \tag{2.25}$$

where

$$LB(x_p) = \min_{\theta \in CS_{1-\alpha}} \{\mu + z_p\sigma\} \quad \text{and} \quad UB(x_p) = \max_{\theta \in CS_{1-\alpha}} \{\mu + z_p\sigma\} \tag{2.26}$$

Applying the sample data in Example 2.1 where (2.3) holds and taking $1 - \alpha = .95$,

$$\hat{\mu} = 27.8, \quad \hat{\sigma} = 9.75914 \quad \text{and} \quad \chi^2_{.95, 2} = 5.9914 \tag{2.27}$$

Confidence intervals of the form (2.25) and (2.26) are found by iteration for various percentiles, and the results are listed in Table 2.1.

To demonstrate the interval estimation of a statistic we utilize the sample coefficient of variation (*CV*) defined by

$$CV = \frac{S}{\bar{X}} = \frac{(n-1)\hat{\sigma}}{n\,\bar{X}} \tag{2.28}$$

TABLE 2.1

Confidence Intervals for Normal Percentiles

Percentile	LB(x$_p$) at θ = (μ, σ)	UB(x$_p$) at θ = (μ, σ)
95th	35.1266 at (22.625, 7.600)	65.4639 at (32.974,19.752)
90th	32.2685 at (21.789, 8.177)	58.3849 at (33.811, 19.175)
75th	26.8273 at (20.103, 9.970)	47.2216 at (35.497, 17.382)
50th	18.9405 at (18.940, 13.676)	36.6596 at (36.659, 13.676)

The prediction set $PS_{1-\alpha}$ fixes θ and consists of $S(X) = (\bar{X}, \hat{\sigma})$ that satisfy (2.24). An approximate $(1 - \alpha)$ 100% prediction interval for the CV is

$$LB(CV) \leq CV = \frac{S}{\bar{X}} \leq UB(CV) \tag{2.29}$$

where the lower and upper bounds are

$$LB(CV) = \min_{S(X) \varepsilon PS_{1-\alpha}} \left\{ \frac{S}{\bar{X}} \right\} \quad \text{and} \quad UB(CV) = \min_{S(X) \varepsilon PS_{1-\alpha}} \left\{ \frac{S}{\bar{X}} \right\} \tag{2.30}$$

If we assume $\mu = 1{,}000$, $\sigma = 100$, and $n = 10$, with the aid of the Excel routine Solver, applying (2.29) in (2.30) yields the 95% PI

$$.04516 \leq CV = \frac{S}{\bar{X}} \leq .15605 \tag{2.31}$$

where $LB(CV) = .04516$ for $S(X) = (1{,}004.931, 45.378)$ and $UB(CV) = .15605$ for $S(X) = (983.314, 153.447)$.

2.2 Sums of Independent Variables

In practice, we often observe a series of independent variables $X_1, ..., X_m$, where the number of variables may be fixed, denoted by m, or stochastic, represented by the discrete random variable N. For example, these could be the future lifetimes of a group of people or the current values of a number of stocks held in a portfolio. Financial and actuarial analysis applications involving the sum or aggregate, denoted S_m, are referred to as a collective risk or aggregate modeling.

In this section the number of variables is a fixed quantity. For positive integer m, the distribution of the aggregate or sum of m random variables is

$$S_m = \sum_{i=1}^{m} X_i \qquad (2.32)$$

The two-variable case, $m = 2$, is now explored and the resulting concepts can be extended to the multiple-variable situation. The distribution of the sum of two independent variables X and Y can be computed using the classical convolution method. The cdf for $S = X + Y$ is defined as

$$F_S(s) = P(S \le s) = P(X + Y \le s) \qquad (2.33)$$

for constant s. Conditioning on $Y = y$, for the fully discrete case we find

$$F_S(s) = \sum_{S_y} P(X \le s - y | Y = y)\, P(Y = y) \qquad (2.34)$$

and for continuous X and Y,

$$F_S(s) = \int_{S_y} P(X \le s - y | Y = y)\, f_y(y)\, dy \qquad (2.35)$$

Further, if X and Y are independent, then the cdfs (2.34) and (2.35) become

$$F_S(s) = \sum_{S_y} F_x(s - y)\, f_y(y) \quad \text{and} \quad F_S(s) = \int_{S_y} F_x(s - y)\, f_y(y)\, dy \qquad (2.36)$$

Here (2.36) gives the formulas for the convolution for the cdfs $F_X(x)$ and $F_Y(y)$, which we denote as F_x*F_y. Taking the differences or derivatives yields the pmf or pdf, respectively. The corresponding pmf in the discrete case and pdf in the continuous case are, respectively,

$$f_S(s) = \sum_{S_y} f_x(s - y)\, f_y(y) \quad \text{and} \quad f_S(s) = \int_{S_y} f_x(s - y)\, f_y(y)\, dy \qquad (2.37)$$

An example of the convolution method in the continuous setting is given in Problem 2.1.

As mentioned, the convolution process can be extended to the situation of $n \ge 3$ independent variables. Let $S_n = X_1 + \cdots + X_n$, where the cdf of X_i is F_i and the cdf of $X_1 + \cdots + X_k$ is $F^{(k)}$ for positive integer $k \le n$. Iteratively, we find the convolution of a cdf by using

$$F^{(i)} = F_i * F^{(i-1)} \tag{2.38}$$

for $i = 2, 3, \ldots, n$. A computational example, analogous to the example in Bowers et al. (1997, p. 35), is now given.

Example 2.6

Let X_1 and X_2 be independent and $S = X_1 + X_2$. The pmfs are $f_i(x)$ and the cdfs are $F_i(x)$ for $i = 1, 2$. The defining probabilities, following (2.34), are given below:

X	$f_1(x)$	$f_2(x)$	$F_1(x)$	$F^{(2)}(x)$	$f^{(2)}(x)$
0	.4	.3	.4	.12	.12
1	.4	.4	.8	.40	.28
2	.1	.1	.9	.63	.23
3	.1	.1	1.0	.78	.15
4	.0	.1	1.0	.91	.13
5	.0	.0	1.0	.97	.06
6	.0	.0	1.0	.99	.02
7	.0	.0	1.0	1.0	.01

Here,

$$F^{(2)}(0) = F_1(0)\, f_2(0) = .4(.3) = .12$$

$$F^{(2)}(1) = F_1(1)\, f_2(0) + F_1(0)\, f_2(1) = .8(.3) + .4(.4) = .40$$

Continuing the pattern, $F^{(2)}(2) = .9(.3) + .8(.4) + .4(.1) = .63$, $F^{(2)}(3) = .78$, $F^{(2)}(4) = .91$, $F^{(2)}(5) = .97$, $F^{(2)}(6) = .99$, and $F^{(2)}(7) = 1$. The pmf of S, $f^{(2)}(x)$, is found by subtracting the consecutive values of $F^{(2)}$. In Problem 2.2, the process is extended to the convolution for a third random variable, computing cdf $F^{(3)}$.

In a theoretical sense, the convolution process can be used to find the distribution of the sum of independent variables for any fixed number of variables. In many practical settings, common in financial and actuarial modeling, several approximations exist to model the distribution of sums of random variables. In Section 2.4, three approximation techniques are presented to model an aggregate sum of independent random variables. This section ends with theoretical distributional considerations of a sum of independent random variables.

In many situations, there exist statistical techniques to find the distribution a linear combination of random variables. We discuss the moment generating function (mgf) approach to this problem. Let a linear combination corresponding to m random variables be $S_m = \Sigma\, a_i X_i$ for constants a_i, where the mgf for X_i is denoted $M_i(t)$ for $1 \le i \le m$. If the random variables are independent the mgf of S_m simplifies as

$$M_{S_m}(t) = E\{\exp(t\,S_m)\} = E\left\{\exp\left(t\sum_{i=1}^{m}a_i X_i\right)\right\} = \prod_{i=1}^{m} M_i(a_i t) \qquad (2.39)$$

The mgf associated with a random variable, if it exists, is unique. Thus, if the mgf of S_m matches the mgf of a known distribution, then the distribution of S must be the distribution associated with the matched mgf. Two illustrative examples follow.

Example 2.7

Let the aggregate claims for two cities, denoted by X_a and X_b, be approximately normal, where $X_a \sim N(100, 20)$ and $X_b \sim N(80, 30)$. We seek the probability that X_a exceeds X_b. To do this we define $Y = X_a - X_b$, and using the form of the normal (1.41) and (2.39), find the mgf of Y:

$$M_Y(t) = \exp(100t + 20t^2/2)\,\exp(-80t + 30t^2/2) = \exp(20t + 50t^2/2)$$

Thus, we observe $Y \sim N(20, 50)$ and we compute

$$(P(Y) > 0) = 1 - \Phi\left(\frac{-20}{50^{1/2}}\right) = \Phi(.2828) = .99766$$

Thus, there is a 99.77% chance that X_a will exceed X_b.

Example 2.8

Independent claim amounts for three types of policies are denoted by X_i for $i = 1, 2, 3$, with association with mgf:

$$M_i(t) = (1 - 2t)^{-i} \quad \text{for } i = 1, 2, 3$$

Let the aggregate claim be $S = X_1 + X_2 + X_3$. Applying (2.39), the mgf of S is

$$M_s(t) = (1 - 2t)^{-6} \qquad (2.40)$$

The mean and variance of S are computed using the mgf. Utilizing (1.40) compute

$$E\{S\} = 12 \quad \text{and} \quad E\{S^2\} = 168$$

and the variance is $\text{Var}\{S\} = 168 - 144 = 24$. Further, based on (2.40) the distribution of the sum follows a gamma distribution with parameters $\alpha = 6$ and $\beta = 2$.

In some important situations, the mgf for sum (2.39) can be simplified. If X_i, $i = 1, 2, \ldots, n$, are iid each with mean μ, standard deviation σ, and mgf $M(t)$, then from (2.39)

$$M_{S_m}(t) = [M(t)]^m \qquad (2.41)$$

This form of the mgf is useful in finding distributions of sums resulting from random samples.

Further, the mgf (2.41) can be used to find the moments of S_m through the derivative formula (1.40). Taking the derivatives of (2.41) and evaluating at $t = 0$ we find

$$E\{S_m\} = m\mu \quad \text{and} \quad \text{Var}\{S_m\} = m\,\sigma^2 \qquad (2.42)$$

Hence, for the sample mean $\bar{X}$, the mean and variance

$$E\{\bar{X}\} = \mu \quad \text{and} \quad \text{Var}\{\bar{X}\} = \frac{\sigma^2}{m} \qquad (2.43)$$

These moments can be used to calculate or approximate relevant probabilities. In the case where the aggregate is modeled by a normal random variable, the cdf is

$$P(S_m \le c) = \Phi\left(\frac{c - m\mu}{m^{1/2}\sigma}\right) \qquad (2.44)$$

for constant c. The next example demonstrates the application of the normal distribution to aggregate variables.

Example 2.9

Let $X_1, \ldots, X_m$ be a random sample, or more specifically iid, from a normal distribution with mean μ and standard deviation σ. From (1.41) and (2.41), the mgf of S_m is

$$M_s(t) = [\exp(t\mu + t^2\sigma^2/2)]_m = \exp(t(m\mu) + t^2(m\,\sigma^2)/2) \qquad (2.45)$$

Since this takes the form of a normally distributed mgf, we conclude S_m is a normal random variable. Further, from (2.45) we realize that S_m has mean and variance given in (2.42) and probabilities can be computed using (2.44).

The mgf technique employed in Example 2.9 to find the distribution of the aggregate is applicable for other distributions, such as Poisson, binomial, and gamma (see Problem 2.3). In the case of other distributions, approximations such as the central limit theorem and Panjer recursion (see Panjer, 1981; Panjer and Wang, 1993) are used to estimate the distribution of the sum or aggregate.

2.3 Order Statistics and Empirical Prediction Intervals

Many statistical techniques are based on information obtained from an arrangement of observations from low to high. In this section, we consider m continuous random variables $X_1, X_2, \ldots, X_m$, where the order statistics arise from an ordered rearrangement of the corresponding observed variables. The order statistics are denoted Y_j or $X_{(j)}$ for $j = 1, 2, \ldots, n$, so that $Y_1 < Y_2 < \ldots < Y_m$. The distribution of the order statistics, or a function of order statistics, is found using mathematical or statistical transformation theory. For a general statistical review of order statistics and their distributions see Hogg et al. (2005, Section 4.6).

Our investigation into order statistics techniques concerns an introduction to the basic theory with two specialized cases that are important in actuarial modeling. Let m continuous random variables be independent. The distributions of the first order statistic, $X_{(1)}$, and the last or mth order statistic, $X_{(m)}$, can be derived using the independence property (1.5). The cdf for the first order statistic is

$$P(X_{(1)} \leq x) = 1 - P(X_{(1)} > x) = 1 - \prod_{i=1}^{m} P(X_i > x) \tag{2.46}$$

for constant x. For the mth order statistic, applying independence and the complement property (1.6) produces the cdf

$$P(X_{(m)} \geq x) = 1 - \prod_{i=1}^{m} P(X_i < x) \tag{2.47}$$

These formulas are central to the computation of probabilities for events defined in terms of the special multiple life statuses explored in Section 7.1.

Example 2.10

Let the future lifetimes of a collection of m people be the independent variables $T_1, T_2, \ldots, T_m$. An economic action, such as paying a life insurance claim upon the first death, is triggered by a function of the n future lifetime variables. In general, the action occurs when, under defined conditions, the status fails to hold. If we have a joint life status, as introduced in Example 1.5, the lifetime of the status is a function of the first order statistic $T_{(1)}$. For constant c, the survivor function of $T_{(1)}$ is

$$P(T_{(1)} > c) = \prod_{i-1}^{m} P(T_i > c) \qquad (2.48)$$

In a similar manner, for the last survivor status the lifetime of the status is a function of the largest order statistic $T_{(m)}$. The survivor function is

$$P(T_{(m)} > c) = 1 - P(T_{(m)} \le c) = 1 - \prod_{i=1}^{m} P(T_i \le c) \qquad (2.49)$$

for $c > 0$. Other combinations of order statistics can be analyzed using the general theory of order statistics.

Financial and actuarial stochastic actions, such as life insurance and stock sales, may be a function of the first or last order statistics. Future lifetime random variables associated with stochastic actions that are functions of multiple lifetimes are discussed in Section 7.1. We end this section with a discussion of two types of prediction intervals based on observed order statistics.

Based on observed order statistics $Y_1 < Y_2 \ldots < Y_m$, nonparametric prediction intervals can be constructed. We present two specific prediction intervals where the first is a symmetric prediction interval (SPI), which utilizes equal tails to compute percentile ranks and straightforward linear interpolation. The percentile ranks for the $(1 - \alpha)100\%$ SPI are given by

$$a = \frac{\alpha}{2}(m+1) \quad \text{and} \quad b = \left(1 - \frac{\alpha}{2}\right)(m+1) \qquad (2.50)$$

The linear interpolation utilizes the function $[d]$, which represents the greatest integer less than or equal to d. For a statistic of interest, denoted by Y, the lower bound is

$$LB(Y) = Y_{[a]} + (a - [a])(Y_{[a]+1} - Y_{[a]}) \qquad (2.51)$$

while the upper bound is

$$UB(Y) = Y_{[b]} + (b - [b])(Y_{[b]+1} - Y_{[b]}) \qquad (2.52)$$

Hence, the $(1 - \alpha)100\%$ SPI for Y is given by

$$LB(Y) \le Y \le UB(Y) \qquad (2.53)$$

The SPI (2.53) has been shown to be effective when applied to normally distributed data.

The second method is based on the bias-correcting percentile method, described by Efron and Tibshirani (1986), and has been demonstrated to be useful when the statistic is not a linear function of an approximatly normal random variable. This method, denoted by BCPI, is a variant of SPI and utilizes the following sample mean and associated empirical distribution value:

$$\bar{Y} = \sum_{i=1}^{m} \frac{Y_i}{m} \quad \text{and} \quad \hat{F}(\bar{Y}) = \sum_{i=1}^{m} I(Y_i \leq \bar{Y}) \Big/ m \tag{2.54}$$

where the indicator function $I(a) = 1$ if a holds and is zero otherwise. The centering quantity and tail areas are defined by

$$z_0 = \Phi^{-1}\left(\frac{\hat{F}(\bar{Y})}{1}\right), \quad \alpha_L = \Phi(2z_0 + z_{(\alpha/2)}) \quad \text{and} \quad \alpha_U = \Phi(2z_0 + z_{(1-\alpha/2)}) \tag{2.55}$$

Based on the tail areas in (2.55), the percentile ranks are defined by

$$a = \alpha_L (m + 1) \quad \text{and} \quad b = \alpha_U (m + 1) \tag{2.56}$$

The BCPI takes the form (2.53) with lower and upper bounds defined by (2.51) and (2.52). If the sample mean coincides with the sample median, both defined in (2.54), then in (2.55) we find $z_0 = 0$ and the BCPI reduces to the symmetric confidence interval or SCI. These methods are demonstrated in the following example.

Example 2.11

In this example, the future lifetime denoted T is approximately normal in distribution, and we use the sample of size 10 given in Example 2.1. For the lifetime variable, a 50% *PI* is desired. First, applying (2.16), a normal distribution-based *PI* takes the form

$$20.991 = 27.8 - .6745(10.095) \leq T \leq 27.8 + .6745(10.095) = 34.609 \tag{2.57}$$

To construct the SPI, from (2.50), the percentile ranks are $a = .25(11) = 2.75$ and $b = .75(11) = 8.25$. The lower and upper bounds, defined by (2.51) and (2.52), are

$$LB(T) = 15 + .75(21 - 15) = 19.6 \quad \text{and} \quad UB(T) = 36 + .25(39 - 36) = 36.75$$

and the 50% SPI is

$$19.6 \leq T \leq 36.75 \tag{2.58}$$

For this example, the BCPI coincides with the SPI, since applying (2.54) and (2.55) we find

$$z_0 = \Phi^{-1}\left(\frac{5}{10}\right) = 0$$

2.4 Approximating Aggregate Distributions

Applications in financial and actuarial modeling often involve a collection or portfolio of financial contracts where the collective is to be analyzed. The insurance company may have a collection of insurance policies or an investor may have a diversified collection of investments. In this section the distribution of the sum or aggregate of random variables is discussed. For moderate sample sizes the convolution method described in Section 2.2 can become cumbersome. There are several approximations to the distribution of aggregate sums of variables used in actuarial science. One method is Panjer recursion, and we refer to Panjer (1981), Panjer and Wang (1993), and Sundt and Jewell (1981) for details and discussion. Another method is the Wilson–Hilferty approximation or the Haldane type A approximation, as presented by Pentikainen (1987). For a discussion of actuarial approaches see Panjer and Willmot (1992, Chapter 6) and Bowers et al. (1997, Chapters 2 and 12). Classical statistical approaches include the Edgeworth expansion (Feller, 1971) or the conjugate density method of Esscher (1932).

In this section, three methods for approximating cumulative probabilities, and hence percentiles, for sums of random variables are presented: the first is the celebrated central limit theorem (CLT) for sums of iid random variables, denoted by CLT; the second is the Haldane type A approximation, as discussed by Pentikainen (1987); and the last is the saddlepoint approximation of Daniels (1954). The latter two approximations can be viewed as extensions of the CLT.

2.4.1 Central Limit Theorem

Much work in theoretical and applied statistics concerns the convergence of a series of random variables. Under various restrictions, limiting distributions have been demonstrated. The Lindeberg–Feller central limit theorem (Feller, 1946) concerns the convergence of a set of independent, but not identically distributed, random variables. Under suitable conditions on the variance of the respective random variables, convergence in distribution of a standardized variable, based on the collective sum of standard normal random variables, is achieved as the sample size increases. Under relaxed conditions, the random variables also are assumed to be independent, and from identical distributions with finite mean and variance, and large sample convergence to normality is assured. This common setting, referred to as the Levy central limit theorem (see Levy, 1954), or simply the central limit theorem, is used throughout this text and is denoted by CLT. In this form, we have m independent variables, $X_1, X_2, ..., X_m$, taken from the same distribution. From this random sample of size m, we compute the sample mean and the aggregate or sum of the random variables denoted by S_m. After proper centering and scaling, the sampling distribution of the sample mean, relating to the aggregate or sum, approaches the normal distribution as the sample size increases. The mean and variance for the sample mean and sum are

$$E\{\bar{X}\} = \mu, \quad \mathrm{Var}\{\bar{X}\} = \frac{\sigma^2}{m}, \quad \text{and} \quad E\{S_m\} = m\mu, \quad \mathrm{Var}\{S_m\} = m\sigma^2$$

The CLT states that for a large sample size, m, the random variable

$$Z = \frac{m^{1/2}(\bar{X} - \mu)}{\sigma} = \frac{S_m - m\mu}{m^{1/2}\sigma} \tag{2.59}$$

is distributed approximately standard normal. From (2.59), percentile, tail probabilities, and prediction intervals can be approximated. The efficiency of any approximation is contingent on many factors. In this case, the shape of the distribution associated with the random variables and the deviation from the assumptions play a role. In any event, the convergence of large sample approximations can be assessed by simulation techniques as discussed in Section 9.2. Further, the statistical consistency of the sample standard deviation, S, allows for the substitution of S for σ in (2.59) in the case of large m.

Example 2.12

The distribution associated with a random sample of size 100 is claimed to have a mean and standard deviation of 65 and 7, respectively. From the sample, we find that

$$\bar{X} = 68$$

The probability that the sample mean is at least 68 is found using (2.59) as

$$P(\bar{X} \geq 65) = P\left(Z \geq \frac{68 - 65}{\dfrac{7}{10}} \right) = 1 - \Phi(4.286) = 0$$

From this we might conclude that the mean is not, as was assumed in the computation, 65, but somewhat larger. This type of investigation lays the groundwork for concepts in statistical hypothesis testing.

In the small sample setting, the central limit theorem can be adjusted to approximate the distribution of the sum of random variables S_m. This has many applications in financial and actuarial modeling. We follow with an introductory example dealing with an insurance claim variable.

Example 2.13

A portfolio consists of $m = 25$ short-term insurance policies. Over the time period for the insurance, $P(\text{claim}) = q = .1$ and the benefit paid, B, has a continuous distribution with mean \$1,000 and standard deviation \$200. This type of policy was discussed in Examples 1.14 and 1.22, where for the ith policy the claim variable is $X_i = I_i B_i$ and the aggregate or sum of the claim variables is S_m, for $m = 25$. Using (1.34) and (1.36), we first find the mean and variance of S_m. For the ith policy the mean and variance are

$$E\{X_i\} = q\, E\{B\} = .1\ (\$1,000) = \$100 \qquad (2.60)$$

and

$$\text{Var}\{X_i\} = q(1 - q)(E(B_i))^2 + q\text{Var}\{B_i\} = .1(.9)(1,000^2) + .1(200^2) = 94,000 \quad (2.61)$$

For the aggregate claim, assuming the individual insurance policies lead to iid claim random variables, using (2.60) and (2.61), the mean and variance are computed as

$$E\{S_m\} = m\, E\{X_i\} = 25(100) = 2,500$$

and

$$\text{Var}\{S_m\} = m\, \text{Var}\{X_i\} = 25(94,000) = 2,350,000$$

The 95th percentile for S_m, in terms of dollars, is approximated using the CLT as

$$\$2,500 + 1.645(2,350,000)^{1/2} = \$5,021.73$$

Hence, the probability of the aggregate claim exceeding \$5,021.73 is about .05. Further, using the CLT approximate, cumulative probabilities, survival probabilities, and prediction intervals can be constructed in a straightforward manner.

Example 2.14

In this example the parameter of interest is the population proportion p. To estimate p we take a random sample of size n and observe the binomial random variable X introduced in Example 1.8 with pmf (1.10). The sample proportion is

$$\hat{p} = \frac{X}{n} \quad \text{where} \quad E\{\hat{p}\} = p \quad \text{and} \quad \text{Var}\{\hat{p}\} = \frac{p(1-p)}{n} \tag{2.62}$$

The central limit theorem holds, where for large sample size n,

$$Z = \frac{\hat{p} - p}{\left(\dfrac{p(1-p)}{n}\right)^{1/2}} \sim N(0,1) \tag{2.63}$$

The random variable (2.63) can be used in the construction of confidence sets and prediction intervals where the pivot is taken to be

$$W(\hat{p}, p) = Z^2 \sim \chi^2_{1.} \tag{2.64}$$

Applying (2.19) based on statistic (2.64), a $(1 - \alpha)100\%$ confidence set for p results in the inequality

$$n(\hat{p} - p)^2 < a\,p(1-p) \tag{2.65}$$

where $\chi^2_{1,1-\alpha} = a$. Solving (2.65) for p gives the confidence interval

$$LB(\mathrm{p}) \le p \le UB(\mathrm{p}) \tag{2.66}$$

where

$$LB(p) = \frac{\hat{p} + \dfrac{a}{2n} - a^{1/2}\left(\dfrac{a}{4n^2} + \dfrac{\hat{p}(1-\hat{p})}{n}\right)^{1/2}}{1 + \dfrac{a}{n}} \tag{2.67}$$

and

$$UB(p) = \frac{\hat{p} + \dfrac{a}{2n} + a^{1/2}\left(\dfrac{a}{4n^2} + \dfrac{\hat{p}(1-\hat{p})}{n}\right)^{1/2}}{1 + \dfrac{a}{n}} \tag{2.68}$$

For large n, assuming a/n terms are negligible, we have the standard confidence interval

$$\hat{p} - z_{1-\alpha}\left(\frac{\hat{p}(1-\hat{p})}{n}\right)^{1/2} \leq p \leq \hat{p} + z_{1-\alpha}\left(\frac{\hat{p}(1-\hat{p})}{n}\right)^{1/2} \tag{2.69}$$

We remark that (2.69) is the typical large sample confidence interval for proportions that is discussed in most standard statistical texts.

2.4.2 Haldane Type A Approximation

Methods other than the CLT have been used to approximate the tail probabilities for aggregate sums of independent random variables. In this section, we discuss the Haldane type A approximation (HAA) for the cdf corresponding to aggregate sums as described by Pentikainen (1987). This method extends the Levy central limit theorem using information from the skewness measure associated with the random variables defined in (1.26).

The aggregate, S_m, is the sum of iid random variables, $X_1, X_2, ..., X_m$, where m is a fixed constant, whose mean and variance are denoted by μ and σ^2. Applying the independence property of the random variables and (2.42), the mean, variance, and central third moment of S_m are given by

$$\mu_s = m\,\mu, \quad \sigma_s^2 = m\sigma^2 \quad \text{and} \quad E\{(S_m - \mu_s)^3\} = m\,E\{(X_i-\mu)^3\} \tag{2.70}$$

The skewness of the aggregate defined by (1.26), utilizing (2.70), is found to be $\mathrm{Sk} = E\{(X_i - \mu)^3\}/(m^{1/2}\sigma^3)$. For a fixed value of s in (2.59), let

$$z = \frac{s - m\mu}{m^{1/2}\sigma}$$

and define $r = \frac{\sigma}{m^{1/2}\mu}$, $h = 1 - \frac{Sk}{3r}$, $\mu(h,r) = 1 - h(1-h)[1 - (2-h)(1-3h)\frac{2}{4}]r^2$, and $\sigma(h,r) = hr[1 - (1-h)(1-3h)\frac{2}{2}]^{1/2}$. The HAA approximation to the cdf of the aggregate sum is

$$P(S_m \leq s) = \Phi\left(\frac{(1+rz)^h - \mu(h,r)}{\sigma(h,r)}\right) \tag{2.71}$$

where $\Phi(a)$ is the standard normal cdf. In many financial and actuarial modeling situations, the aggregate consists of nonnegative random variables,

such as claim values and future lifetimes of stock holding, and approximation (2.71) applies with $h > 0$, $\mu(h, r) > 0$, and $\sigma(h, r) > 0$.

Example 2.15

Consider the portfolio of 25 insurance policies with associated claim variables as given in Example 2.13, where the proportion of claims is $q = .1$
and the claim benefit variable B has an approximate normal distribution,
specifically $B \sim N(\mu_B = \$1,000,\ \sigma_B^2 = \$40,000)$. We wish to approximate the
probability the aggregate exceeds $\$5,000$. Using the CLT with approxi-
mate standard normal random variable (2.59),

$$P(S_{25} > 5000) = 1 - \Phi\left(\frac{5,000 - 2,500}{1532.97}\right) = 1 - \Phi(1.63082) = .05146$$

To apply HAA, we first find a form for the third central moment in this
situation. Noting the claim variable is $X = I\,B$, the central third moment,
similar to the derivation of the variance in (1.36), is

$$E\{(X - \mu)^3\} = q\,E\{(B - \mu_B)^3\} + 3q(1 - q)\text{Var}\{B\}\mu_B + q(1 - 2q)(1 - q)\,\mu_B^3 \qquad (2.72)$$

The normality of B and the computations of Example 2.13 imply $E\{(X - \mu_x)^3\} = 82,800,000$ and $Sk = .574604$. Further, $r = .613188$, $h = .6876413$, $\mu(h, r) = .9392614$, and $\sigma(h, r) = .4082815$, and the HAA approximation (2.71)
is used to compute

$$P(S_{25} > 5000) = 1 - \Phi\left(\frac{(1 + .6132(1.6308))^{.6876} - .9392614}{.4083}\right) = .0500$$

We note the CLT and HAA computations are quite close. The exact probability can be obtained using simulation methods as discussed in Section 9.2.
Using simulation, the probability is found to be .06616. Hence, both approximations have a large relative error of about $(.06616 - .05)/.06616 = .24425$.

2.4.3 Saddlepoint Approximation

In the previous two sections, approximations for the cdf of sums of iid random variables depend only on theoretical moments. Here also are approximation methods that utilize information drawn from the form of the mgf of
the individual random variables. Since their introduction by Daniels (1954),
saddlepoint approximations (SPAs) have been used to approximate tail probabilities corresponding to sums of independent random variable. Field and
Ronchetti (1990) give an in-depth discussion of the accuracy of saddlepoint
approximations, where the approximation is shown to be accurate for sample

sizes as small as one. Further, in the case of data from a normal distribution, the SPA reduces to the CLT approximation. For this reason, the SPA can be viewed as an extension to the CLT in the case of small sample sizes. Saddlepoint approximations have been applied to a variety of situations, examples of which are discussed in the articles by Goutis and Casella (1999), Huzurbazar (1999), Butler and Sutton (1998), Tsuchiya and Konishi (1997), and Wood et al. (1993).

In the simplest setting, there are independent identically distributed random variables, $X_1, \ldots, X_m$, where m is fixed and the aggregate or sum is denoted S_m. Unlike other saddlepoint approximation developments that utilize the cumulants of hypothesized distributions, this discussion is based on the associated moments. The moment generating function of X_i is assumed to exist and is denoted by $M_X(\beta)$, where $E\{X_i\} = \mu$ and $\text{Var}\{X_i\} = \sigma^2$. The corresponding moment generating function for $Z = (X - \mu)/\sigma$ is

$$M_Z(\beta) = M_X\left(\frac{\beta}{\sigma}\right)\exp(-\beta\mu/\sigma) \tag{2.73}$$

For a fixed value of t, let β satisfy

$$M_Z^{(1)}(\beta) = t\, M_Z(\beta) \tag{2.74}$$

To solve for β in (2.74), a numerical algorithm, such as the Solver routine in Excel, may be required. Further, let

$$c = \exp(\beta t)/M_Z(\beta) \quad \text{and} \quad \alpha^2 = \frac{M_Z^{(2)}(\beta)}{M_Z(\beta)} - t^2 \tag{2.75}$$

For constant s, the SPA for the cdf of the aggregate, as given in Field and Ronchetti (1990, p. 37), takes the form

$$P(S_m \le s) = \Phi(\text{sgn}(t)(2m\ln(c))^{1/2}) - c^{-m}\frac{\dfrac{1}{\beta\alpha} - \dfrac{\text{sgn}(t)}{(2\ln(c))^{1/2}}}{(2m\pi)^{1/2}} \tag{2.76}$$

where $t = (s - m\mu)/(m\sigma)$ and $\text{sgn}(t)$ takes the value $+1$ if $t > 0$, -1 if $t < 0$, and 0 if $t = 0$. Approximate percentiles for S_m, leading to prediction intervals, can be constructed by inverting (2.76) for a fixed probability.

In an application, for a chosen s and resulting t the value β, defined by (2.74), is computed using a numerical method, such as Newton's Method or the secant method (see Stewart (1995, p. 170) or Burden and Faires (1997, Chapter 2). The Solver program in Excel can be utilized for this computation (see Problem 2.11).

Example 2.16

In this example the SPA is demonstrated and the results are compared
with the CLT and HAA approximations. The individual iid random variables are exponential with pdf given by $f(x) = (1/\theta) \exp(-x/\theta)$, for support
$S = (0, \infty)$, so that the mean is $\mu = \theta$, the variance $\sigma^2 = \theta^2$, and the mgf is
$M_i(\beta) = (1 - \beta\theta)^{-1}$. For a fixed value of s, we find $t = (s - m\theta)/(m\theta)$, and solving (2.74) for β, we compute (2.75). Applying the SPA to the exponential
distribution, we find the required constants are

$$\beta = \frac{t}{1+t}, \quad c = (1-\beta)\exp[\beta(1+t)] \quad \text{and} \quad \alpha^2 = \frac{2\beta}{(1-\beta)^2} + 1 - t^2 \quad (2.77)$$

In this case, the distribution of the aggregate is known to be gamma (see
Problem 2.3c). For different sample sizes, the exact percentile points computed using the gamma distribution with parameters $\alpha = m$ and $\theta = 1$ are
found. The cdf probabilities associated with these percentile points using
the CLT, HAA, and SPA are given in Table 2.2 for sample sizes of $m = 1$
and $m = 2$.

From Table 2.2, we see that the HAA and SPA outperform the CLT. This
is to be expected, since these approximations use more information, specifically, information about the skewness of the random variables. The CLT is
most efficient in the case of symmetric random variables. The SPA is the
most efficient method with respect to skewed distributions, efficient percentile approximations for the exponential distribution even for sample sizes of
one and two.

In general, the SPA requires computation of (2.73), (2.74), and (2.75), which
requires the form of the mgf. In some settings the mgf requirement may be
partly relaxed. This is the case in Section 2.5.2, where a saddlepoint approximation for the cdf of Poisson compound sums is constructed based on specified moments. The SPA has applications in financial and actuarial modeling
and has been extended to the case of life table data with uniform distributions within each year by Borowiak (2001).

TABLE 2.2

CLT, HAA, and SPA Percentile Approximations

			Sample Size					
		$m = 1$				$m = 2$		
Percentile	.99	.95	.90	.75	.99	.95	.90	.75
CLT	.9998	.9770	.9036	.6504	.9995	.9738	.9092	.6879
HAA	.9900	.9513	.9023	.7512	.9899	.9506	.9012	.7511
SPA	.9900	.9498	.8997	.7502	.9900	.9499	.8998	.7499

2.5 Compound Aggregate Variables

Generally speaking, a compound random variable is a random variable whose stochastic structure consists of multiple random pieces. In this context, we explore the distribution of the sum of iid random variables where the number of random variables is itself a random variable. Statistical properties, techniques, and applications associated with these compound random variables are presented using the concepts and formulas of previous sections. Applications of these types of compound random variables in the actuarial and financial fields include investment portfolio analysis and collective risk modeling (see Bowers et al., 1997, Chapter 12).

2.5.1 Expectations of Compound Aggregate Variables

Modifying the structure of the aggregate variable of Section 2.4, the random variables $X_1, X_2, \ldots, X_N$ are assumed to be iid where the number of random variables, N, is a discrete random variable. Let $E\{X_i\} = \mu_1$, $E\{X_i^2\} = \mu_2$, and $\text{Var}\{X_i\} = \sigma^2$. The random variable of interest is the compound aggregate

$$S_N = \sum_{i=1}^{N} X_i \tag{2.78}$$

where the pdf of N is given by $P(N = n)$, which has support S_N. The mean and variance of (2.78) can be found using statistical conditioning arguments. Assume $X_1, \ldots, X_N$ and N are independent and the joint mpf is given by $f(n, s_N) = P(N = n) f(s_n \mid N = n)$, where $f(s_n \mid N = n)$ is the pdf of S_n. Applying this joint mpf, the expectation of the aggregate is

$$E\{S_N\} = \sum_{S_N} P(N = n) E\{S_n \mid N = n\} = \mu_1 E\{N\} \tag{2.79}$$

Further, conditioning on N, the variance is found to be

$$\text{Var}\{S_N\} = \mu_1^2 \, \text{Var}\{N\} + \sigma^2 \, E\{N\} \tag{2.80}$$

These formulas are used to construct statistical inference such as confidence and prediction intervals. The derivations of (2.79) and (2.80) are considered in Problem 2.8.

The mgf corresponding to the compound aggregate S_N also can be found using a conditioning argument. Letting the mgf of X_i be $M(t)$ for $i = 1, \ldots, N$, conditioning on N, the mgf of (2.78) is

$$M_{S_N}(t) = E_N\{E\{\exp(tS_N) \mid N\}\} = \sum_{S_N} P(N = n)(M(t))^n \tag{2.81}$$

Following the techniques of Section 1.4, the mean (2.79) and the variance (2.80) can be found from the mgf (2.81) by taking the usual derivatives (see Problem 2.8). Two illustrative examples follow where the second describes the celebrated compound Poisson process.

Example 2.17

Let N have the geometric distribution with the form of the pmf given in Example 1.19, where $P(N = n) = pq^n$ for $n = 0, 1, ...$ Applying the summation formula (1.31) to (2.81) the mgf is found to be

$$M_{S_N}(t) = \sum_{n=0}^{\infty} p(qM(t))^n = \frac{p}{1 - qM(t)} \qquad (2.82)$$

Taking the derivative of (2.82) and following (1.40) produces the mean $E\{S_N\} = pq\mu_1/(1 - q)^2$. The random variable associated with the mgf of (2.82) is referred to as a compound geometric random variable.

Example 2.18

A group of insurance policies produces N claims where N is modeled by a Poisson random variable introduced in Example 1.9 with parameter λ. The distribution of S_N is said to be a compound Poisson random variable. Since $E\{N\} = \text{Var}\{N\} = \lambda$, utilizing (2.79) and (2.80) we find the mean and variance of the aggregate are

$$E\{S_N\} = \mu_1 \lambda \quad \text{and} \quad \text{Var}\{S_N\} = \mu_2 \lambda \qquad (2.83)$$

Following (2.81) (see Problem 1.4) the mgf of S_N is

$$M_{S_N}(t) = \exp(-\lambda) \sum_{n=0}^{\infty} \frac{(\lambda M(t))^n}{n!} = \exp(\lambda(M(t) - 1)) \qquad (2.84)$$

This mgf is used in the next section to validate and construct limiting distributions for the compound Poisson random variable. These distributions are employed in actuarial collective risk models, and we refer the reader to Bowers et al. (1997, Chapter 12).

2.5.2 Limiting Distributions for Compound Aggregate Variables

Limiting distributions exist for compound distributions, such as the compound Poisson and compound negative binomial, and we refer the reader to Bowers et al. (1997, Sections 12.5 and 12.6). In this section, we discuss limiting distribution approximations for the compound Poisson distribution. The first directly utilizes the CLT, while the subsequent two approximations are variants of the saddlepoint approximation. Other approximation approaches exist, such as the discretizing method given by Panjer (1981).

The aggregate S_N is composed of iid random variables and a stochastic number N that follows a Poisson distribution. Let the first two moments be denoted $E\{X_i\} = \mu_i$ for $i = 1, 2$, the moments of S_N follow (2.83), and the standardized variable be

$$Z_N = \frac{S_N - \lambda\mu_1}{(\lambda\mu_2)^{1/2}} \qquad (2.85)$$

Utilizing (2.84), the mgf of Z_N can be written as

$$M_Z(t) = E\{\exp(t\, Z_N)\} = \exp(-t\, \lambda\mu_1/(\lambda\mu_2)^{1/2})\; M_{S_N}\, (t/(\lambda\mu_2)^{1/2}) \qquad (2.86)$$

This development assumes the mgf of X_i, denoted $M(t)$, exists, and using a Taylor series expansion (see Problem 1.4) we have the approximation

$$M(t) = 1 + \frac{\mu_1 t}{1!} + \frac{\mu_2 t^2}{2!} + \frac{\mu_3 t^3}{3!} + \dots \qquad (2.87)$$

Putting (2.84) and (2.87) into (2.86) yields the mgf of (2.85),

$$M_Z(t) = \exp(t^2/2 + o(\lambda)) \qquad (2.88)$$

where $o(\lambda)$ are terms that approach zero as λ approaches infinity. Hence, as λ approaches infinity, $M_Z(t)$ approaches the mgf of the standard normal distribution $\exp(t^2/2)$. By the continuity theorem, as discussed in Hogg et al. (2005, p. 216), Z_N, defined by (2.85), converges in distribution to a standard normal random variable as λ approaches infinity. This is demonstrated through Example 2.19.

Example 2.19

Let the amounts of accident claims, X_i, be iid with mean $\mu = 100$ and variance $\sigma^2 = 100$ and the number of claims, N, be Poisson with mean $\lambda = 50$. For the aggregate claim variable S_N, from (2.83), $E\{S_N\} = 5,000$ and $Var\{S_N\} = 1,000,000$. The approximate probability the sum of the claims is less than 7,000 utilizing (2.85) and the CLT is

$$P(Z \le (7,000 - 5,000)/1,000) = \Phi(2) = .9772$$

Further, applying (2.16), a 95% prediction interval, using $z_{.975} = 1.96$, for S_N is

$$5000 - 1.96(1,000) = 3,040 \le S_N \le 5,000 + 1.96(1,000) = 6,960$$

Prediction limits for the aggregate claim S_N are \$3,040 on the low side and \$6,960 on the high side, and their accuracy depends on the appropriateness of the CLT.

The saddlepoint approximation approach can also be applied to the compound Poisson distribution. If the distribution of the iid individual random variable X_i is specified, the SPA technique can be directly applied. Utilizing (2.84) the mgf of (2.85) is

$$M_Z(\beta) = \exp(-\beta \, \lambda^{1/2} \, \mu_1/\mu_2^{1/2} + \lambda \, (M(\beta/((\lambda\mu_2)^{1/2}) - 1)) \qquad (2.89)$$

where the $M(\alpha)$ denotes the mgf corresponding to X_i for $i \geq 1$. For a chosen distribution for X_i with mgf $M(\alpha)$, the SPA can be computed. Following the SPA as outlined in Section 2.4.3, for any $t = (s - \lambda\mu_1)/(\lambda\mu_2)^{1/2}$, the cdf approximation of S_N, modifying (2.76), is

$$P(S_n \leq s) = \Phi(\mathrm{sgn}(t)(2\ln(c))^{1/2}) - \frac{1}{c(2\Pi)^{1/2}} \left[\frac{1}{\beta\sigma} - \frac{\mathrm{sgn}(t)}{(2\ln(c))^{1/2}} \right] \qquad (2.90)$$

An application of the SPA for aggregate sums using Excel is outlined in Problem 2.11.

In applications, the individual distributions, and hence the mgf $M_x(\alpha)$, may not be known. In this case, we introduce a moment-based saddlepoint approximation approach. A three-moment approximation is obtained by first letting

$$H_Z(\beta) = \ln[M_Z(\beta)] = -\beta \frac{\mu_S}{\sigma_S} + \lambda M\left(\frac{\beta}{\sigma_S}\right) - 1 \qquad (2.91)$$

Function (2.91) is now approximated using a truncated Taylor series. Computing $H_Z(0) = H_Z^{(1)}(0) = 0$, $H_Z^{(2)}(0) = 1$, and $H_Z^{(3)}(0) = \mu_3 \beta^3/(6 \, \mu_2^{3/2} \, \lambda^{1/2})$, a three-moment approximation is

$$H_Z(\beta) = \frac{\beta^2}{2} + \frac{\mu_3\beta^3}{6\mu_2^{3/2}\lambda^{1/2}} \qquad (2.92)$$

Taking the natural log of (2.92) produces the approximation to (2.89):

$$M_Z(\beta) = \exp(\beta^2/2 + \mu_3 \, \beta^3/(6 \, \lambda^{1/2} \, \mu_2^{3/2})) \qquad (2.93)$$

To construct a three-moment saddlepoint approximation for the cdf of the aggregate, the first two derivatives of (2.93) are required and, utilizing the compound Poisson distribution, are found to be

$$M_Z^{(1)}(\beta) = M_Z(\beta)\left(\beta + \frac{\mu_3\beta^2}{2\mu_2^{3/2}\lambda^{1/2}}\right) \tag{2.94}$$

and

$$M_Z^{(2)}(\beta) = M_Z^{(1)}(\beta)M_Z^{(1)}(\beta)/M_Z(\beta) + M_Z(\beta)\left[1 + \frac{\mu_3\beta}{\mu_2^{3/2}\lambda^{1/2}}\right] \tag{2.95}$$

Following the saddlepoint approximation approach in Section 2.4.3, for $t = (s - \lambda\mu_1)/(\lambda^{1/2}\mu_2^{1/2})$ the β value solves relation (2.74), which simplifies to

$$t = \beta + \frac{\mu_3\beta^2}{2\mu_2^{3/2}\lambda^{1/2}} \tag{2.96}$$

The resulting positive value is

$$\beta = \left(\frac{\lambda^{1/2}\mu_2^{3/2}}{\mu_3}\right)\left[\left(1 + \frac{2t\mu_3}{\lambda^{1/2}\mu_2^{3/2}}\right)^{1/2} - 1\right] \tag{2.97}$$

The three-moment saddlepoint approximation utilizing (2.92) to (2.97) is denoted by MSPA. This approach used to model the cdf is outlined in Problems 2.12, 2.13, and 2.14. In the following example both the SPA and MSPA methods are demonstrated.

Example 2.20

In this example, we consider the compound Poisson distribution with parameter λ, where X_i is distributed as an exponential random variable, with pdf given in (1.16), and parameter $\theta > 0$. Thus, $\mu_1 = \theta$, $\mu_2 = 2\theta^2$, and $\mu_3 = 6\theta^3$, so that, from (2.83), $\mu_s = \lambda\theta$ and $\sigma_s = (2\lambda)^{1/2}\theta$. For a demonstration, let $\lambda = 10$ and $\theta = 1$; the probability the aggregate sum is at most 15 is explored. The exact probability can be found by conditioning on $N = n$ and using the fact that, for a fixed n, S_n is distributed as a gamma random variable with parameters $\alpha = n$ and $\beta = 1$ (see Problem 2.3c). The exact probability is computed as

$$P(S_N \leq 15) = \sum_{n=0}^{\infty} P(N = n)P(S_n \leq 15) = .86584 \tag{2.98}$$

TABLE 2.3

Compound Poisson Approximations

Approximation	$P(S_N \leq 15)$	β	$M_z(\beta)$	$M_z^{(1)}(\beta)$	$M_z^{(2)}(\beta)$
SPA	.86566	.82065	1.51046	1.68875	4.66296
MSPA	.86093	.86631	1.56508	1.74981	4.43096

The CLT and saddlepoint approximation methods of this section are now applied. Here $s = 15$ and, from (2.85), $t = (s - \lambda\theta)/((2\lambda)^{1/2}\theta) = (15 - 10)/(20^{1/2}) = 1.118034$, so the CLT approximation gives $\Phi(1.11803) = .86822$. The SPA uses the mgf of the individual variables $M_x(\beta/\sigma_s) = (1 - \beta/(2\lambda)^{1/2})^{-1}$, where (2.89) becomes

$$M_Z(\beta) = \exp\left[-\beta(\lambda/2)^{1/2} + \lambda\left(1 - \frac{\beta}{(2\lambda)^{1/2}} \right)^{-1} - 1 \right) \right] \qquad (2.99)$$

The derivatives of (2.99) are computed as

$$M_Z^{(1)}(\beta) = M_Z(\beta)\exp(-(\lambda/2)^{1/2} + (\lambda/2)^{1/2}(1 - \beta/(2\lambda)^{1/2})^{-2}) \qquad (2.100)$$

and

$$M_Z^{(2)}(\beta) = M_Z(\beta)\exp([(\lambda/2)^{1/2}((1 - \beta/(2\lambda)^{1/2})^{-2} - 1)]^2 + (1 - \beta/(2\lambda)^{1/2})^{-3}) \qquad (2.101)$$

For a fixed value of t using (2.99) and (2.100), the value of β solving (2.74) takes the form

$$\beta = (2\lambda)^{1/2}\left[1 - \left(\frac{\lambda}{2}\right)^{1/4}\left(t + \left(\frac{\lambda}{2}\right)^{1/2} \right)^{-1/2} \right] \qquad (2.102)$$

Computing yields $\beta = .82065$, $c = 1.657155$, $\sigma = 1.355403$, and $(2 \ln(c))^{1/2} = 1.005089$. The SPA-required computations (2.99) to (2.102), along with the MSPA approximation to (2.98), are listed in Table 2.3.

From Table 2.3 we observe that all three approximations give efficient estimators of the true probability. This is due to the quick convergence of this standardized compound Poisson random variable with parameters $\lambda = 10$ and $\theta = 1$.

2.6 Regression Modeling

One of the most widely utilized statistical techniques is that of linear regression. Linear regression applied to a collection of variables can demonstrate relationships among the variables and model predictive structures. In this section, we present an introduction to simple linear regression with applications in the modeling and analysis of financial systems. This exposition is not meant to be a comprehensive discussion but to give the flavor of the interaction of regression and financial estimation. For an introduction to the theory and application of linear regression modeling we refer to Myers (1990) and Draper and Smith (1998).

In simple linear regression modeling there are two variables of interest. The independent or predictor variable, denoted by X, impacts upon the dependent or response variable, denoted Y. The response is considered a random variable and the observed data take the form of ordered pairs (x_j, y_j), for $j = 1, 2, \ldots, n$. A linear relationship between the variables is assumed, and the simple linear regression model is

$$y_j = \beta_0 + \beta_1 x_j + e_j \quad \text{for} \quad 1 \leq j \leq n \tag{2.103}$$

where β_0 and β_1 are intercept and slope parameters and e_j is the error term. For estimation purposes, the error e_j is assumed to be iid from a continuous distribution that has mean zero and constant variance for $j = 1, 2, \ldots, n$. Classical statistical inference procedures, such as hypothesis testing of parameters and confidence intervals for parameters, require the additional assumption that the errors are normally distributed with constant variance.

For observed data sets, such as stock prices or insurance claims over time, the linear regression model is applied and statistical inference procedures, such as response prediction and prediction intervals, yield practical insights. Model diagnostics used to check the modeling assumptions verify the applicability of the regression analysis. For example, a plot of the data points referred to as a scatter plot can be utilized to demonstrate the linearity assumption (demonstrated in Example 2.21). The normality assumption can be investigated through the technique of probability plotting (discussed in Section 2.8.1).

Model (2.103), containing one predictor variable, is referred to as a simple linear regression model. In general, other predictor variables may be added to the model, resulting in a multiple linear regression model. In the case of multiple linear regression, topics of variable selection, influence, colinearity, and testing become important. In this section we apply only simple linear regression models and standard regression inference techniques to financial and actuarial data.

2.6.1 Least Squares Estimation

To estimate the parameters β_0 and β_1, the mathematical method of least squares has desirable statistical properties and is commonly employed. Under general regularity conditions, Jennrich (1969) has shown that general least squares estimators are asymptotically normal as the sample size increases. In this method, the sum of squares error is used as a measure for fit of the estimated line to the data and is given by

$$Q = \sum_{j=1}^{n} e_j^2 = \sum_{j=1}^{n} (y_j - \beta_0 - \beta_1 x_j)^2 \qquad (2.104)$$

Taking the partial derivatives of (2.104), the least squares estimators are found and they take the form

$$\hat{\beta}_1 = \frac{S_{xy}}{S_{xx}} \quad \text{and} \quad \hat{\beta}_0 = \bar{y} - \hat{\beta}_1 \bar{x} \qquad (2.105)$$

where $\bar{y} = \sum \frac{y_j}{n}$, $\bar{x} = \sum \frac{x_j}{n}$, $S_{xy} = \sum(y_j - \bar{y})(x_j - \bar{x})$, and $S_{xx} = \sum(x_j - \bar{x})^2$.

Applying the least squares estimators (2.105) to the linear relationship, the estimated or fitted regression line is

$$\hat{y} = \hat{\beta}_0 + \hat{\beta}_1 x \qquad (2.106)$$

One of the primary applications of (2.106) is the prediction of y at future values of x. The accuracy of such a prediction depends on the accuracy of the model, through modeling shape, and efficiency of estimation.

One of the oldest measures of the linear fit of the estimated regression model is the correlation coefficient denoted ρ. This parameter arises where both X and Y are random variables and the random vector (X, Y) comes from the bivariate normal distribution (see Hogg et al., 2005, p. 174). The parameter ρ measures the linear association between X and Y. Theoretically, $-1 \le \rho \le 1$, with proximity to -1 or $+1$ indicating a linear relationship. It is a mathematical fact that $\rho^2 = 1$ if and only if the probability Y is a linear function of X with probability one (see Rohatgi, 1976, p. 175). The sample estimate of this parameter is called the sample correlation coefficient and is computed as

$$r = \frac{S_{xy}}{(S_{xx} S_{yy})^{1/2}} \qquad (2.107)$$

where $S_{yy} = \sum y_j^2 - (\sum y_j)^2/n$. Here, r measures the linear relationship between X and Y, where $-1 \le r \le 1$, and the closer $|r|$ is to 1, the closer the points lie to a straight line. Another useful diagnostic statistic connected to the correlation coefficient is the coefficient of determination defined by r^2. The statistic (2.107), used to assess the accuracy of the estimated regression line (2.106), can be computed using Excel, as in Problem 2.15. An example of applying linear regression to modeling stock prices follows:

TABLE 2.4

Stock Price over 1 Year

Month	1	2	3	4	5	6	7	8	9	10	11	12
Price	10.5	11.31	12.75	12.63	12.17	12.56	12.17	12.56	14.69	15.13	12.75	13.44

Example 2.21

The price of a stock is recorded at the start of each month for a year and is listed in Table 2.4. The goal is the prediction of the stock price over time where we apply the regression of X, the month, on Y, the stock price. The least squares (2.105) are found to be

$$\hat{\beta}_0 = 11.0617 \quad \text{and} \quad \hat{\beta}_1 = .255$$

and the estimated regression line (2.106) is

$$\hat{y} = 11.0617 + .255\,x$$

A scatter plot of the data listed in Table 2.4 is given in Figure 2.1, along with the estimated least squares line. We see the data are increasing over time, but the linear fit of the data is somewhat questionable. The estimated slope, given by .255, indicates the price of the stock is increasing over time. Further, the sample correlation coefficient is found to be $r = .724$, indicating a linear relationship. Based on this regression model, we would expect the price of the stock after 6 months to be $11.0617 + 6(.255) = 12.59$. Analysis questions about the efficiencies of these types of point estimates arise, and we now consider statistical inference topics associated with regression models.

There are many statistical programs that deal with linear regression analysis. For an observed data set, Excel can be used to draw a scatter plot, compute correlations, and run standard linear regression analysis. An example of using Excel in this context is outlined in Problem 2.15.

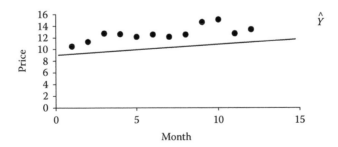

FIGURE 2.1
Regression of stock prices.

2.6.2 Regression Model-Based Inference

Desirable properties of the least squares estimators, such as unbiasedness and minimum variance, are well known and do not rely on the form of the distribution of the error term. For a review of theoretical topics in linear models, we refer to Searle (1971) and Myers and Milton (1998). For inference techniques, we require that the error terms are normally distributed with zero mean and common standard deviation σ, i.e., in (2.103) we assume $e_i \sim$ iid $N(0,\sigma^2)$, for $i = 1, 2, \ldots, n$. Under this assumption, inference procedures rely on the resulting normality of the least squares estimators. Further, when normality of the errors is assumed, the least squares estimators coincide with the maximum likelihood estimators.

Normality of the error term in (2.103) implies that many statistics used in statistical analysis also have normal distributions. In particular, the least squares estimator of the slope and the predicted response associated with fixed $X = x_0$ are both normal random variables, where

$$\hat{\beta}_1 \sim N\left(\beta_1, \frac{\sigma^2}{S_{XX}}\right) \quad \text{and} \quad \hat{y} \sim N\left(\beta_0 + x_0\beta_1, \sigma^2\left(\frac{1}{n} + \frac{(x_0 - x)^2}{S_{xx}}\right)\right) \quad (2.108)$$

An estimator of σ^2, based on the fitted model (2.106) and independent of the random variables in (2.108), is

$$S^2 = \frac{\sum_{i=1}^n (y_i - \hat{y}_i)^2}{n-2}$$

where we denote $S = (S^2)^{1/2}$ and associated degrees of freedom, since two parameters are estimated, is $n - 2$. Statistical distribution theory can be used to construct confidence intervals by constructing pivots from (2.108) in the form of a t random variable. A $(1 - \alpha)100\%$ confidence interval for β is

$$\hat{\beta} - t_{(1-\alpha/2, n-2)} \frac{S}{(S_{xx})^{1/2}} \le \beta \le \hat{\beta} + t_{(1-\alpha/2, n-2)} \frac{S}{(S_{xx})^{1/2}} \quad (2.109)$$

where a $(1 - \alpha)100$th percentile point for a t random variable with degrees of freedom $n - 2$ is denoted by $t_{(1-\alpha, n-2)}$. In a similar manner, a prediction interval for the expected value of Y at $X = x_0$, defined by $y_0 = \beta_0 + \beta_1 x_0$, takes the form

$$\hat{y}_0 - t_{(1-\alpha/2, n-2)} S\left(\frac{1}{n} + \frac{(x_0 - \bar{x})^2}{S_{xx}}\right)^{1/2} \le y_0 \le \hat{y}_0 + t_{(1-\alpha/2, n-2)} S\left(\frac{1}{n} + \frac{(x_0 - \bar{x})^2}{S_{xx}}\right)^{1/2} \quad (2.110)$$

Other confidence intervals, such as for a corresponding response associated with a value of x, can be constructed.

The width of the confidence intervals can be used to judge the efficiency of the estimation where more accuracy is signified by shorter confidence intervals. Different factors can adversely affect the width of the interval. The choice confidence coefficient and sample size has an effect, with larger samples resulting in better accuracy. Further, utilizing (2.110), the farther x_0 is from the dependent variable's mean, the wider the interval.

Example 2.22

Consider the stock prices and regression model in Example 2.21, where $s^2 = .846332$ and $s_{xx} = 143$. The 95% confidence interval (2.109) is computed as $.083971 \leq \beta \leq .42679$. This indicates the price for the stock is increasing over time, but the slope of the increase is not certain, as seen by the large width of the confidence interval for β. Further, the estimated price in 6 months or at future time $x = 13$ using (2.106) is 12.59. Computing a 95% confidence interval for the estimated price from (2.110) yields $10.15 \leq y \leq 15.02$. Again, a large-width confidence interval implies that the estimated stock price after 6 months is uncertain.

2.7 Autoregressive Systems

In the practice of modeling financial and actuarial systems, dependent random variables play an important role. In applications, we may observe a series of random variables $X_1, X_2, \ldots$, where the individual random variables are dependent. Variables such as interest and financial return rates are often modeled using dependent variable systems. There are dependent variable techniques, such as time series, moving average, and auto regressive modeling. In this section, an introduction to dependent modeling is presented where one such modeling procedure, namely, autoregressive modeling of order 1, is discussed. For a discussion of dependent random variable models we refer to Box and Jenkins (1976).

In a general autoregressive system of order j, with observed $x_1, x_2, \ldots$, the relation between variables is defined by

$$x_t = \mu_j + \sum_{i=1}^{j} \phi_i (x_{t-i} - \mu_{t-i}) + e_t \tag{2.111}$$

where μ_j and ϕ_j are constants and the error terms e_j are independent random variables for $j \geq 1$. It is clear from (2.111) that the random variables x_j are dependent. The order of the autoregressive system designates the dependent structure of the data, and our discussion focuses on an autoregressive system of order 1, taking $j \geq 2$.

An autoregressive process or system of order 1, denoted $AR(1)$, follows (2.111), and taking $j \geq 1$ can be written as

$$x_1 = \mu_1 + e_1 \quad \text{and} \quad x_j = \mu_j + \phi(x_{j-1} - \mu_{j-1}) + e_j \quad \text{for} \quad j \geq 2 \qquad (2.112)$$

where the error terms e_j are independent and from the same distribution, with zero mean and variance $\text{Var}\{e_j\} = \sigma^2$. From (2.112), we solve iteratively to find

$$x_j = \mu_j + \sum_{i=1}^{j} \phi^{j-i} e_i \qquad (2.113)$$

From this, the mean and variance of X_i are given by

$$E\{X_i\} = \mu_i \quad \text{and} \quad \text{Var}\{X_i\} = \sigma^2 (1 - \phi^{2j})/(1 - \phi^2) \qquad (2.114)$$

The moments given in (2.114) are functions of unknown parameters that need to be estimated before these formulas can be utilized. After estimation, approximate inference techniques, such as point estimation and confidence intervals, can be applied.

Alternative conditions to (2.112) on the observed sequence of random variables lead to moving average-dependent variable models. In moving average modeling conditions relating the observed variables, $X_j - \mu_j$, and error terms are imposed for $j \geq 1$. A mixture of autoregressive and moving average models leads to ARMA and integrated ARIMA models. For an introduction and exposition to these techniques see Box and Jenkins (1976, Chapter 3).

In an $AR(1)$ system we estimate the parameters ϕ and σ^2 using the method of least squares. Let $Z_i = X_i - \mu_i$ for $i \geq 1$. In matrix notation, (2.113) is written as

$$\mathbf{Z} = \mathbf{P}\,\mathbf{e} \qquad (2.115)$$

where $\mathbf{e} = (e_1, \dots, e_n)'$ and $\mathbf{P}$ is comprised of elements p_{ij} where

$$p_{ij} = \begin{cases} 0 & \text{for} \quad i < j \\ 1 & \text{for} \quad i = j \\ \phi^{i-j} & \text{for} \quad i > j \end{cases} \qquad (2.116)$$

The inverse of $\mathbf{P}$, denoted $\mathbf{P}^{-1}$, exists and is comprised of elements

$$q_{ij} = \begin{cases} 1 & \text{for} \quad i = j \\ -\phi & \text{for} \quad i = j+1, j = 1, 2, \dots, n-1 \\ 0 & \text{for} \quad \text{otherwise} \end{cases} \qquad (2.117)$$

In the absence of a presumed trend in the means, the means are estimated by $\hat{\mu}_j$ for $1 \le j \le n$. Unknown parameters are estimated using least squares applied to

$$Z_i = X_i - \hat{\mu}_i$$

The least squares point estimate for ϕ is found to be

$$\hat{\phi} = \frac{\sum_{i=1}^{n-1} Z_i Z_{i+1}}{\sum_{i=1}^{n} Z_i^2} \qquad (2.118)$$

Using (2.115), (2.117), and (2.118), we compute $\hat{e} = P^{-1}Z$ and write the least squares estimator of σ^2 as

$$\hat{\phi}^2 = \frac{\sum_{i=1}^{n} \hat{e}_1^2 - \frac{\left(\sum_{i=1}^{n} \hat{e}_i\right)^2}{n}}{n-1} \qquad (2.119)$$

In the computation of (2.119), it is useful to note that

$$\sum_{i=1}^{n} e_i = (1-\phi) \sum_{i=1}^{n} Z_i + \phi Z_n \qquad (2.120)$$

and

$$\sum_{i=1}^{n} e_i^2 = (1+\hat{\phi}^2) \sum_{i=1}^{n} Z_i^2 - \hat{\phi}^2 Z_1^2 - 2\hat{\phi} \sum_{i=2}^{n} Z_i Z_{i-1} \qquad (2.121)$$

The distributional properties of the $AR(1)$ fitted model are not easy to assess, and in some cases, using the asymptotic normality of the least squares estimators, approximate inference can be constructed.

An ad hoc approximate prediction interval for a new response can be formed using the $AR(1)$ model when the means at the individual locations are treated as fixed. The estimators (2.118) and (2.119) are used to produce the interval estimate for a new value X_{n+m}.

$$\hat{\mu}_{n+m} + z_{(1-\alpha/2)}\hat{\sigma}\sqrt{\frac{1-\hat{\phi}^{2(m+n)}}{1-\hat{\phi}^2}} \le X_{n+m} \le \hat{\mu}_{n+m} + z_{(1-\alpha/2)}\hat{\sigma}\sqrt{\frac{1-\hat{\phi}^{2(m+n)}}{1-\hat{\phi}^2}} \qquad (2.122)$$

We remark that to employ these $AR(1)$, an assumed model of mean values, μ_j for $j \geq 1$, is required. The efficiency of prediction interval (2.122) can be evaluated using simulation techniques such as those discussed in Section 9.2. An example of $AR(1)$ modeling is now given.

Example 2.23

In this example the $AR(1)$ model is applied to the stock price data given in Table 2.4. The means at the different locations are fixed at $\mu_i = 11 + .2i$ for $i \geq 1$. Utilizing formulas (2.118) to (2.121), we compute the estimates:

$$\hat{\phi} = .37448 \quad \text{and} \quad \hat{\sigma}^2 = .86206$$

The 95% prediction interval (2.122) at a future time corresponding to 18 months computes as $10.77405 \leq x_{18} \leq 14.69928$. Comparing these findings with the least squares analysis in Example 2.22, we observe that in this instance $AR(1)$ resulted in a slightly larger variance point estimate and a wider resulting confidence interval.

Autoregressive models can be used in conjunction with collective risk models. The aggregate sum is composed of dependent individual random variables as defined by (2.111). This area is open for new approaches.

2.8 Model Diagnostics

In financial and actuarial applications, the underlying structure of the stochastic components plays an important role. In practice, the underlying stochastic structure is modeled by choosing appropriate discrete, continuous, or mixed distributions. Actuarial examples include the stochastic modeling of insurance claim amounts, status lifetimes, or future stock prices where the proper choice of an underlying statistical distribution is critical. Current references on statistical model selection and verification are Klugman et al. (2008) and Myers (2008). In this section, assessment methods for assumed statistical distributions for empirical data are presented that are both graphical and analytic in nature and can be adapted for discrete and continuous distributions.

The efficiency of an assumed continuous distribution can be examined through graphical and analytical methods. The graphical technique referred to as probability plotting is outlined and applied to typical actuarial and financial settings. In particular, probability plots for the Weibull, Pareto, and lognormal are demonstrated through numerical examples. Extending the concept of probability plotting, an analytic measure based on the statistical generalized least squares used for continuous-type model assessment is proposed. General formulas are given and applied to empirical data sets through examples.

In some financial and actuarial modeling settings the underlying stochastic structure may be discrete in nature. In particular, for target populations of individual deaths, counts may only be reported for yearly intervals. This is the basis of life and mortality tables examined in Chapter 5. In statistical reliability modeling this is referred to as interval data. A diagnostic measure for a proposed discrete distribution applied to observed interval data based on a generalized least squares approach is proposed. The proposed measure used to assess the discrete distribution efficiency is demonstrated in an example where the best from a set of competing discrete distributions is chosen.

2.8.1 Probability Plotting

In probability plotting applications a proposed continuous statistical distribution for observed data is assessed graphically. In general, the proposed statistical distribution is a function of parameter vector $\theta = (\theta_1, \theta_2, ..., \theta_m)$, and the cdf is denoted by $F(x, \theta)$. As demonstrated in Rohatgi (1976, p. 203), the distribution of observed cdf is uniform on the unit interval. Based on observed data, such as realized failure times or past insurance claim amounts, the observed order statistics, as discussed in Section 2.3, are $y_1 \leq y_2 \leq ... \leq y_n$. Suggestions for the choice of nonparametric cdf quantities associated with the order statistics have appeared in the literature. Johnson (1951) and Benard and Bos-Levenbach (1953) suggested a median-based approach, but we utilize the more popular mean-based nonparametric cdf quantities of Hahn and Shapiro (1967), defined by

$$F_i^* = \frac{i}{1+n} \quad \text{for} \quad i = 1, 2, ..., n \tag{2.123}$$

The main concept in probability plotting is to equate the values of (2.123) to their assumed cdf values:

$$F_i^* = F(y_i, \theta) \quad \text{for} \quad i = 1, 2, ..., n \tag{2.124}$$

where deviations in (2.124) indicate distributional inadequacies. Monotonic transformations applied to (2.124) are used to form a linear regression model in terms of independent variable $w(F_i^*)$ and dependent variable $h(y_i)$ for $1 \leq i \leq n$. The linear regression model (2.103) becomes

$$h(y_i) = \beta_0 + \beta_1 w(F_i^*) + e_i \tag{2.125}$$

for error term e_i for $i = 1, 2, ..., n$. The slope parameter β_1 and intercept parameter β_0 are related to the parameter components in θ. The least squares estimators are found and used to form regression estimators for the unknown parameters.

If the proposed statistical distribution models the observed data efficiently, then the transformed data $(w(F_i^*), h(y_i))$, for $1 = 1, 2, \ldots, n$, form an approximate linear relationship and (2.124) holds with small deviations. The scatter plot the transformed data give is referred to as the probability plot, and measures of linear relationship such as correlation or coefficient of variation can be used to assess the linearity. This is demonstrated in Example 2.24.

Example 2.24

A collection of automobile claims are examined where we test to see if the normal distribution can be utilized with the unknown parameters μ and σ^2. The observed order statistics are $y_1 \leq y_2 \leq \ldots \leq y_n$. Relation (2.124) is

$$F_i^* = \Phi\left(\frac{y_i - \mu}{\sigma}\right) \quad \text{leading to} \quad y_i = \mu + \sigma \Phi^{-1}(F_i^*) \qquad (2.126)$$

Thus, the linear regression model (2.125) is used where

$$h(y_i) = y_i, \quad w_i = w(F_i^*) = \Phi^{-1}(F_i^*), \quad \beta_0 = \mu, \text{ and } \beta_1 = \sigma \qquad (2.127)$$

The observed order statistics, as well as the transformed quantities (2.127), are listed in Table 2.5.

Utilizing the transformed data in Table 2.5, the least squares are found, producing the regression estimators of the parameters defined in (2.127):

$$\hat{\mu} = 2.04 \quad \text{and} \quad \hat{\sigma} = .463$$

The sample correlation (2.107) is found to be $r = .975$ and the probability plot corresponding to the proposed normal distribution is displayed in Figure 2.2.

The probability plot in Figure 2.2 deviates from linearity to a somewhat minor degree. Therefore, other statistical distributions might be investigated to model the claim amounts.

TABLE 2.5

Probability Plotting of Claim Data

Index	1	2	3	4	5	6	7	8	9	10
Order statistics	1.5	1.7	1.8	1.8	1.9	2.0	2.1	2.3	2.5	2.8
$h(y_i)$	1.5	1.7	1.8	1.8	1.9	2.0	2.1	2.3	2.5	2.8
$w(F_i^*)$	−1.335	−.908	−.605	−.349	−.114	.114	.349	.605	.908	1.335

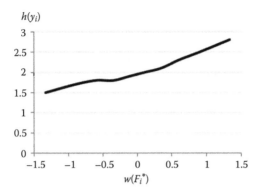

FIGURE 2.2
Normal probability plot.

In financial and actuarial applications, typical distributions used to fit empirical data, such as stock prices or insurance claim amounts, include nonnegative and heavy tailed distributions. In particular, we consider probability plotting techniques for the nonnegative distributions discussed in Section 1.6, namely, the Pareto, lognormal, and Weibull with respective cdfs of the forms (1.46), (1.51), and (1.54). For these distributions, the required transformations dictated by the mathematical structure in (2.124) are listed in Table 2.6.

An Excel application outlining these procedures for the nonnegative random variables of Section 1.6 is discussed in Problem 2.17. The application of these distributions to a data set is demonstrated in the next example.

Example 2.25

The probability plotting approach is used to explore claim data of Example 2.24, listed in Table 2.5, in terms of an underlying Pareto, lognormal, and Weibull distribution. Computed quantities utilized in Table 2.6 transformations are listed in Table 2.7.

TABLE 2.6

Probability Plot Variables for Pareto, Lognormal, and Weibull

Distribution	$h(y_i)$	$w(F_i^*)$	β_0	β_1
Pareto	$\ln(y_i)$	$-\ln(1 - F_i^*)$	$\ln(\beta)$	$1/\alpha$
Lognormal	$\ln(y_i)$	$\Phi^{-1}(F_i^*)$	α	β
Weibull	$\ln(y_i)$	$\ln(-\ln(1 - F_i^*))$	$\ln(\beta)$	$1/\alpha$

TABLE 2.7

Probability Plotting Transformation Quantities

Index i	1	2	3	4	5	6	7	8	9	10
$\ln(y_i)$	.405	.531	.588	.588	.642	.693	.742	.833	.916	1.03
$-\ln(1 - F_i^*)$	.095	.201	.318	.452	.606	.788	1.012	1.299	1.705	2.398
$\Phi^{-1}(F_i^*)$	−1.335	−.908	−.605	−.349	−.114	.114	.349	.605	.908	1.335

The least squares estimators are found for the proposed distributions, and the transformed regression parameters estimates are computed. These parameter estimates along with the observed correlation are given in Table 2.8.

From Table 2.8, based on the correlation measure, the best linear fit results from an underlying lognormal distribution. The probability plots corresponding to the Pareto and lognormal distributions are given in Figure 2.3. From Figure 2.3 we observe that the probability plot associated with the lognormal distribution is more linear than that of the Pareto distribution. Probability plots are useful in application, but choosing between competing statistical distributions with similar plots may require an analytic assessment measure.

TABLE 2.8

Probability Plot Comparison for Pareto, Lognormal, and Weibull

Computations	Distributions			
	Normal	Pareto	Lognormal	Weibull
Correlation r	.975	.978	.988	.965
Regression parameter	$\mu = 2.04$	$\alpha = 3.984$	$\alpha = .697$	$\alpha = 5.513$
Estimates	$\sigma = .463$	$\beta = 1.606$	$\beta = .224$	$\beta = 2.196$

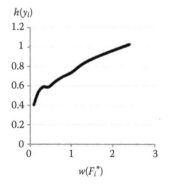 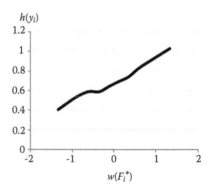

FIGURE 2.3
Probability plots for Pareto (left panel) and lognormal (right panel).

2.8.2 Generalized Least Squares Diagnostic

This section presents a quantitative diagnostic measure of the usefulness of a proposed continuous statistical distribution to observe data based on the statistical concept of generalized least squares. The mathematical forms presented in the probability plotting approach of Section 2.8.1 are combined with the theory of the distribution of quadratic forms in normal variables. The proposed measure is a quadratic form in the difference between the assumed and empirical cumulative distribution functions at the order statistics values. In application, the diagnostic measurement can be utilized to choose between competing distributions. We refer to Graybill (1976, p. 34) and Searle (1971) for useful statistical background

The assumed continuous statistical distribution is defined in terms of the cdf $F(x, \theta)$. Applying statistical theory, the distribution of F_i^*, defined by (2.123), for $i = 1, 2, \ldots, n$, coincides with the distribution of the nth order statistics, based on iid uniform random variables on the unit interval. The mean and variance under a true model are

$$E\{F_i^*\} = F(y_i, \theta) = F_i \tag{2.128}$$

and

$$\operatorname{Var}\{F_i^*\} = \frac{i(n - i + 1)}{(n + 1)^2 (n + 2)} \tag{2.129}$$

for $i = 1, 2, \ldots, n$. Further, the covariance for F_i^* and F_j^* is

$$\operatorname{Cov}\{F_i^*, F_j^*\} = \frac{i(n - j + 1)}{(n + 1)^2 (n + 2)} \tag{2.130}$$

for $1 \leq i < j \leq n$. The derivation of (2.128), (2.129), and (2.130) is considered in Rohatgi (1976, p. 311). Thus, the empirical cdf vector $(F_1^*, F_2^*, \ldots, F_n^*)$ has mean vector $(F_1, F_2, \ldots, F_n)$ with components (2.128) and dispersion matrix Σ with elements defined by (2.129) and (2.130). The inverse of Σ, denoted by V, exists and is composed of nonzero entries defined by

$$v_{ii} = \frac{2(n + 1)^2 (n + 2)}{n + 1} \quad \text{for} \quad 1 \leq i \leq n \tag{2.131}$$

$$v_{i, i+1} = \frac{-(n + 1)^2 (n + 2)}{n + 1} \quad \text{for} \quad 1 \leq i \leq n - 1 \tag{2.132}$$

and

$$v_{i+1, i} = \frac{-(n + 1)^2 (n + 2)}{n + 1} \quad \text{for} \quad 2 \leq i \leq n \tag{2.133}$$

To demonstrate inverse elements (2.131), (2.132), and (2.133), it is enough to show that the product Σ **V** gives the identity matrix.

An analytic measure of the efficiency of the proposed continuous distribution is based on the statistical generalized least squares concept. The generalized least squares measure (GLSM) is defined by the quadratic form

$$Q = (F_1^* - F_1, F_2^* - F_2, ..., F_n^* - F_n) \mathbf{V} (F_1^* - F_1, F_2^* - F_2, ..., F_n^* - F_n)^T \quad (2.134)$$

where for row vector **a** the transpose is denoted by $\mathbf{a}^T$. Using (2.131), (2.132), and (2.133), (2.134) becomes

$$Q = \left[\frac{2(n+1)^2(n+2)}{n+1} \right] \left[\sum_{i=1}^{n} (F_i^* - F_i)^2 + \sum_{i=1}^{n-1} (F_i^* - F_i)(F_{i+1}^* - F_{i+1}) \right] \quad (2.135)$$

Analytic measure (2.125) can be used to choose between competing distributions as demonstrated in the following example. The Excel computation and application of (2.135) is outlined in Problem 2.18.

Example 2.26

For the observed data of Example 2.24 given in Table 2.4, the GLSM defined by (2.135) is utilized to select between the normal, Pareto, lognormal, and Weibull distributions. The results, using computations in Examples 2.24 and 2.25, are given in Table 2.9.

From Table 2.9, we observe the Pareto and lognormal distributions yield a much lower GLSM, indicating superior fit, thus eliminating the normal and Weibull from consideration. The slightly lower GLSM indicates a better fit for the lognormal distribution over the Pareto. This finding is consistent with the probability plotting results of Figure 2.3.

2.8.3 Interval Data Diagnostic

Observed data may be discrete in nature where frequencies corresponding to stochastic events over predefined intervals, such as future time intervals, are recorded. In reliability and statistics this is referred to as interval data. In actuarial science, interval data play a central role with the construction and verification of mortality and life tables (see Section 5.7). A diagnostic measure for interval data used to assess the efficiency of an assumed underlying

TABLE 2.9

GLSM for Proposed Distributions

Distribution	Normal	Pareto	Lognormal	Weibull
GLSM	13.0527	6.5472	6.5182	18.8536

statistical distribution, based on the generalized least squares concept, is presented and demonstrated.

For the general interval data, the support of the underlying variable of interest is partitioned into disjoint sets, $E_i = [i, i+1)$, where the probability associated with E_i is denoted $p_i = P(E_i)$, for $i = 1, 2, ..., m$. Based on a random sample of size n, interval data consist of observing the frequencies associated with the disjoint sets. Discrete random variables associated with the observed frequencies are

$$X(i) = \text{Number of observations in } E_i \qquad (2.136)$$

for $i = 1, 2, ..., m$, subject to the constraints

$$\sum_{i=0}^{m} p_i = 1 \quad \text{and} \quad \sum_{i=0}^{m} X(i) = n \qquad (2.137)$$

hold. In particular, we consider interval data and resulting diagnostic measures useful in actuarial science applications.

In common actuarial science applications life table construction the range of individual lifetimes is divided into separate years. Thus, $E_i = [i, i+1)$ for $i = 0, 1, ..., m$. The random vector $(X(1), X(2), ..., X(m))$ is multinomial and the point estimators for p_j are the relative frequencies

$$\hat{p}_i = \frac{X(i)}{n} \quad \text{where} \quad E\{\hat{p}_i\} = p_i \qquad (2.138)$$

for $i = 0, 1, ..., m$. The variance and covariance are

$$\text{Var}\{\hat{p}_i\} = p_i \frac{1-p_i}{n} \quad \text{and} \quad \text{Cov}\{\hat{p}_i, \hat{p}_j\} = -\frac{p_i p_j}{n} \qquad (2.139)$$

The dispersion matrix for the random vector $(\hat{p}_1, \hat{p}_2, ..., \hat{p}_m)$ is denoted by Σp with elements defined by (2.137) and (2.138). The inverse of Σ_p is denoted by Vp and takes the form

$$\underset{-p}{V} = \left(\frac{n}{p_0}\right)\underline{1}\,\underline{1}^t + \text{diag}\left(\frac{n}{p_1}, \frac{n}{p_2}, ..., \frac{n}{p_n}\right) \qquad (2.140)$$

where $\mathbf{1}$ denotes a column vector consisting of ones. The validity of the inverse (2.140) can be demonstrated by showing the product of Σ_p and V_p yields the identity matrix. An alternative derivation is given in Hogg et al. (2005, p. 347).

In application, a proposed statistical model for the interval data defines the probabilities $p_i = P(E_i)$ for $i = 0, 1, ..., m$. This may result from an assumed discrete statistical distribution or actuarial life table. Analogous to the approach of Section 2.8.2, an assessment measurement for a discrete distribution based on the GLSM is constructed using (2.140). The GLSM for discrete distributions, similar to (2.134), is

$$W = (\hat{p}_1 - p_1, \hat{p}_2 - p_2, ..., \hat{p}_n - p_n)\, \mathbf{V}\,(\hat{p}_1 - p_1, \hat{p}_2 - p_2, ..., \hat{p}_n - p_n)^T \quad (2.141)$$

Applying form (2.140) we find that (2.141) reduces to

$$W = n\left[\sum_{i=0}^{n} \frac{(\hat{p}_i - p_i)^2}{p_i}\right] \quad (2.142)$$

The proposed interval data diagnostic measure is demonstrated in the following example where a collection of competing discrete distributions are compared for an observed data set. Statistic (2.142) is the classic chi squared goodness of fit test statistic. Under a correct model, the asymptotic distribution of (2.142) is a chi square variable, but is applicable only when none of the proportions are small. This generally is not the case in life table applications.

Example 2.27

The observations of the multinomial variable given in Table 1.1 are considered and an underlying discrete distribution is desired. Here the mean of K is calculated as $64/25 = 2.56$. A collection of three discrete random variables, namely, the geometric, Poisson, and binomial random variables, with theoretical means given by q/p, λ, and np, are considered. Equating the theoretical means with 2.56, crude parameter estimates result and we find the competing distributions are the geometric with $q = .7$, Poisson with $\lambda = 2.7$, and binomial with $n = 5$ and $p = .5$. The observed relative frequencies and the theoretical probabilities for the discrete distributions are listed in Table 2.10.

The GLSM is used to select between the three competing discrete distributions, and the values of (2.141), computed using (2.142), are listed in Table 2.11.

TABLE 2.10

Empirical and Discrete Distributions

Observed Data		Proposed Distributions		
$J = j$	$X(j)/n$	Geometric	Poisson	Binomial
0	.08	.3000	.0672	.0313
1	.16	.2100	.1815	.1563
2	.20	.1470	.2450	.3125
3	.32	.1029	.2205	.3125
4	.16	.0720	.1488	.1563
≥ 5	.08	.1681	.1371	.1930

TABLE 2.11

GLSM for Discrete Data

Distribution	Geometric	Poisson	Binomial
Q	.8040	.0828	.1930

From Table 2.11 we see that the Poisson distribution is superior to the other two with a smaller GLSM defined by Q.

Problems

2.1. Let X be continuous uniform on $[0, 1]$ independent of Y continuous uniform on $[0, 3]$ and $S = X + Y$.

  a. Use (2.36) to compute $F_s(s)$.

  b. Compute the pdf $f_s(s)$ given by (2.37).

2.2. In Example 2.7 we add a third independent random variable with pdf defined by $f_3(x) = .2$ for $x = 0, 1, 2, 3$, and 4.

  a. Using (2.38) find the cdf $F_s(s)$ as defined by (2.33).

  b. Compute the pmf $f_s(s)$ given by (2.37).

2.3. For a collection of iid random variables let the aggregate S_m be as defined in (2.32). Applying (2.41) and the uniqueness of the mgf find the pmf or pdf for S_m where the distribution of X_i is given by (a) the Poisson pmf, $f(x) = \exp(-\lambda)\lambda^x/x!$ for $S = \{0, 1, ...\}$, and $\lambda > 0$, (b) the binomial pmf, $f(x) = [n!/(x!(n - x)!)]p^x (1 - p)^{n-x}$ for $S = \{0, 1, ..., n\}$ and $0 < p < 1$, and (c) the gamma pdf, $f(x) = [1/(\Gamma(\alpha)\beta^\alpha)]x^{\alpha-1} \exp(-x/\beta)$ for $S = (0, \infty)$ and $\alpha > 0$ and $\beta > 0$.

2.4. For an insurance policy the amount of the claims, denoted B, are distributed approximately normal with mean $1,000 and standard deviation $300. The probability of a claim is .05 and the claim variable is $X = I B$. Find (a) the mean and variance of X and (b) the 99th percentile of X.

2.5. A portfolio consists of 30 independent insurance policies of the type defined in Problem 2.4 where the aggregate claim variable is S_{30}.

  a. Find the mean and variance of S_{30}.

  b. Approximate the probability that S_{30} would not exceed 3,000 using the CLT.

  c. Using the CLT compute a 95% prediction interval corresponding to S_{30}.

2.6. The SPA is to be applied to the aggregate sum of Problem 2.5.

   a. For the claim variable $X = I B$ give the form of the mgf.

   b. For $Z = (X - \mu_x)/\sigma_x$, where μ_x and σ_x^2 are found in Problem 2.4, using (2.73) find formulas for $M_z(\beta)$, $M_z^{(1)}(\beta)$, and $M_z^{(2)}(\beta)$.

2.7. A portfolio of stock values is denoted by X_i for $i \geq 1$. Let the aggregate of values be S_N, where due to future transactions N is a Poisson random variable with mean $\lambda = 20$. The distribution of the stock values is not known, but we estimate $\mu_1 = 1$, $\mu_2 = 1.5$, and $\mu_3 = 3$.

   a. Compute the mean and standard deviation of the aggregate.

   b. Using the CLT, approximate the probability that the portfolio will be valued at more than 25.

   c. Using the CLT, approximate the 95th percentile for the aggregate.

2.8. Consider the compound random variable S_N. Use the conditioning formulas (1.86) and (1.87), substituting N and X for $w(x)$ and $v(y)$, respectively, to show (2.79) and (2.80) hold. Further, show the mgf takes the form of (2.81) and the moments can be computed using this mgf.

2.9. It is claimed that for a type of life insurance three risk categories, A, B, and C, are in a ratio of 1:2:3. A survey of policies found 5 of A, 12 of B, and 13 of C. Apply Q given by (2.142) to evaluate this claim.

Excel Problems

2.10. A random sample from a normal distribution is 1,000, 1,200, 800, 750, 220, 330, 410, and 2,000.

   a. What are the mean, median, and standard deviation of the sample?

   b. Compute the 99% confidence interval for the mean given by (2.12).

   c. Compute the 99% confidence interval for the mean based on the T distribution where $t_{1-.005}$ with degrees of freedom is 7.
   Excel: Data, Data Analysis, Descriptive Statistics, Input Range A1:A8, NORMSINV(.995) $= z_{.995}$, TINV(.005,7) $= t_{1-.005,7}$

2.11. We use SPA outlined in Problem 2.6 for the setting of Problems 2.4 and 2.5 to approximate $P(S_{30} > 3,000)$. Here $s = 3,000$ and $t = (3,000 - 30(50))/(30(228.035)) = .219265$.

a. Give the mean and standard deviation for S_N.

b. Find β using Solver.

c. Compute c and α^2 following (2.75).

d. Compute $P(S_{30} \leq 3{,}000)$ using (2.76) and then $P(S_{30} > 3{,}000)$. Excel: Set formulas for $M_z(\beta)$, $M_z^{(1)}(\beta)$, and $M_z^{(2)}(\beta)$ from Problem 2.6 and Data, Solver to solve (2.74).

2.12. MSPA given by (2.90) is used to approximate the aggregate, which is at most 25, where $\lambda = 20$, $\mu_1 = 1$, $\mu_2 = 1.5$, and $\mu_3 = 3$.

a. Give $E\{S_N\}$ and $[Var\{S_N\}]^{1/2}$.

b. Give s, t, and using (2.97), β.

c. Using (2.93), (2.94), and (2.95) find $M_z(\beta)$, $M_z^{(1)}(\beta)$, and $M_z^{(2)}(\beta)$.

d. Compute c and α^2 following (2.75).

e. Compute $P(S_N \leq 25)$ using (2.90). Excel: Utilize basic formulas to compute $M_z(\beta)$, $M_z^{(1)}(\beta)$, and $M_z^{(2)}(\beta)$.

2.13. A compound Poisson distribution with $\lambda = 5$ is considered where the individuals are independent exponential with parameter $\theta = 1$. Compute $P(S_N \leq s)$, where the parameters are $\lambda = 5$, $\mu_1 = 1$, $\mu_2 = 2$, $\mu_3 = 6$, and $s = 10$, (a) using MSPA (2.90) applying (2.93) to (2.97), (b) using the CLT, and (c) exactly. Excel: Problem 2.12 extension: NORMSDIST, POISSON, GAMMADIST.

2.14. Based on the MSPA Excel formulas in Problem 2.12, find the 90th percentile for parameters $\lambda = 20$, $\mu_1 = 1$, $\mu_2 = 1.5$, and $\mu_3 = 3$. Excel Problem 2.12 extension: Data, Solver, setting Target Cell Equal to .90.

2.15. The Dow Jones Index is observed over 12 months with the following ending values in terms of 1,000 yield results:

Month	1	2	3	4	5	6	7	8	9	10	11	12
Index	1.07	1.09	1.05	.96	1.08	1.10	1.06	1.05	.99	.91	.92	.98

The linear regression with the month as the predictor variable and the index as the response is considered.

a. Draw a scatter plot and comment on the linearity.

b. Run the regression and give the correlation, the slope and intercept estimators, and a 95% confidence interval for the slope.

c. Estimate the price index at the end of the next month.

d. Is the index increasing or decreasing?

e. Using (2.110) find a 95% confidence interval for the expected index at month 13.

Excel: Chart Wizard, Scatter and Data, Data Analysis.

2.16. The AR(1) techniques are applied to Problem 2.13, where $\mu_j = 1.021667$ (mean).

a. Using (2.118) and (2.119) compute estimates for φ and σ^2.

b. Compute a 95% confidence interval for the index at month 13 using (2.122).

Excel: Problem 2.15 extension: Basic operations.

2.17. The probability plotting regressions are applied to the data of Table 2.7.

a. For Pareto give the correlation, regression estimates, and parameter estimates.

b. For lognormal give the correlation, regression estimates, and parameter estimates.

c. For Weibull give the correlation, regression estimates, and parameter estimates.

Excel: NORMSINV and, Data, Data Analysis, Regression.

2.18. Diagnostic Q defined by (2.135) is computed for the data and setting of Problems 2.15 and 2.17.

a. Compute Q for the lognormal.

b. Compute Q for the Weibull.

Excel: Problem 1.33 extension: Basic operations.

Solutions

2.1. a. $F_s(s) = s^3/6$ for $0 \le s < 1$, $s^2/6 - (s-1)^2/6$ for $1 \le s < 3$, $s - 3/2 - (s-1)^2/6$ for $3 \le s < 4$, and 1 for $s \ge 4$.

b. $f_s(s) = s^2/2$ for $0 < s \le 1$, $1/3$ for $1 < s < 3$, and $4/3 - s/3$ for $3 < s < 4$.

2.3. a. Problem 1.9 gives $M_x(t) = \exp(\lambda(e^t - 1))$ so $M_s(t) = \exp(m\lambda(e^t - 1))$ and S_m is Poisson with mean $m\lambda$.

b. $M_x(t) = (1 - p - pe^t)^n$ so $M_s(t) = (1 - p - pe^t)^{mn}$ and S_m is binomial with parameters mn and p.

c. $M_x(t) = (1 - \beta t)^{-\alpha}$ so $M_s(t) = (1 - \beta t)^{-m\alpha}$ and S_m is gamma with parameters $m\alpha$ and β.

2.4. a. First (1.34) gives $E\{X\} = .05(1,000) = 50$ and (1.36) gives $Var\{X\} = .05(300^2) + .05(.95)(1,000^2) = 52,000$.

b. Using pdf (1.20) implies $.01 = .05(1 - \Phi((B_{.99} - 1,000)/300)$ so $B_{.99} = 1,000 + (300) z_{.8} = 1,252.486$.

2.5. a. Following Example 2.13 $E\{S_{30}\} = (30)(50) = 1{,}500$ and $\mathrm{Var}\{S_{30}\} = 30(52{,}000) = 1{,}560{,}000$.

    b. $1 - \Phi((300 - 1{,}500)/1{,}249) = 1 - \Phi(1.200) = .1150$.

    c. $1{,}500 - 1.96(1{,}249) \le S_{30} \le 1{,}500 + 1.96(1{,}249)$.

2.6. a. $M_x(t) = 1 - q + q\,\exp(\mu t + t^2\sigma^2/2)$.

    b. $M_Z(\beta) = (1 - q)\exp(-\beta_x/\sigma_x) + q\exp[\beta(\mu - \mu_x)/\sigma_x + \sigma^2\beta^2/(2\sigma_x^2)]$, $M_Z^{(1)}(\beta) = -(1 - q)(\mu_x/\sigma_x)\exp(-\beta_x/\sigma_x) + q[(\mu - \mu_x)/\sigma_x + \sigma^2\beta/\sigma_x^2]\exp[\beta(\mu - \mu_x)/\sigma_x + \sigma^2\beta^2/(2\sigma_x^2)]$, and $M_Z^{(2)}(\beta) = (1 - q)(\mu_x/\sigma_x)^2\exp(-\beta_x/\sigma_x) + (\sigma^2/\sigma_x^2)\,q\,\exp[\beta(\mu - \mu_x)/\sigma_x + \sigma^2\beta^2/(2\sigma_x^2)] + q[(\mu - \mu_x)/\sigma_x + \sigma^2\beta/\sigma_x^2]^2\exp[\beta(\mu - \mu_x)/\sigma_x + \sigma^2\beta^2/(2\sigma_x^2)]$.

2.7. a. $E\{S_N\} = 20$, $[\mathrm{Var}\{S_N\}]^{1/2} = 5.4773$.

    b. $\Phi((20 - 25)/5.4773)$.

    c. $20 + 1.645(5.47726) = 19.01$.

2.9. The claim implies $P(A) = 1/6$, $P(B) = 2/6$, and $P(C) = 3/6$, while the sample proportions are 5/30, 12/30, and 13/30. Computing gives $Q = 2/3$. Applying the chi square goodness of fit test Q the 95% percentile for a chi square random variable with degrees of freedom 2 is 5.991, indicating the claim is not disproved.

2.10. a. 838.75, 775, 578.1606.

    b. $312.23 \le \mu \le 1365.27$.

    c. $15.2 \le \mu \le 1662.38$.

2.11. b. Using $\mu = 1{,}000$, $\sigma = 300$, $\mu_x = 50m$ and $\sigma_x^2 = 52{,}000$, we find $\beta = .15229$, $M_z(\beta) = 1.02488$, $M_z^{(1)}(\beta) = .22472$, and $M_z^{(2)}(\beta) = 2.09851$.

    c. $c = 1.033956$, $\alpha = 1.41403$.

    d. $P(S_{30} \le 3{,}000) = .9006275$, so $P(S_{30} > 3{,}000) = .099172$.

2.12. a. $E\{S_N\} = 20$, $[\mathrm{Var}\{S_N\}]^{1/2} = 5.4772$.

    b. $s = 25$, $t = .91287$, and $\beta = .79692$.

    c. $M_z(\beta) = 1.416717$, $M_z^{(1)}(\beta) = 1.293278$, and $M_z^{(2)}(\beta) = 3.00957$.

    d. $c = 1.46103$, $\alpha = 1.136219$.

    e. $P(S_N \le 25) = .820077$.

2.13. a. .92333.

    b. .94308.

2.14. $s_{90} = 27.2503$.

2.15. a. Scatter plot is somewhat linear with a small negative slope.

    b. $r = .6575$, intercept estimate $= 1{,}101$, slope estimate $= -.012$ and $-.02199 < \beta_1 < -.00234$.

    c. $1.101 - 13(.012) = .945$.

    d. Decreasing since the confidence interval for the slope contains only negative values.

2.16. a. .5469, .05525.

2.17. a. $r = .9777$, regression estimates .4739 and .2510, parameter estimates 1.6063 and 3.9835.

      b. $r = .9885$, regression estimates .6967 and .2238, parameter estimates .6967 and .2238.

      c. $r = .9645$, regression estimates. 7865 and .1813, parameter estimates 2.1958 and 5.5141.

2.18. a. $Q = 6.4623$.

      b. $Q = 11.1341$.

3

Financial Computational Models

Financial or actuarial models are used to quantify and analyze future financial actions. These actions may be contingent on many factors, such as time, price, and speculation, and the resulting models fall into one of two main types. In the first, the actions are deterministic and are completely defined in terms of their form and timing. This is true for monthly mortgage payments that continue for a fixed number of years where the interest rate may be either fixed or variable. In the second type, the financial action itself may be initiated by a stochastic event. Examples of stochastic financial actions include the payment of a benefit associated with an insurance policy at the time of death of the policyholder or the purchase of a stock at the time its value exceeds a predefined price. Following actuarial science nomenclature, the collection of economic actions with defined future time conditions is referred to as a status model. Thus, following the previous descriptions, the two basic types of status models are deterministic and stochastic.

The value of an investment or series of payments depends on numerous factors that include the interest rate or the return rate and the length of time of the investment. Typically, financial analyses containing one or more actions are functions of a future time random variable denoted by T, where $T \geq 0$. These actions are evaluated in terms of some reference time fixed before the analysis and may correspond to a future time or the present time. The present value function is an evaluation at initial time $t = 0$ associated with financial actions at future time T and is denoted $PV(T)$. The future value is analysis computed at time T and is given by $FV(T)$. If these models possess a nonstochastic financial action time component, they are generally referred to as deterministic status models and include the compounding of interest, growth of stock holdings, evaluation of annuities consisting of a series of monetary payments, and computation of combination economic actions over fixed time periods. In this context the interest or return rates may be either fixed or stochastic.

Classically there are two types of financial actions that dominate financial and actuarial analysis. In the first, a single monetary action, such as an investment at a specified interest or estimated return rate, is analyzed. For simplicity we refer to both interest rates and investment return rates by the single term of financial rates, or simply rates. In the second, there is a series of monetary payments that may differ in amount, or may be equal or level, referred to as an annuity. These models are analyzed at both the present and general future time reference points. For a review of deterministic interest and annuity models we refer to Kellison (2009).

In this chapter financial computations associated with deterministic status models are discussed. Basic concepts and formulas for future and present value functions associated with single monetary units are presented in Section 3.1. Single investment or interest models with various forms of growth or compounding are discussed in Sections 3.1.1, 3.1.2, and Section 3.1.3. Annuity models are discussed for both the discrete payment, in Section 3.2.1, and continuous payment, in Section 3.2.2, structures. The chapter concludes with an extension of these models to stochastic financial rate models in Section 3.3. Both interest and annuity models are analyzed in the case of stochastic rates.

3.1 Fixed Financial Rate Models

Money can be invested with financial intermediaries, such as a bank, where the receiver pays the investor for the right to utilize the invested capital or the stock market in the hopes that its value will increase. An amount of money, called the principal, earns additional value, referred to as interest or investment return, over time. The additional monetary worth of an investment grows through time as a function of an interest or a return rate. In this section these rates are considered nonstochastic and may be modeled as either a discrete or a continuous function over future time. Formulas for both the present value and the future value of the principal are given, and we start with a discussion of required basic financial computations.

3.1.1 Financial Rate-Based Calculations

The financial rate, whether from bank interest or speculative investment, is often defined in terms of yearly percentages. For principal F_0 the financial rate associated with 1 year is denoted i and the amount of interest earned is the product iF_0. Thus, after 1 year the future value is the sum of the principal plus the interest:

$$FV(1) = F_0 (1 + i) \qquad (3.1)$$

Formula (3.1) is an example of simple yearly interest, and if the time length differs from 1, the formula is altered. If the lengths of the time periods are equal and correspond to a fraction of a year, such as $(1/m)$th of a year, the period rate is computed as $i^{(m)}/m$. For example, if $m = 12$, the time period is in terms of months, while yearly quarters are indicated by $m = 4$.

Modeling monetary growth can be extended to arbitrary discrete time periods defined by the partition $0 = w_0 < w_1 < \ldots < w_n$ yielding disjoint time

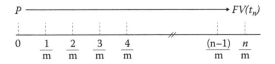

FIGURE 3.1
Future value of an annuity-immediate.

intervals $(w_{j-1}, w_j]$, and the rate during the *j*th interval is assumed fixed at i_j. If the time intervals correspond to $(1/m)$th of a year, the equal-width time periods are $E_{j/m} = ((j-1)/m, j/m]$ for $j \geq 1$. The future value of the investment after multiple time periods found by repeated use of (3.1) and is referred to as compounding interest. After *n* time periods the future time is given by $t_n = n/m$ years and the future value of the initial investment of F_0 is

$$FV(t_n) = F_0 \prod_{j=1}^{n} (1+i_j) \qquad (3.2)$$

The graph of this function is given in Figure 3.1.

This formula is general in nature and can be adapted to common interest time frame situations as discussed later and in Problem 3.15.

For the equal-length discrete time period model consisting of *m* periods per year with a constant interest rate, the nominal annual rate is $i^{(m)}$. Hence, the rate per period is $i^{(m)}/m$ and the time corresponding to *n* periods is denoted t_n. Applying (3.2) the future value at time t_n is

$$FV(t_n) = F_0 \left(1 + \frac{i^{(m)}}{m} \right)^n \qquad (3.3)$$

In the monetary growth structure (3.3) interest is applied to both principal and interest amounts in the previous time period and is referred to as compound interest. Examples are now given that demonstrate the application of these formulas and the financial growth of an initial investment over time.

Example 3.1

Let an initial investment or principal be $F_0 = \$100$ and the nominal annual interest rate be $i^{(m)} = .12$. Suppose we compound interest monthly, so that the periods are 1 month long and in yearly units the future time is denoted $t_j = j/12$. In addition, the monthly or period interest rate is $i^{(m)}/m = .12/12 = .01$. The future value after 1 year, or 12 periods, using (3.3) is

$$FV(1) = \$100 \, (1.01)^{12} = \$112.68$$

From this we note that the investment growth, in terms of simple interest (3.1), is 12.68%. Comparing monthly interest to yearly interest, an increase of .6% in the future value is realized.

Example 3.2

A bank investment of $500 earns interest that is compounded quarterly where the nominal annual interest rate is $i^{(4)} = .08$. The quarterly rate is .08/4 = .02, and based on (3.3), the future value after 1 year is

$$FV(1) = \$500(1.02)^4 = \$541.22$$

In a financial investment the realized quarterly rates may vary. For example, let the quarterly rates be .01, .015, .025, and .03, noting the mean of these rates is .02. In this case, using (3.2), the future value is

$$FV(1) = \$500(1.01)(1.015)(1.025)(1.03) = \$541.15$$

The future value of the investment differs slightly from the computation using the mean rate, but the order of the differing rates is immaterial.

The rate of compounding, in terms of time period length, affects the future value calculation. Compounding interest or investment growth continuously over a fixed time is achieved by letting the number of equal-length periods approach infinity. To find the limiting future value, we let m approach infinity in formula (3.3). The continuous rate is often called the force of interest, and is denoted by δ, and at time $T = t$ the future value takes the form

$$FV(t) = F_0 \exp(\delta t) \tag{3.4}$$

Future value (3.4) is based on a continuous future time structure and not discrete time periods. The different rate settings, demonstrated in Example 3.3 and explored in Problems 3.1 and 3.15, result in slightly different future value computations.

Example 3.3

The future value of $100 after 1 year is calculated under various rate settings where the annual rate is specified only as .12. Using (3.3) and (3.4) we find the following future values:

1. Simple interest: $FV(1) = \$100(1 + .12) = \$112.00$
2. Compound monthly: $FV(1) = \$100(1 + .01)^{12} = \$112.68$
3. Compound continuously: $FV(1) = \$100\exp[(1)(.12)] = \$112.75$

The rate of compounding is directly related to the future value of the investment, with compounding continuously yielding the largest future value and simple interest producing the smallest yield.

In the analysis of financial and actuarial models, direct comparison of financial growth based on annual rates and continuous growth is useful. Equating (3.1) and (3.4) yields relations between discrete and continuous rates given by

$$i = \exp(\delta) - 1 \quad \text{and} \quad \delta = \ln(1 + i) \tag{3.5}$$

We refer to the general discrete model, where the intervals correspond to $(1/m)$th of a year with interest rate $r_{j/m}$, and we relate the continuous-type rates for discrete rates for period $E_{j/m}$ denoted by $\omega_{j/m}$ as

$$r_{j/m} = \exp(\omega_{j/m}) - 1 \quad \text{and} \quad \omega_{j/m} = \log(1 + r_{j/m}) \tag{3.6}$$

for $j \geq 1$. These relations play an important role in associations that exist among computations concerning differing financial and actuarial models. To compare interest and investment rates from different sources, the effective annual interest rate is often utilized and an illustrative example follows.

Example 3.4

An investment offer claims a continuous rate δ over a specified time. Using (3.5) the effective yearly interest rate can be computed for any continuous rate. A comparison of these rates is given in Table 3.1, where force of interest δ and the associated effective annual rate i are listed. Here i is higher than the corresponding continuous rate and can be utilized to make financial comparisons.

The time frame associated with the evaluation of a future financial action can be arbitrarily specified. Financial actions, such as investments or insurance payments, valued at F_t at future time $t > 0$ may be analyzed at the initial time $t = 0$. This is referred to as the present value and in the discrete time period setting, assuming compound interest, is found by inverting formula (3.2). The present value of a payment of F_t at future time t_n, corresponding to n future time periods, with associated rates i_j for $j = 1, 2, \ldots, n$ is

$$PV(t_n) = F_t \prod_{j=1}^{n} (1 + i_j)^{-1} \tag{3.7}$$

Figure 3.2 symbolizes this computation.

TABLE 3.1

Continuous Percentage Rate and Annual Rate Comparison

δ	.10	.12	.14	.16	.18	.20	.22
i	10.5	12.7	15.0	17.4	19.7	22.1	24.6

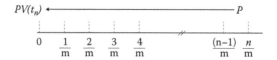

FIGURE 3.2
Present value of an annuity-immediate.

This is demonstrated in Example 3.5 where the rates are not level or constant.

Example 3.5

Over a 6-month period the monthly rates, i_j for $j = 1, ..., 6$, are .02, .03, .01, .04, .02, and .06. The present value of one unit, using (3.7), is found to be

$$PV(t_6) = (1.02)^{-1}(1.03)^{-1}(1.01)^{-1}(1.04)^{-1}(1.02)^{-1}(1.06)^{-1} = .838112$$

If we use the mean interest rate of .03 the present value of one unit over 6 months is $PV(t_6) = (1.03)^{-6} = .837484$, and the mean rate is found to under-value the present value computation.

Analogous to future value computations, other present value formulas are constructed. Consider the discrete time period setting where the periods are $(1/m)$th of a year and the period interest rate is level corresponding to a nominal annual interest rate $i^{(m)}$. The present value associated with an investment valued at P at future time t_n corresponding to n periods is

$$PV(t_n) = P\left(1 + \frac{i^{(m)}}{m}\right)^{-n} \tag{3.8}$$

Further, if the financial growth is compounded continuously at continuous rate δ the present value of a payment of F_t at future time t is found inverting (3.4) as

$$PV(t) = F_t \exp(-t\delta) \tag{3.9}$$

The relationships associated with present value computations are explored in Example 3.6.

Example 3.6

The present value of an investment is desired when after 5 years the investment is valued at $1,000. Here the yearly interest rate is specified only as a constant at .12. The present value after 5 years is compared under various interest schemes:

1. Simple interest: $PV(5) = \$1,000 \ (1 + 5(.12))^{(-1)} = \$625.00$
2. Compound monthly: $PV(5) = \$1,000 \ (1 + .12/12)^{-60} = \$550.45$
3. Compound continuously: $PV(5) = \$1,000 \ \exp(-.12(5)) = \$548.81$

As one would expect, we note that the greatest effect on the present value function occurs with continuous compounding where only an initial investment of $548.81 is required to produce the future value of $1,000.

It is sometimes useful to associate these calculations in terms of the time line. For a monetary investment the future value goes forward in time and the investment increases in value. The opposite is true for a present value of an investment where the time line proceeds backward and the value of the investment decreases. In the next section the discrete time period model is formalized.

3.1.2 General Period Discrete Rate Models

In this section we explore the general discrete rate models for both the future and present value functions that form a basis in terms of concepts and formulas for models described in later chapters. The future value of an investment of F_0, given by (3.2), can be rewritten using the technique of compounding period rates. Over time period $E_{j/m}$ the continuous rate, introduced in (3.4), is denoted ω_j for $j \geq 1$. If the principal is compounded m times a year, then over the time period $(0, w_n]$, where $w_n = n/m$, the cumulative interest rate δ_n and the accumulation function (AF) reflecting the increase in investment value are given by

$$\delta_n = \sum_{j=1}^{n} \omega_j \quad \text{and} \quad AF(t_n) = \exp(\delta_n) \tag{3.10}$$

Based on these notations the future value after n periods or at time t_n is

$$FV(t_n) = F_0 \, AF(t_n) \tag{3.11}$$

This notation is useful in various discrete time period settings. As before, the continuous-time structure is defined in terms of a limiting extension of the discrete time period model case as the time period shrinks.

In both the discrete time period and the continuous-time structures the present value function is constructed utilizing a discount function. Inverting (3.11) for future time t_n, a discount function defined over $(0, t_n]$, $DF(t_n)$, reflects a decrease in value and is defined as

$$DF(t_n) = \exp(-\delta_n) \tag{3.12}$$

The present value function corresponding to payment of F_t at future time $t = t_n$ is

$$PV(t_n) = F_t\, DF(t_n) \qquad (3.13)$$

This formula is a restating of (3.7), where the return rates may vary from period to period, and is useful in various settings, specifically when the rates are stochastic in nature.

3.1.3 Continuous-Rate Models

General continuous-rate models are now introduced where the rate is a function of future time t and is denoted ω_t. The future value and the present value functions are extensions of the continuous compounding rate formulas for discrete time periods. In the simplest case the continuous interest rate ω_t is nonnegative and integrable for $t > 0$ and the cumulative rate over $(0, t]$ is defined by

$$\delta_t = \int_0^t \omega_u\, du \qquad (3.14)$$

Analogous to the discrete time period models of Section 3.1.2 the accumulation and discount functions reflect the time effect on monetary values and over $(0, t]$ are given by

$$AF(t) = \exp(\delta_t) \quad \text{and} \quad DF(t) = \exp(-\delta_t) \qquad (3.15)$$

Following (3.11) and (3.13) the future value and present value functions corresponding to future time t are

$$FV(t) = F_0\, AF(t) \quad \text{and} \quad PV(t) = F_t\, DF(t) \qquad (3.16)$$

These formulas are robust in that they can be utilized to model and analyze a variety of financial contracts and settings. This is demonstrated in the next example and discussed in Problem 3.5.

Example 3.7

A principal of \$100 is invested for 1 year where the return rate over the year is estimated to be increasing linearly from .10 to .14. Hence $\omega_t = .04t + .10$ and the cumulative rate, from (3.14), is

$$\delta_1 = \int_0^1 \omega_u\, du = (.02t^2 + .10t)\big|_{t=1} = .12$$

Applying (3.15) and (3.16) gives accumulation function $AF(1) = \exp(.12) =$ 1.127497 and future value $FV(1) = \$100 \exp(.12) = \$112.75$. Coincidentally, this quantity matches with the compound continuous future value computed in Example 3.3.

3.2 Fixed-Rate Annuities

An annuity is a series of payments where the time between the payments is called the period. The value of the jth payment is denoted by π_j, for $j = 1, 2, \ldots$. The future value of the annuity is often called the amount of the annuity and represents the sum of the future values of the separate payments. In many financial and actuarial settings the payments are level or equal denoted by π. In an ordinary or annuity-immediate, payments are made at the end of time periods, while in an annuity-due, payments are made at the start of each time period. In this section the financial rates are considered nonstochastic and are fixed over future time periods. For a mathematical discussion of annuities we refer to Guthrie and Lemon (2004).

3.2.1 Discrete Annuity Models

The discrete time period setting of Section 3.1.2 holds where the time line is partitioned into disjoint time periods $E_{j/m} = ((j - 1)/m, j/m]$ and the period interest rate is r_j, for $j \geq 1$. Initially consider an annuity where payments, π_j, are made at the end of each period or at $t_j = j/m$, for $j \geq 1$. The amount or future value of the annuity at time t_n is the sum of the future values of the separate payments and is represented in Figure 3.3.

Utilizing (3.2) the future value takes the form

$$FV(t_n) = \pi_n + \sum_{s=1}^{n-1} \pi_s \prod_{j=s+1}^{n} (1 + r_j) \tag{3.17}$$

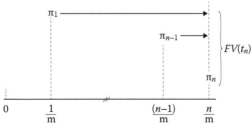

FIGURE 3.3
Future value of an annuity immediate.

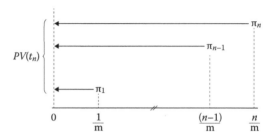

FIGURE 3.4
Present value of an annuity immediate.

Formulas similar to (3.17) are used to model other financial situations, in particular in the actuarial setting when payments may be made at the start of each period (see Problem 3.16).

In common financial and actuarial settings the structure of (3.17) is often simplified. For an ordinary annuity consisting of equal payments π made at the end of each period, we assume the monthly time period model where the year is divided into m equal-length time periods. The period rates are assumed to be all the same, given by $r = i^{(m)}/m$, and future value at time t_n, (3.17), reduces, using the summation formula (1.31), to

$$FV(t_n) = \pi \sum_{j=0}^{n-1} (1+r)^j = \pi \frac{(1+r)^n - 1}{r} \tag{3.18}$$

Formula (3.18) has many applications, such as house mortgages, loan payments, and the evaluation of future pension benefits, and is demonstrated in Problem 3.3 and the following example.

Example 3.8

A deposit of $150 is made at the end of each month for 4 years where the nominal annual interest rate is 3.6%. The interest rate per month is $r = .036/12 = .003$ and the future value of the annuity at $t_{48} = 4$ years from (3.18) is

$$FV(4) = \$150 \frac{(1.003)^{48} - 1}{.003} = \$7,731.76$$

The total future value of the payments, without interest, is $7,200, which is approximately 7% less than the interest enhanced or loaded monetary sum.

The formula for the present value of a discrete annuity can be written in terms that utilize the cumulative rate defined in (3.10). Letting the annuity payments be equal, for integers $s < n$, the accumulation function over the time interval between t_s and t_n is defined to be

$$AF(t_s, t_n) = \exp(\delta_n - \delta_s) \tag{3.19}$$

where δ_n is given by (3.10). The future value of the annuity at time t_n, given by (3.17), becomes

$$FV(t_n) = \pi \left(1 + \sum_{j=1}^{n-1} AF(t_j, t_n) \right) \tag{3.20}$$

These formulas are general in nature and can be specified to fit standard financial rate problems, such as life annuities and pension plans, discussed later in Sections 6.5 and 7.3, respectively.

As in the case of bank interest or principal investments, the aggregate value of the annuity payments may be projected backward in time to $t = 0$ to find the present value of an annuity. The present value of an annuity-immediate up to time t_n is the sum of the present values of the separate payments, π_j for $j \geq 1$, made at the end of each time period as represented in Figure 3.2.

Applying (3.7) the present value

$$PV(t_n) = \sum_{s=1}^{n} \pi_s \prod_{j=l}^{s} (1 + r_j)^{-j} \tag{3.21}$$

where r_j is the jth period financial rate. In the common setting where equal payments, π, are made at the end of each $(1/m)$th of a year, (3.21) can be simplified. If the period rates are all $r = i^{(m)}/m$, the present value of the annuity up through time t_n can be written as

$$PV(t_n) = \pi \sum_{j=l}^{n} (1 + r)^{-j} = \pi \frac{1 - (1 + r)^{-n}}{r} \tag{3.22}$$

Again, this formula is a result of the summation formula (1.31) and an example of its application now follows.

Example 3.9

A series of payments of \$150 made at the end of each month is made in an account for 5 years where the nominal annual interest rate is fixed at $i = .036$. Thus, $r = .036/12 = .003$, $\pi = \$150$, and $t_{60} = 5$, and utilizing (3.22),

$$PV(5) = \$150 \frac{1 - (1.003)^{-60}}{.003} = \$8,225.23$$

The sum of the payments without interest is \$9,000, and we observe a decrease of $774.77/9,000 = 8.6\%$. The lower value of the present value calculation in the presence of interest is due to the growth of money over time.

The formulas for the present value of an annuity can be written in terms of the discount function (3.15). If the payments are level at π, the present value of the ordinary annuity or annuity-immediate up to time $t = t_n$, from (3.21), is

$$PV(t_n) = \pi \sum_{j=1}^{n} DF(t_j) \tag{3.23}$$

If the period payments vary, adjustments to (3.23) are required.

In actuarial calculations, such as life insurance and life annuity expectations, the annuities-due, where payments valued at one unit are made at the start of each year for k years, play a central role. If the yearly interest rate is i, the future value of the annuity at the end of year k is computed using standard summation (see Problem 1.3) and is denoted

$$\ddot{s}_{\overline{k|}} = \sum_{r=1}^{k} (1+i)^r = \frac{(1+i)^k - 1}{d} \tag{3.24}$$

where $d = i/(1 + i)$. Similarly, the present value of unit payments made at the start of each year for k years is

$$\ddot{a}_{\overline{k|}} = \sum_{r=0}^{k-1} (1+i)^{-r} = \frac{1 - (1+i)^{-(k)}}{d} \tag{3.25}$$

Through these formulas the effect of time is evident and the notations given in (3.24) and (3.25) are given for completeness, as they are commonly used in actuarial science texts and literature.

3.2.2 Continuous Annuity Models

Similar to the continuous-rate models of Section 3.1.3, annuity models can be constructed from a continuous-time point of view. Continuous annuity models, of which there are two types, are extensions of the basic discrete time period annuity models. The first type is a discrete time period model with continuous-rate structure. In this case payments follow a discrete pattern, corresponding to interval time periods, and the rate is modeled by a continuous function. In the second, the more traditional treatment, payments are made continuously through time periods at time-sensitive rates. In all cases, as presented in Section 3.1.2, the cumulative rate is defined by (3.14), and for times between s and t, for $s < t$, the accumulation function is given by

$$AF(s,t) = \exp\left(\int_s^t \omega_u \, du \right) \tag{3.26}$$

In a discrete time period annuity the future value is computed by applying (3.26) to (3.20). An example is given in Example 3.10.

Example 3.10

In a level annuity $1,000 is invested at the end of each quarter for 1 year. The continuous rate is estimated to be increasing over the next year according to the function $\delta_t = .02t + .04$ for $0 \le t \le 1$. From (3.26) the accumulation function corresponding to future times $s < t$ is computed as

$$AF(s,t) = \exp\left(\int_s^t (.02u + .04)du\right) = \exp(.01(t^2 - s^2) + .04(t - s))$$

and the future value of the annuity after 1 year, utilizing (3.20), is computed as

$$FV(1) = \$1,000\left(1 + \sum_{j=1}^{3} AF\left(\frac{j}{4}, 1\right)\right) = \$1,000(4.0825209) = \$4,082.52$$

In these constructions, we note the potential flexibility in the structure of the accumulation function, where a variety of shapes for financial growth can be modeled.

The second type of continuous annuity is considered where the payments themselves are considered to be continuous in nature. For an annuity with continuous payments, denoted at time t by π_t, the future value and present value associated with future time t take the form

$$FV(t) = \int_0^t \pi_s AF(s, t)\, ds \quad \text{and} \quad PV(t) = \int_0^t \pi_t\, DF(t)\, dt \qquad (3.27)$$

where formulas (3.14), (3.15), and (3.26) are utilized. In the case of level continuous payments, which are often used in the analysis of life annuity premiums, the payments are constant with respect to time, where $\pi_t = \pi$ for $t \ge 0$ and formulas (3.27) are suitably adjusted. In Problem 3.7 the constant payments and rate situation is explored.

Example 3.11

We estimate the annuity structure presented in Example 3.10 given as $\delta_t = .02t + .04$ for $0 \le t \le 1$, where the annuity payments are level continuous payments associated with a yearly value of $\pi = \$4,000$. Utilizing (3.27) the present value becomes

$$PV(1) = \$4000 \int_0^1 \exp(-.01s^2 - .04s)\, ds = \$4530.37$$

The integral in the calculation is evaluated by first completing the square inside the exponential function and then utilizing the normal distribution. Further, present value computations associated with arbitrary future time t are easily made by adjusting the integral limits.

3.3 Stochastic Rate Models

In practice it is often the case that the return rates for investments or bank interest rates corresponding to loans or annuities are stochastic in nature. In these cases the rates are modeled as random variables with associated distributions. The effect of varying interest and return rates on financial and actuarial computations has been the focus of much work. The effects of interest rates on surrender features of insurance have been explored by Grosen and Jorgensen (1997). Further, examples of modeling the force of interest using stochastic processes and time series models are given in Nielsen and Sandmann (1995), Frees (1990), and Panjer and Bellhouse (1980).

As in the previous sections, there are two basic cases for the rate random variables designated by discrete and continuous time period models. In the discrete case years are partitioned into m equal time periods. The rate random variable $\omega_{j/m}$ over the jth interval has mean and variance $\gamma_{j/m}$ and $\beta_{j/m}^2$ for $j \geq 1$. In the continuous setting the mean and variance of the random rate ω_u are integrable functions given by γ_u and β_u^2 for $u \geq 0$. Basic statistical concepts and measures, including percentiles, expectations, and prediction intervals, are utilized to analyze present value and future value functions in the presence of stochastic rates.

Following the approach of Kellison (2009, Chapter 12) the normal distribution plays a central role in the analysis of stochastic rates models. This approach is equivalent to modeling of stochastic rates using the lognormal random variable introduced in Section 1.6.2 (see Problem 3.14). To assess the applicability of modeling rates with the normal distribution, graphical assessments, such as probability or hazard plots, may be employed. For a review of the usage of these plots in the selection of probability distributions for empirical data see Nelson (2005, Chapter 3) and Tobias and Trindade (1995, Chapter 6). The topic of applying probability plots to empirical data is explored in Section 2.8.1.

3.3.1 Discrete Stochastic Rate Model

In this section, the approach of Borowiak (1999b) is utilized to model stochastic rates in the discrete time interval setting. The equal-width time periods E_j are associated with rates ω_j that are considered random variables with mean γ_j and variance β_j^2 for $j \geq 1$. The rates are assumed to follow a specified distribution, such as the normal or uniform distribution. In particular, if $\ln(1+i)$ is normally distributed, probability computations and prediction intervals for future and present value functions exist. This is introduced in the following investment example.

Example 3.12

An investment of F_0 is made for n periods where the interest is compounded according to monthly time periods. The period rates are assumed to be independent and identically distributed normal random variables. This distributional assumption implies that the cumulative rate (3.10) is a normal random variable where

$$\delta_n \sim N(\mu_n, \sigma_n^2) \quad \text{where } \mu_n = \sum_{i=1}^{n} \gamma_i \quad \text{and} \quad \sigma_n^2 = \sum_{i=1}^{n} \beta_j^2 \qquad (3.28)$$

For constant c, the probability that the future value exceeds c is found utilizing the standard normal random variable, as in (1.19), and is

$$P(FV(t_n) > c) = P(F_0 \exp(\psi_n) > c) = 1 - \Phi\left(\frac{\ln\left(\dfrac{c}{F_0}\right) - \mu_n}{\sigma_n}\right) \qquad (3.29)$$

where $\Phi(c)$ is the standard normal distribution function. Further, applying the normality in (3.28) a $(1 - \alpha)100\%$ prediction interval (PI) for ψ_n following (2.16) is $(\mu_n - z_{1-\alpha/2}\,\sigma_n, \mu_n + z_{1-\alpha/2}\,\sigma_n)$ and directly yields the PI for the future value at time t_n:

$$P\exp(\mu_n - z_{1-\alpha/2}\sigma_n) \le FV(t_n) \le P\exp(\mu_n + z_{1-\alpha/2}\sigma_n) \qquad (3.30)$$

The reader is left to form prediction intervals for the present value function, and this construction is developed in Problem 3.8. For example, if $F_0 = \$10,000$, $m = 12$, $\gamma_j = .01$, and $\beta_j = .01$, for all $j \ge 1$, then after 5 years the probability of the future value exceeding \$20,000 following (3.30) is

$$P(FV(5) > 20,000) = 1 - \Phi\left(\frac{\left(\ln\left(\dfrac{2}{1}\right) - 60(.01)\right)}{(60^{1/2})(.01)}\right) = 1 - \Phi(1.21) = .1132$$

From this we interpret that there is an 11.32% chance that the future value after 5 years will exceed \$20,000. Also, following (3.30) a 95% PI for the future value after 5 years is

$$\$15,655 = \$10,000 \exp(.6 - 1.96(.006)^{1/2}) \le FV(5) \le \$10,000\exp(.6 + 1.96(.006)^{1/2})$$

$$= \$21,208$$

As with any statistical interval estimate, statistically reasonable values for the future value are described in the prediction interval.

Expectations and variances for future and present value functions can be computed following basic statistical techniques. The expected value of the rate and discount functions in (3.15) is

$$E\{AF(t_n)\} = E\{\exp(\delta_n)\} \quad \text{and} \quad E\{DF(t_n)\} = E\{\exp(-\delta_n)\} \tag{3.31}$$

Applying (3.16) the expected future and present values corresponding to principal P and time t_n are

$$E\{FV(t_n)\} = F_o\, E\{AF(t_n)\} \quad \text{and} \quad E\{PV(t_n)\} = F_t\, E\{DF(t_n)\} \tag{3.32}$$

Expectation (3.32) can sometimes be streamlined using the moment generating function (mgf) (see Section 1.4). In the discrete monthly period model, consisting of m equal time periods, if the rates corresponding to the sequential time periods are independent and identically distributed (iid) random variables with corresponding mgf $M_m(t)$, then (3.31) becomes

$$E\{AF(t_n)\} = [M_m(1)]^n \quad \text{and} \quad E\{DF(t_n)\} = [M_m(-1)]^n \tag{3.33}$$

From (3.33) we note that the expected future and present values are computed using mgf evaluated at 1 and -1 and are

$$E\{FV(t_n)\} = P\,[M_m(1)]^n \quad \text{and} \quad E\{PV(t_n)\} = P\,[M_m(-1)]^n \tag{3.34}$$

A variety of probability distributions and corresponding moment generating functions can be applied to model the financial rates. A set of examples using the normal distribution to model the financial rates, demonstrating concepts and formulas, are now explored. The use of the uniform distribution is considered in Problem 3.10.

Example 3.13

The investment in Example 3.12 is considered where the forces of interest are assumed to be iid normal random variables with period mean and variance given by, respectively, γ and β^2. Using (3.34) and the normal mgf (1.41), the expected accumulation function is

$$E\{AF(t_n)\} = [M_m(1)]^n = [\exp(\gamma + (1/2)\,\beta^2)]^n = \exp(n\gamma + n\beta^2/2) \tag{3.35}$$

In particular, if \$10,000 earns interest compounded monthly, find the expected future value after 5 years where, in terms of each month, $\gamma = .01$ and $\beta = .01$, and the expected accumulation function is

$$E\{AF(5)\} = \exp(60((.01) + (1/2)(60)(.01^2)) = 1.82759$$

From (3.32) the expected future value is

$$E\{FV(5)\} = \$10,000(1.82759) = \$18,275.34$$

As anticipated, the expected future value is within the prediction bounds computed in Example 3.12.

Example 3.14

In this example we apply an $AR(1)$ system introduced in Section 2.7 to a collection of rates over time. Utilizing the notations of that section, based on n periods $\delta_n = S_n$, we have

$$\delta_n \sim N(\mu_n, \sigma_n^2) \quad \text{where} \quad \mu_n = \sum_{i=1}^{n} \gamma_i \quad \text{and} \quad \sigma_n^2 = \sigma^2 \sum_{i=1}^{n} \frac{(1 - \phi^{n-i+1})^2}{(1 - \phi)^2}$$

(3.36)

Using these formulas, statistical calculations, such as survival probabilities and prediction intervals given in (3.29) and (3.30), can be computed based on hypothetical parameter values. This procedure is extended to empirical data in Example 3.16.

Example 3.15

The discrete monthly period model consisting of m equal time periods per year holds. The expected present value of a payment in t years is computed where the forces of interest are assumed to be iid normal random variables with monthly period values $\gamma_j = \gamma$ and $\beta_j^2 = \beta^2$ for all $j \geq 1$. Using the normal mgf (1.41), the expected present value associated with future time $t = t_n$ is

$$E\{PV(t_n)\} = F_t \exp(n(-\gamma + (1/2)\beta^2)) \tag{3.37}$$

where $n = tm$. For example, if $m = 12$, $P = \$1,000$, $\gamma = \beta = .01$, and $t = 10$, then $n = 120$, and using (3.37),

$$E\{PV(10)\} = \$1,000 \exp(120 \, (-.01 + .01^2/2)) = \$303.01$$

We note that if the force of interest is fixed at .12 per year, we find $PV(10) = \$301.19$.

In practice, distribution parameters, such as means and variances, are estimated from observed data. Statistical formulas are often applied to observed data by treating the point estimates as the fixed constants and substituting computed values for unknown parameters. This is theoretically sound with statistical estimates that converge to their associated estimated parameters as the sample size gets large. The reliability of these substituted estimator procedures can be evaluated using simulation methods (see Section 9.2). In Example 3.16 empirical data are considered and the method of substituting estimators is demonstrated.

Example 3.16

At the end of each of the periods utilizing (3.10) and (3.11), the value of an initial investment of $X_0 = F_0$ takes the form $X_j = FV(j) = F_0 \exp(\delta_j)$ for $j = 1, 2, \ldots, n$. Thus, for period E_j, the estimated instantaneous return rate is computed as

$$\omega_{j+1} = \ln\left(\frac{X_{j+1}}{X_j}\right) \quad \text{for} \quad j = 0, 1, \ldots \tag{3.38}$$

and we let

$$Z_j = \omega_j - \bar{\omega} \quad \text{for} \quad j \geq 1$$

where $\bar{\omega}$ is the sample mean. For $n = 11$ these observed values are given in Table 3.2. The future value of the investment is to be estimated at time $t = 12$. The rates are assumed to be normal with a constant mean, and we apply, one at a time, both the independent rate and $AR(1)$ models.

In the independent rate model the point estimates of the parameters are the sample mean and variance of ω_j given by

$$\bar{\omega} = .006996 \quad \text{and} \quad \hat{\sigma}^2 = \frac{\sum Z_i^2}{n-1} = .011237$$

The estimated expected accumulation at time $t = 12$ is

$$E\{AF(12)\} = \exp(12(.006996) + (1/2)(12)(.011237)) = 1.163432$$

and the estimated expected future value (3.32), based on initial value 12.5, becomes

$$E\{FV(12)\} = 12.5\,(1.163432) = 14.5429$$

TABLE 3.2

Empirical Return Rate

Time	0	1	2	3	4	5	6	7	8	9	10	11
X_j	12.5	10.5	11.5	10.7	12.5	12.5	12.2	12.5	14.7	14.2	12.7	13.5
ω_j	−.174	.091	−.072	.156	.000	−.024	.024	.162	−.035	−.112	.062	
Z_j	−.181	.084	−.079	.148	−.007	−.081	.017	.155	−.042	−.119	.054	

Using the $AR(1)$ approach the least squares point estimates are computed using (2.118) and (2.119). Letting $W = P^{-1} Z$ we find

$$\hat{\phi} = \sum Z_{i+1} \frac{Z_i}{\sum Z_i^2} = -.2124887 \quad \text{and} \quad \hat{\sigma}^2 = .0100199$$

Utilizing the formulas in (3.36), $\hat{\mu}_{12} = 12(.006996) = .083952$ and $\hat{\sigma}_{12}^2 = .0367907$. The estimate of the expected accumulation function is

$$E\{AF(12)\} = \exp(.083952 + (1/2)(.0367907)) = 1.107768$$

and the expected present value estimate is

$$E\{PV(12)\} = 12.5(1.107768) = 13.8471$$

We remark that for both the iid and $AR(1)$ models, prediction intervals can be constructed using asymptotic statistics, relying on the asymptotic normality of the least squares estimators.

In the case of increasing or decreasing trends in financial rates, statistical model fitting may be applied. If the changes in the financial rates are linear with time, then a linear regression model can be utilized. An Excel example is outlined in Problems 3.18 and 3.19 and Example 3.17. If the financial rate is not linear in time, then transformations or other regression techniques, such as nonlinear regression, are possible.

Example 3.17

The investment data of Example 3.16 given in Table 3.2 are considered where the rates are assumed to be from a normal distribution. The future value at $t = 12$ is to be estimated. To do this we apply regression techniques (see Section 2.6) defining the response variable as $Y_j = \ln(\frac{X_j}{X_0})$ and predictor variable $J = j$ for $j = 1, 2, \ldots, 12$. The least squares estimators (2.105) are found to be

$$\hat{\beta}_0 = -.15919 \quad \text{and} \quad \hat{\beta}_1 = 0.025698$$

Utilizing the squares line (2.106), the point estimates for time 12 are

$$\hat{\delta}_j = -.15919 + .025698j \quad \text{and} \quad \hat{\delta}_{12} = .14918$$

This result indicates an increasing trend in the return rates. In fact, for this regression the sample correlation coefficient is found to be $r = .8106$, indicating a linear fit. The estimated value at time $t = 12$ is

$$\hat{F}V(12) = 12.5 \exp(.14918) = 14.511$$

Prediction intervals for future values can be computed by first computing a confidence interval for δ_j and applying (3.10) and (3.11).

3.3.2 Continuous Stochastic Rate Models

This section presents a general continuous-rate approach, based on stochastic normal rates, to model present and future values. For a fixed future time t, cumulative rate, δ_t, is considered a random variable where the mean and variance are based on integrable functions γ_t and β_t for $t \geq 0$. Based on future time t, the mean and variance of δ_t are

$$\mu_t = \int_0^t \gamma_u \, du \quad \text{and} \quad \sigma_t^2 = \int_0^t \beta_u^2 \, du \tag{3.39}$$

Applying a specific probability distribution to the cumulative rate ψ_t allows the calculation of survival probabilities, percentiles, and expectations for future and present value functions.

Example 3.18

The investment over time with normally distributed rates, discussed in Example 3.12, is altered so that the cumulative rate is continuous and follows a normal distribution. Hence,

$$\delta_t \sim N(\mu_t, \sigma_t^2) \quad \text{where} \quad \mu_t = \int_0^t \gamma_u du \quad \text{and} \quad \sigma_t^2 = \int_0^t \beta_u^2 \, du \tag{3.40}$$

Utilizing (3.40), many statistical and probabilistic computations are possible. For example, the probability the future value exceeds c is

$$P(FV(t) > c) = 1 - \Phi\left(\frac{\ln\left(\dfrac{c}{F_0}\right) - \mu_t}{\sigma_t}\right) \tag{3.41}$$

A $(1 - \alpha)100\%$ prediction interval for the future value constructed using standard normal percentiles is

$$F_0 \exp(\mu_t - z_{1-\alpha/2}\,\sigma_t) \leq FV(t) \leq F_0 \exp(\mu_t + z_{1-\alpha/2}\,\sigma_t) \tag{3.42}$$

The construction of counterparts of (3.41) and (3.42) in the case of the present value function is left to the reader in Problem 3.11.

General formulas for the expected future and present value functions given in (3.31) and (3.32) are adapted to this setting of continuous stochastic rate structure. The rates are assumed to be normal random variables following (3.40) so that the expected accumulation function associated with future time t is

$$E\{AF(t)\} = E\{\exp(\delta_t)\} = \exp(\mu_t + (1/2)\,\sigma_t^2) \qquad (3.43)$$

while the expected discount function is

$$E\{DF(t)\} = \exp(-\mu_t + (1/2)\,\sigma_t^2) \qquad (3.44)$$

Analogous to the discrete case, the expected future and present value functions are

$$E\{FV(t)\} = P\,E\{AF(t)\} \quad \text{and} \quad E\{PV(t)\} = P\,E\{DF(t)\} \qquad (3.45)$$

These expectation formulas are utilized when the shape of the mean and variance functions can be modeled, as the following two examples show.

Example 3.19

A principal of $100 is invested for 1 year, where the mean associated with the rate is defined by the integrable function $\gamma_u = .04u + .10$ and the variance is defined by $\beta_u^2 = .002(1 + .001u)$. Using the structure in (3.39) the mean and variance for 1 year are

$$\mu_1 = \int_0^1 (.04u + .1)\,du = .12 \quad \text{and} \quad \sigma_1^2 = \int_0^1 (.002(1 + .001u))\,du = 1.0025$$

Applying the normal distribution to model the accumulation computation (3.43) is

$$E\{AF(1)\} = \exp(.12 + (1/2)(1.0025)) = 1.86125$$

and the expected future value follows from (3.45) and is computed as $186.13.

Example 3.20

In this example, a continuous rate follows a normal distribution through time where the mean and variance are constant. From (3.39), this construction implies the mean and variance over time t are $\mu_t = t\mu$ and $\sigma_t^2 = t\sigma^2$, where the yearly mean and variance are μ and σ^2, respectively. Statistical formulas (3.39) through (3.42) hold, as do present value function counterparts. For example, the probability that the present value, associated with future time t, is at most c is

$$P(PV(t) \le c) = 1 - \Phi\left(\frac{-\ln\left(\dfrac{c}{F_0}\right) - t\mu}{t^{1/2}\sigma}\right) \qquad (3.46)$$

From (3.46), the $(1 - \alpha)100\%$ percentile for this present value function is given by

$$PV(t)_{1-\alpha} = F_0 \exp(-t\mu + z_{1-\alpha} t^{1/2} \sigma) \qquad (3.47)$$

Further, following formulas (3.43), (3.44), and (3.45), the expectation of the future value and present value functions associated with $T = t$ are

$$E\{FV(t)\} = F_0 \exp(t(\mu + \sigma^2/2)) \quad \text{and} \quad E\{PV(t)\} = F_t \exp(t(-\mu + \sigma^2/2)) \qquad (3.48)$$

In particular, consider a monetary sum of \$1,000 that is required in 18 months. How much money do we need to invest now to have confidence that the future obligation will be satisfied? The yearly investment return rate is estimated to be between 6 and 14%. The estimated rate is centered at $i = 10\%$, so that the continuous-rate point estimated is taken to be $\ln(1.1) = .0953$, with a likely range of $\ln(1.14) - \ln(1.06) = .07277$. Using a 2 standard deviation interval estimate, we estimate the yearly quantities $\mu = .0953$ and $\sigma = .07277/4 = .0182$. The expected required investment from (3.48) is

$$E\{PV(1.5)\} = \$1,000 \exp(1.5(-.0953 + .0179^2/2)) = \$1,000(.866577) = \$866.58$$

A more conservative estimate results from using the 95th percentile for the present value. From (3.47)

$$PV(1.5)_{.95} = \$1,000 \exp(-1.5(.0953) + 1.645(1.5^{1/2})(.0179)) = \$898.62$$

The difference in these two values is due to the skewness of the present value function and the probability requirement.

3.3.3 Discrete Stochastic Annuity Models

The case of discrete time period ordinary annuities, consisting of payments made at the start of each time period, with level or constant payments, is the topic for this section. Other annuities with varying payments can be modeled by adjustments to the formulas that follow. For annuities with level payments of π made at the start of each time interval, the future and present values are given by

$$FV(t_n) = \pi \sum_{j=0}^{n-1} AF(t_j, t_n) \quad \text{and} \quad PV(t_n) = \pi \sum_{j=0}^{n-1} DF(t_j) \tag{3.49}$$

In the annuity model, setting the computation of probabilities leading to percentiles and prediction intervals is not straightforward. These statistical computations can most easily be handled by statistical approximation techniques or statistical simulations as discussed in Section 9.2.

Assuming (3.10) holds where the cumulative interest rate is a normal random variable, prediction intervals can be computed. For the future value associated with future time t_n, defined by (3.49), a $(1 - \alpha)100\%$ *PI* is given by

$$\pi \sum_{j=0}^{n-1} \exp(j\gamma - z_{1-\alpha/2} j^{1/2}\beta) \le FV(t_n) \le \pi \sum_{j=0}^{n-1} \exp(j\gamma + z_{1-\alpha/2} j^{1/2}\beta) \tag{3.50}$$

To derive (3.50), accumulation function (3.19) and percentile (3.47) are required.

For discrete time period, annuities expectation formulas for the future and present value functions can be derived. For an ordinary discrete annuity, the expected future and present values are, respectively,

$$E\{FV(t_n)\} = \pi \left(1 + \sum_{j=l}^{n-1} E\{AF(t_j, t_n)\} \right) \quad \text{and} \quad E\{PV(t_n)\} = \pi \left(\sum_{j=1}^{n-1} E\{DF(t_j)\} \right) \tag{3.51}$$

In the case of iid rates, the formulas in (3.51) can be written in terms of the corresponding mgfs similar to that of (3.33) and are left to the reader. A computational example for the expectation for the present value of an annuity, using the normal mgf, follows.

Example 3.21

An annuity-due continues for n periods where equal payments of π are made at the start of each period. Employing (3.51), the expected present value associated with future time t_n is

$$E\{PV(t_n)\} = \pi \sum_{j=1}^{n} [M_m(-1)]^j = \pi \frac{M_m(-1) - M_m(-1)^{n+1}}{1 - M_m(-1)} \tag{3.52}$$

where $M_m(-1) = \exp(-\mu_m + (1/2)\sigma_m^2)$. In particular, consider payments of $\$1,000$ made at the start of each month for 25 years where the interest rates are assumed to be iid normal with yearly mean $\mu = .08$ and yearly variance $\sigma^2 = .024$. Hence, $\mu_{12} = .006666$ and $\sigma_{12}^2 = .002$, and we find $M(-1)$ $= .994349$ and $E\{PV(25)\} = \$1,000(143.631538) = \$143,631.15$. We note that if the rate is deterministic, then $\beta_j = 0$, $M(-1) = .9933555$ and $E\{PV(25)\} = \$1,000(129.26773) = \$129,167.73$.

From our examples and computations, we observe that the stochastic nature of the financial rates influences the annuity more than the interest. This is due to the propagation of the stochastic effect present in formulas such as (3.50) and (3.51).

3.3.4 Continuous Stochastic Annuity Models

This chapter ends with a discussion of a stochastic annuity structure applied to continuous-type annuities. For continuous annuities with constant payment π, the future and present values, associated with future time t, are defined by

$$FV(t) = \pi \int_0^t AF(t,s)\, ds \quad \text{and} \quad PV(t) = \pi \int_0^t DF(s) \tag{3.53}$$

As in the case of discrete stochastic annuities, percentiles and prediction intervals can be computed by simulation and approximation methods. The formulation of prediction intervals is now discussed assuming stochastic normal rates.

Prediction intervals for the future and present value functions in the continuous model can be found similar to the discrete setting when the rates are assumed to be normal random variables. Assuming yearly mean μ and variance σ^2, using (3.40) with constant mean and variance functions, a $(1-\alpha)100\%$ prediction interval for the future value function takes the form

$$\pi \int_0^t \exp(s\mu - z_{1-\alpha/2} s^{1/2} \sigma)\, ds \leq FV(t) \leq \pi \int_0^t \exp(s\mu + z_{1-\alpha/2} s^{1/2} \sigma)\, ds \tag{3.54}$$

In applications, the evaluation of (3.54) likely will require numerical integration procedures or packages.

The expected future and present values for continuous stochastic annuities can be computed assuming an underlying distribution for rates. Based on the constructions in (3.53), for future time t these expectation formulas are

$$E\{FV(t)\} = \pi \int_0^t E\{AF(t,s)\}\, ds \quad \text{and} \quad E\{PV(t)\} = \pi \int_0^t E\{DF(s)\}\, ds \tag{3.55}$$

As mentioned before, numerical methods and approximations can be used to evaluate (3.55) for specific cases.

Example 3.22

A continuous annuity is considered where the rates are iid and $\mu_t = t\mu$ and $\sigma_t^2 = t\sigma^2$ for $t > 0$. If we assume the rates are distributed as normal random variables, then $E\{DF(t)\} = \exp(-t(\mu - (1/2)\sigma^2)$ and the expected present value is

$$E\{PV(t)\} = \pi \int_0^t \exp(-s(\mu - (1/2)\sigma^2))\, ds = \pi(1 - \exp(-t(\mu - (1/2)\sigma^2))/(\mu - (1/2)\sigma^2)$$

Letting $\pi = \$12{,}000$, yearly parameters $\mu = .08$ and $\sigma^2 = .024$, and applying (3.55), we compute EPV(25) = $\$12{,}000(12.01935) = \$144{,}232.32$. Since the payments are made in a continuous manner, the expected present value is more than the discrete time period counterpart.

Problems

3.1. If $950 is invested in a bank account where the effective annual interest rate is 6% using (3.3), find the future value after 3 years if the interest is compounded (a) yearly, (b) monthly, and (c) continuously.

3.2. a. If the effective annual interest or return rate is 5%, what is the corresponding continuous compounding rate?

   b. If the yearly interest rate of 6% is compounded monthly, what is the corresponding effective yearly interest rate?

3.3. At the end of every month a deposit of $100 is made in a bank account where the annual interest rate is 6%. How much money is in the account after (a) 5 years, (b) 10 years, and (c) 20 years?

3.4. A university wants to endow a professor position where $100,000 is required annually. Money is deposited in a trust and only the interest or growth is spent. How much money is needed if it can be invested at an effective annual rate of 7.5%?

3.5. Over a 2-year period the return rate on an investment grows from 4% to 10% in a linear fashion.

   a. Find the δ_t function in (3.14).

   b. What is the future value of an investment of $2,000 after 18 months?

3.6. Consider a continuous annuity with unit premium and present value given by (3.27) where AF $(0, t)$ is given in Example 3.10. Show $PV(t) = F_t \, (2\pi 50)^{1/2} \exp(.04)[\Phi((t + 2)/50^{1/2}) - \Phi(2/50^{1/2})]$.

3.7. After 5 years a payment of \$10,000 is due where the annual interest is $i = .06$. To finance this payment an annuity with level payments of π per year is made. Find the value of π if (a) payments are made at the start of each month, and (b) payments are made continuously, from a bank account, throughout the 5 years.

3.8. Using (3.12), (3.13), and (3.28) construct the formula for a $(1-\alpha)100\%$ prediction interval for the present value function associated with amount F_t at future time $t = t_n$ assuming a discrete stochastic rate structure based on monthly equal-width time periods.

3.9. A sum of \$5,000 is invested where the yearly return rate, δ, follows Example 3.20 and is considered a normal random variable with yearly mean .08 and standard deviation .03. After 5 years find (a) the expected future value, (b) the probability the future value is at most \$8,000, and (c) a 95% prediction interval for the future value.

3.10. For an investment, return rates are independent and random, taking values between .06 and .12. Considering the accumulation as uniform random variables, after 5 years find (a) the expected accumulation function (3.33), (b) the expected discount function (3.33), and (c) the expected future value function based on an investment of \$5,000.

3.11. Consider the continuous stochastic rate structure given in (3.39) and (3.40). For the present value function give counterpart formulas to (3.41) and (3.42), namely, (a) $P(PV(t) > c)$ for $c > 0$ and (b) a $(1 - \alpha)100\%$ prediction interval for the present value function associated with future time $T = t$.

3.12. You are to receive a lump sum payment of \$20,000 in 10 years. You can invest money where the return rate is distributed normal with mean .08 and standard deviation .025. For this future payment find (a) the expected present value and (b) a 95th percentile for the present value. What does this value mean?

3.13. An annuity-due making payments at the start of each month is conducted for 1 year. If the payments are for \$1,000 and the investment rate is a random variable, compute the expected present value after 1 year using (3.52) for (a) iid normal investment rates where the yearly mean is 8% with a standard deviation of .03, and (b) iid uniformly distributed yearly investment rates between 6 and 12%.

3.14. For rate function (3.10) let $Y_j = \exp(\delta_j)$ for $1 \le j \le n$.

    a. Give an expression for $AF(t_n)$ in terms of Y_j.

    b. If $\delta_j \sim N(\mu_j, \sigma_j^2)$ use (1.51) to show Y_j is a lognormal random variable as presented in Section 1.47 for $1 \le j \le n$.

Excel Problems

3.15. We invest \$500 where over a 6-month period the annual rates are given as

Month	1	2	3	4	5	6
Annual rate	.05	.06	.06	.07	.07	.09

    a. Give the value of the rate and accumulation function over the 6-month period.

    b. Compute the future value after 6 months.

    c. If \$800 were required to pay a note due at the end of the 6-month period, how much money would need to be invested at the beginning?

    Excel: PRODUCT.

3.16. Annuity payments π_j are made at the start of each month where the return rates for 6 months are given in Problem 3.15. The payments are \$200 for 2 months, \$250 for 2 months, and \$300 for 2 months. What is the future value of the annuity after the 6 months?

    Excel: Basic operations and Function.

3.17. Under the setting and results of Problem 3.6 compute the present value of 4,000 after 5 years.

    Excel: NORMSDIST.

3.18. The opening stock prices for 14 weeks are found to be

Week	1	2	3	4	5	6	7	8	9	10	11	12	13	14
Price	6.2	6.5	5.9	6.3	6.8	7.2	7.0	7.3	7.9	8.1	8.2	7.3	8.5	8.7

We follow the procedure of Example 3.17, where $X = $ Price and $\delta_j = \ln(X_j/X_1)$ for $j = 1, 2, \ldots, 13$, and we consider the regression of j on δ_j.

    a. Give the estimated linear regression equation for δ_j.

    b. Estimate the price at the end of the 14th week.

    Excel: Data, Data Analysis, Regression.

3.19. Apply the approach of Example 3.16 on the data of Problem 3.18.

a. Compute $\bar{\omega}$ and $\hat{\sigma}^2$.

b. Estimate the price at the start of week 14.
Excel: Basic functions, SUMPRODUCT.

Solutions

3.1. a. $950(1.06)^3 = 1{,}131.47$.

b. $950(1.005)^{36} = 1{,}136.85$.

c. $950\exp(.18) = 1{,}137.36$.

3.2. a. (3.5) gives $\delta = \ln(1.05) = .04879$.

b. $1.06 = (1+i/12)^{12}$ gives $i = 12[1.06^{1/12} - 1] = .05841$.

3.3. (3.18) gives (a) 6,977, (b) 16,387.93, and (c) 46,204.09.

3.4. From $100{,}000 = .075\,P$ we need $P = 1{,}333{,}333$.

3.5. a. $\omega_t = a + bt$ and $\omega_0 = .04$ and $\omega_2 = .1$ implies $\omega_t = .04 + .03t$.

b. (3.15) and (3.16) give $AV(1.5) = .098285$ and $FV(1.5) = 2196.57$.

3.6. We integrate $\exp(.01s^2 + .04s)$ by completing the square as $.01(s+2)^2 - .04$. The present value formula then uses the normal pdf.

3.7. a. $r = .06/12$, $n = 60$, and $FV(5) = (\pi/12)(1 + r)[(1 + r)^n - 1]/r = 70.11{,}888\pi/12$, so $\pi = 10{,}000/5.84324 = 1711.38$.

b. (3.5) gives $\delta = \ln(1.06)$, and using (3.26) and (3.27), $FV(5) = 5.80456\pi$ and $\pi = 10{,}000/5.80456 = 1{,}722.78$.

3.8. $PV(t_n) = F_t \exp(-\delta_n)$ and $F_t \exp(-\mu_n - z_{1-\alpha/2}\,\sigma_n) \le PV(t_n) \le F_t \exp(-\mu_n + z_{1-\alpha/2}\,\sigma_n)$.

3.9. a. $E\{FV(5)\} = 5{,}000\exp(5(.08 + .03^2/2)) = 7{,}475.93$.

b. $P(FV(5) \le 8/5) = \Phi((\ln(8/5) - 5(.08))/(5^{1/2}(.03))) = .851654$.

c. $5{,}000\exp(5(.08) - 1.969(5^{1/2}).03) \le FV(5) \le 5{,}000\exp(5(.08) + 1.969(5^{1/2}).03)$ gives $6{,}540.13 \le FV(5d) \le 8{,}507.25$.

3.10. The mgf of δ is $M_\delta(t) = (\exp(.12t) - \exp(.06t))/(.06t)$.

a. $M_\delta(1)^5 = (1.0943)^5 = 1.5695$.

b. $M_\delta(-1)^5 = (.9140)^5 = .6378$.

c. $E\{V(5)\} = 5{,}000(1.569489) = 7{,}847.77$.

3.11. a. $P(P\exp(-\delta_t) > c) = \Phi((\ln(F_t/c) - \mu_t)/\sigma_t)$.

b. $F_t \exp(-\mu_t - z_{1-\alpha/2}\,\sigma_t) \le PV(t) \le F_t \exp(-\mu_t + z_{1-\alpha/2}\,\sigma_t)$.

3.12. a. (3.48) gives $E\{PV(10)\} = 20{,}000 \exp(-10(.08) + 10(.025^2)/2) = 9014.71$.

b. (2.65) gives $PV(10)_{.95} = 20{,}000 \exp(-10(.08) + 1.645(10^{1/2})(.025)) = 10{,}234.67$.

3.13. a. $M_m(-1) = \exp(-(.08/12) + (1/2)(.03^2)/12) = .99339$ so $E\{PV(1)\} = 1{,}000(11.496) = 11{,}496.92$.

b. $M_m(-1) = .9140677$ and $E\{PV(1)\} = 1{,}000(7.018286) = 7{,}018.29$.

3.14. a. $AF(t_n) = Y_1 Y_2 \ldots Y_n$.

3.15. a. Using (3.2), (3.10), and (3.11), $\delta_n = \Sigma \ln(1 + r_j) = .033238$, $AF(1/2) = \exp(.033238) = 1.033796$.

b. Using (3.12), $FV(1/2) = 1.03379(500) = 516.90$.

c. $PV(1/2) = 800/1.03379 = 773.85$.

3.16. $FV(1/2) = 1{,}529.84$.

3.17. Here $F_t = 4{,}000$, $t = 5$, and $PV(5) = 16{,}791.21$.

3.18. a. $\delta_j = -.03031 + .02792j$.

b. $-.03031 + .02792(14) = .3606$, $FV(14) = 6.2 \exp(.3606) = 8.8922$.

3.19. a. .022704 and .00524.

b. $E\{FV(14)\} = 6.2 \exp(14(.022704) + (1/2)(14)(.00524)) = 8.838218$.

4

Deterministic Status Models

In financial and actuarial analysis a status model is a collection of future economic actions based on a set of well-defined conditions. In deterministic status models the future times associated with the actions are contractually fixed. If one or more of the future times are random, the model is referred to as a stochastic status model. In both cases, the status model may contain unknown parameters or constants to be determined. Financial computations, such as those discussed in Chapter 3, are utilized in relevant analysis. A typical example includes house payments where the cost of the house is financed by a series of future payments. The status model defines the terms of the amount and time structure of the payments, and conditions change when the loan is paid off and the payments end. The amount of the individual payments is determined from present value functions constructed based on variables such as the amount of the loan and value of the interest rate.

In this chapter the analysis of deterministic status models is based on a general loss function or loss model approach. The loss function, introduced in Section 4.1, represents a difference between expenditures and revenue and is dependent on the conditions of the status. Loss models for deterministic and stochastic rates are described, respectively, in Sections 4.1.1 and 4.1.2. Two basic criteria for analyzing loss models are discussed. The first is a risk function approach, discussed in Section 4.2.1, based on the expectation of the loss function. The second utilizes percentiles associated with the loss function and is discussed in Section 4.2.2. In practice both approaches lead to the determination of relevant economic quantities. Analysis of single-risk models, where either the revenue or expenditure is fixed, results in formulas for insurance pricing, investment pricing, and option pricing, discussed, respectively, in Sections 4.3.1, 4.3.2, and 4.3.4. This approach is extended to the analysis of aggregate models in Section 4.4. The analyses of both fixed and random number of variables aggregate models are introduced. The chapter ends with an introduction to the theory and analysis of the stochastic surplus model in Section 4.5.

4.1 Basic Loss Model

From the perspective of an investor, insurance company, or lending institution, to analyze future economic actions we consider a basic loss model. For these actions at future time t the lender or insurer may have future values of expenditures, denoted FVE(t), that are offset by future values of revenues, denoted FVR(t). A general loss model is composed of a loss function, the difference between the future values of expenditures and the revenues, and associated conditions and actions. In a like manner, taking the initial time as a reference point, the present values of the revenue and the expenditure associated with future time t are denoted PVR(t) and PVE(t), respectively. Hence, the loss function can be defined with respect to the two basic time frames, namely, future and present values, and are

$$LF(t) = FVE(t) - FVR(t) \quad \text{or} \quad LF(0) = PVE(t) - PVR(t) \tag{4.1}$$

The choice of the expenditures and revenue functions in (4.1) depends on the context of the modeling to be done, and in financial and actuarial applications the present value approach is the most common. In this construction the lender or insurer makes a profit associated with actions at future time t when $LF(t) < 0$. If the financial structure is completely deterministic in nature, then the structure given by (4.1) is directly used for analysis. Applications utilizing loss models in financial or actuarial settings include the analysis of house and automobile loans and insurance, and the computation of their premiums.

4.1.1 Deterministic Loss Models

In deterministic loss models, all the components in loss function (4.1) are assumed fixed, lacking any random component. This deterministic structure includes the future time when status conditions change, initiating one or more economic actions. At future time t an equilibrium state holds if there is a balance between the expenditures and the revenues. Applying the initial time as a reference point (4.1) implies

$$LF(0) = 0 \quad \text{or} \quad PVE(t) = PVR(t) \tag{4.2}$$

The equivalence principal (*EP*) dictates that unknown constants in the loss model are chosen so that the conditions and equations in (4.2) hold. Applications of loss models and *EP* analysis include fixed interest rate loan payment computations that are demonstrated in a series of examples.

Example 4.1

A payment of F_t is due in t years and is offset by level or equal annuity payments of π made at the end of each period m times a year. The status model defines the loan time period, and the status fails after a total time of $n = mt$. If the interest is compounded m times a year and the effective rate of interest is r, the expenditure PVE(n) is the present value of F_n

associated with future time $t = t_n$. Further, the present value of the revenue, PVR(n), is the sum of the present values of the annuity payments. Using (3.8) and (3.22), the loss function is

$$LF(n) = F_n(1+r)^{-n} - \pi \frac{1-(1+r)^{-n}}{r} \qquad (4.3)$$

Applying *EP* (4.2) to (4.3), the premium payment π reduces to

$$\pi = \frac{F_n r}{(1+r)^n - 1} \qquad (4.4)$$

In particular, an amount of \$10,000 is needed in 5 years where interest is compounded monthly and the nominal annual interest rate is 12%. How much money does the person have to save each month to meet this requirement? Here $r = .12/12 = .01$ and $n = 12(5) = 60$. From (4.4), the person needs to save

$$\pi = \frac{\$10,000(.01)}{(1.01)^{60} - 1} = \$122.44$$

each month. In this setting the interest rate, the payment amount, and the length of time involved determine the required premium payments.

Example 4.2

A house is purchased for \$120,000 where the nominal annual interest rate on a 30-year loan is 9%. How large a down payment, D, is required so that the monthly mortgage payments are only \$900.00? The monthly interest rate is $r = .09/12 = .0075$, the number of payments is $n = 12(30) = 360$, and PVE(t) = \$120,000. The present value of the revenue is PVR(t) = D + PVP(t), where PVP(t) represents the present value of the premium payments (3.22). Applying (3.2) and solving for D,

$$D = \$120,000 - \frac{\$900[1-(1.0075)^{-360}]}{.0075} = \$8,146.32$$

In application, any cost related to the loan must be included into the expense term. Models that include various types of expenses, fixed and variable, are explored in the context of life insurance in Section 6.11.

Example 4.3

How many end of the month payments of \$225 are needed to pay off a loan of \$10,000 if the nominal annual interest rate is 8%? Here $r = .08/12$

$= .0067$, $PVE(t) = \$10,000$, $\pi = 225$, and $PVR(t)$ is given by (3.22). Solving for n in (4.2) yields

$$n = \frac{-\ln(.70222)}{\ln(1.0067)} = 52.17$$

Hence, there are 52 full payments and one partial payment consisting of $\$225(.17) = \$38.25$.

The previous examples illustrate the applications of *EP* on deterministic loss models to solve for unknown constants. More examples are discussed in Problems 4.1 to 4.3. This analysis technique is the basis for more complex financial and actuarial analysis presented in later discussions. Section 4.1.2 explores deterministic status models where the rate, such as an interest rate or an investment return rate, is stochastic with an associated probability distribution.

4.1.2 Stochastic Rate Models

In applications, a deterministic status model may produce a loss model where loss function (4.1) contains stochastic components. This is the case in a house mortgage computation where the interest rate is variable and modeled by an appropriate random variable. In this section, deterministic status models where rates are stochastic are explored. Two demonstrational examples assuming normally distributed rates follow, one utilizing the future value function and the other the present value function, as discussed in Chapter 3.

Example 4.4

An investor wants to compare two types of investments. The first invest-ment is a fixed-rate annuity where level payments, π, are made at the start of each month for n years. In the second, a one-time investment of value F_0 is made where the return rate is considered a normal random variable with yearly mean μ and variance σ^2, where normality structure (3.40) holds, so that over time t, $\delta_t \sim N(\mu_t = t\mu, \sigma_t^2 = t\sigma^2)$. A comparison of these economic actions by relating the values at future time $t = n$ years of the investment and the annuity is conducted. For the annuity formula (3.18) applies, while the investment computation employs (3.10) and (3.11). At future time $t = t_n$, corresponding to n equal periods, we find the competing future values

$$\text{FVR}(t_n) = \frac{\pi((1+r)^n - 1)}{r} \quad \text{and} \quad \text{FVE}(t) = F_0 \exp(\delta_t) \quad\quad (4.5)$$

Based on the definitions in (4.5), loss function (4.1) implies the investment outperforms the annuity if the loss function is positive, and utilizing the normality assumption,

$$P(LF(t_n) > 0) = 1 - \Phi\left(\frac{\ln\left(\frac{FVR(t_n)}{F_0}\right) - t_n\mu}{t_n^{1/2}\sigma} \right) \quad (4.6)$$

Probability (4.6) is a direct measurement of the likelihood that the investment outperforms the fixed-rate annuity. The magnitude of the difference in the competing economic actions is estimated by taking the expectation of the loss function. The expected loss, using the moment generating function (mgf) of the normal random variable given in (1.41), takes the form

$$E\{LF(t_n)\} = F_0 \exp(t(\mu + \sigma^2/2)) - FVR(t_n) \quad (4.7)$$

In particular, the fixed-rate annuity for $t = 10$, $m = 12$, $\pi = \$100$, and $r = .06/12 = .005$, from (3.18), gives future value $FVR(10) = \$16{,}387.934$. In the stochastic investment, let $F_0 = \$9{,}000$, and over each year let the normal parameters be $\mu = .06$ and $\sigma = .02$. Using (4.6) we find

$$P(LF(10) > 0) = 1 - \Phi(-.0174) = 1 - .495716 = .504284$$

Hence, 50.43% of the time the stochastic investment would outperform the fixed-rate competitor. Further, applying (4.7), the expectation of the loss function is

$$E\{LF(10)\} = \$9{,}000 \exp(10(.06 + .02^2/2)) - 16{,}387.94 = \$43.97$$

Depending on accepted risk tolerance, the expected difference in the future values of these investments is minimal, but both analyses indicate, marginally, that the stochastic investment should be taken over the fixed-rate annuity.

Example 4.5

An investor has a choice between a sum of money, F_0, now or a payment of F_t at future time t. If the money can be invested where the cumulative rate (3.14) is the continuous normal random variable, specifically $\delta_t \sim N(t\mu, t\sigma^2)$, and (3.15) and (3.16) hold, which does the investor take? In other words, what future value F_t would entice the investor to decline the immediate amount F_0? To answer, the present value loss function (4.1) is constructed, taking $PVE(t) = F_0$ and $PVR(t) = F_t \exp(-\delta_t)$:

$$LF(t) = F_0 - F_t \exp(-\delta_t) \quad (4.8)$$

One solution to find F_t is to take the expectation of (4.8) and apply the EP. Using the mgf associated with $\delta_t \sim N(t\mu, t\sigma^2)$, given in (1.41), the expectation of (4.8) is

$$E\{LF(t)\} = F_0 - F_t \exp(-t(\mu - \sigma^2/2)) \qquad (4.9)$$

A second technique involves probability computations based on the realization of a positive loss function. A desirable value of F_t produces a sufficiently small positive loss probability. Applying the normal distribution of rates, the probability of a positive loss is

$$P(LF) > 0) = 1 - \Phi\left(\frac{\ln\left(\dfrac{F_t}{F_0} - t\mu\right)}{t^{1/2}\sigma}\right) \qquad (4.10)$$

In particular, if $F_0 = \$1,000$, the yearly parameters are $\mu = .08$, $\sigma = .025$, and $t = 3$, and setting (4.9) equal to zero yields

$$F_t = \$1,000 \exp(-3(.08) + 3(.025^2)/2) = \$1,270.06$$

Hence, any future amount over $\$1,270.06$ yields a greater present value expectation over the immediate $\$1,000$. The probability of a positive loss associated with $F_t = \$1,270.06$ using (4.10) is

$$P(LF > 0) = 1 - \Phi\left(\frac{\ln(1.27) - .24}{3^{1/2}(.025)}\right) = 1 - .490945 = .501055$$

This indicates the investor may desire a greater future value amount over that obtained by the EP loss function expectation approach to ensure a greater present value over the immediate sum.

The previous examples introduce two techniques, one based on expectation and the other on probability assessments, to evaluate stochastic loss models. In the next section, specific criteria are formally presented to evaluate and estimate deterministic status models, based on stochastic loss models, and aid in the selection of rival financial and actuarial strategies.

4.2 Stochastic Loss Criterion

To analyze a deterministic status model, a loss model, with associated loss function (4.1), is constructed based on present value functions where one or both contain a stochastic component. Two main approaches are utilized to analyze stochastic loss models leading to statistical inferences, such as estimation and prediction intervals. The first approach, broadly referred to as risk analysis, bases inferences and decisions on the expectation of the loss function. The second approach bases analyses on the percentile calculations associated with the stochastic loss function, thereby restricting the likelihood of adverse loss realizations. These techniques or criteria for analyzing stochastic loss models are presented in the next two sections. Other criteria have been proposed, such as the exponential utility function as presented in Bowers et al. (1997, Chapter 1).

4.2.1 Risk Criteria

In decision theoretic-based statistics a loss function is constructed based on possible actions, realities, and probabilities where the expectation is referred to as the risk. This nomenclature is adopted in the context of stochastic loss models where the risk is defined as the expectation of the stochastic loss function (4.1) and is

$$R(t) = E\{PVE(t) - PVR(t)\} = E\{PVE(t)\} - E\{PVR(t)\} \tag{4.11}$$

The equilibrium principle (*EP*), introduced in Section 4.1.1, is extended to stochastic loss models and implies that all unknown constants yield zero risk as defined by (4.11), so that

$$E\{PVE(t)\} = E\{PVR(t)\} \tag{4.12}$$

Other criteria exist in the analysis of stochastic loss models, such as the application of an exponential utility weight function as discussed by Gerber (1976, 1979).

The general risk criteria (*RC*), applied in conjunction with *EP*, dictates that (4.12) holds and is utilized as a tool for unknown constant evaluation. This approach is a standard in financial and actuarial analysis, and for a statistical development of the *EP*, we refer to Lukacs (1948). An investment example now follows that demonstrates the application of the risk criteria.

Example 4.6

Amount F_0 is invested for 4 years where the return rate is assumed to be a normal random variable following Example 4.5, where $\delta_t \sim N(t\mu, t\sigma^2)$ for future time $t > 0$. If the goal is to have a sum of F_t after t years, what amount do you need to invest? In this case, the expected expenditure is $E\{PVE(t)\} = F_0$, while the present value of the revenue is $PVR(t) = F_t \exp(-\delta_t)$. Applying the *RC* through (4.12) yields

$$F_0 = E\{PVR(t)\} = F_t \exp(-t(\mu - \sigma^2/2)) \tag{4.13}$$

In particular, if $F_t = \$1,000$, $t = 4$, $\mu = .12$, and $\sigma = .06$, then, from (4.13),

$$F_0 = E\{PVR(4)\} = \$1,000 \ \exp(-4(.12) + .5(4)(.06^2)) = \$623.25$$

Hence, $623.25 is required for the investment using *RC*. Statistically, based on many identical investments of this type, the mean principal investment that meets the conditions is $623.25.

As with all *RC* applications, estimated constants or parameters, referred to as statistical point estimators, contain no statistical measure of their reliability. In the next section, a percentile criterion associated with a probabilistic measure of reliability is explored.

4.2.2 Percentile Criteria

An alternative criterion to the *RC* to assess financial and actuarial models is based on percentile analysis of associated stochastic loss models. The $(1 - \alpha)100$th percentile, as described in Section 2.1.3, associated with the loss function in (4.1) is denoted $lf(t)_{1-\alpha}$ and satisfies

$$1 - \alpha = P(LF(t) \leq lf(t)_{1-\alpha}) \tag{4.14}$$

Based on the distribution of the rate, percentiles for stochastic loss function can be found leading to statistical inferences that include prediction intervals. Unknown constants can be chosen that, to a desired probability, loss amounts are limited.

Percentiles defined in (4.14) can be used not only to estimate unknown constants, but also to select among possible financial and actuarial decisions. In a percentile criterion, unknown constants and actions are chosen so that the probability of a positive loss function is a small specified amount. One used in applications is the 25th percentile criteria, denoted by $PC(.25)$, where, in (3.20), $\alpha = .25$ with $lf(t)_{.75} = 0$. Therefore, $PC(.25)$ defines constants in the loss model so that a positive loss function occurs only 25% of the time.

Example 4.7

Consider the setting of Example 4.6 where the goal is to achieve amount S over time t based on investment F_0 and the cumulative rate $\delta_t \sim N(t\mu, t\sigma^2)$. In this case, $PC(.25)$ is applied with $E\{PVR(t)\} = F_0$ and $PVE(t) = F_t \exp(-\delta_t)$ and a positive loss occurs. The present value of the target investment amount exceeds F_0 and (4.14) becomes

$$.25 = P(F_t \exp(-\delta_t) > F_0) = P\left(\delta_t < \ln\left(\frac{F_t}{F_0}\right)\right) = \Phi\left(\frac{\ln\left(\dfrac{F_t}{F_0}\right) - t\mu}{t^{1/2}\sigma}\right)$$

Equating the 25th percentile for the standard normal random variable $z_{.25} = -z_{.75}$ in the above equation, we solve for the $PC(.25)$ estimate of P given by

$$P_{.25} = F_t \exp(-t\,\mu + z_{.75}\, t^{1/2}\sigma) \qquad (4.15)$$

In the context of Example 4.6, $F_t = \$1,000$, $\mu = .12$, and $\sigma = .06$, and from (4.15),

$$P_{.25} = \$1,000 \exp(-4\,(.12) + .6745(.12)) = \$670.95$$

where $z_{.75} = .6745$. Thus, $F_0 = \$670.95$ produces a positive loss 25% of the time under identical conditions and the percentile approach is observed to be more conservative than the risk method, producing a larger required investment.

4.3 Single-Risk Models

Single-risk models take loss functions of the form (4.1) where one of the components, either $PVE(t)$ or $PVR(t)$, is not future time dependent. This is a common loss function structure in financial and actuarial applications, including the computation of house mortgages with stochastic rates. In this section, the rate is a continuous random variable, and either the RC or $PC(.25)$, introduced in Sections 4.2.1 and 4.2.2, are employed in the analysis.

In an insurance setting, an insurance company may have a future obligation, such as life insurance over a fixed time period, financed by one payment called a net single premium. In an investment setting, such as the stock market, the revenue depends on the stochastic return compared to a fixed monetary amount called the net single revenue. To encompass both financial and actuarial settings applying the RC to a single-risk model through formula (4.12), the net single value (NSV) is defined as either

$$NSV = E\{PVE(t)\} \quad \text{or} \quad NSV = E\{PVR(t)\} \qquad (4.16)$$

This approach is central to the pricing of future economic actions such as life insurance and investments, and in particular, we explore insurance pricing, investment pricing, and option pricing.

4.3.1 Insurance Pricing

In an insurance setting, the company incurs more expenses than just the payment of the direct benefit. Other expenses, both fixed and variable, may be added to the overall cost (see Section 6.11). For this reason the net single value of the direct insurance claims may be less than the expected value of the direct and indirect expenses, and the difference is referred to as a loading denoted by LD. The gross premium, denoted G, finances possible future administrative expenses, and thus exceeds the expected insurance payment. If the present value of the claim is taken to be $PVE(t)$, applying RC the gross premium is related to the loading by

$$LD = G - E\{PVE(t)\} \tag{4.17}$$

In this case, the policyholder, because of the loaded premium, may receive a dividend. For simplicity we consider a short time period, ignore the effect of interest, and set the claim amount X equal to the present value of expenditures. For constant k, $0 < k < 1$, the dividend is a function of kG and is defined by

$$D_{kG} = \begin{cases} kG - X & \text{if } X \leq kG \\ 0 & \text{if } X \geq kG \end{cases} \tag{4.18}$$

If claim amount X is a continuous random variable with pdf $f(x)$, the expected dividend from (4.18) is

$$E\{D_{kG}\} = kG\, P(X < kG) - \int_0^{kG} x\, f(x)\, dx \tag{4.19}$$

The structure of formula (4.19) for expected dividends includes computations similar to computations associated with an insurance policy containing a deductible claim amount. Deductible insurance is discussed in the next two examples, and the analysis of a collection of such policies is addressed in Section 4.4. The force of interest must be included in these computations when the length of time under consideration is not minimal. Further, Excel computations for these computations are discussed in Problems 4.22 and 4.23.

Example 4.8

In the case of an accident, an insurance policy pays benefit B where the actual loss is Y, where $B \leq Y$. Here, Y is assumed to be a continuous random variable with pdf $f(y)$ and the probability of an accident is q. If $Y = B$, then the claim variable is $X = I$ (claim)Y, where the indicator function is defined as I(claim) $= 1$ if there is a claim and is zero otherwise (see Example 1.14). Applying the single-risk structure with EP expectation (1.34) produces the net single value

$$NSV = E\{X\} = q\, E\{B\} = q\, E\{Y\} \tag{4.20}$$

The variance of the claim amount is found using (1.35) and (1.36). In the event (4.20) is too large, forcing a larger than desirable premium, NSV may be lowered by introducing a deductible structure. In this case, the claim payments do not start until the loss exceeds a deductible amount d and are decreased by the deductible amount. For deductible d, the amount of the benefit paid, B, takes the form

$$B = \begin{cases} 0 & \text{if } Y \leq d \\ Y - d & \text{if } Y \geq d \end{cases} \tag{4.21}$$

The claim variable with fixed deductible d is $X = I(\text{claim})B$, where (4.21) holds. The deductible increases the probability of a zero benefit and decreases the height of the continuous pdf portion. Applying the RC to the single-risk model, the expectation of (4.21) is

$$\text{NSV} = q \left[\int_d^\infty yf(y)\,dy - dP(Y \geq d) \right] \tag{4.22}$$

For any chosen distribution of loss amounts, designated by the pdf $f(y)$, the expectation (4.22) can be computed. Utilizing the NSV as the insurance premium, the loss function for deductible insurance takes the form $L = I(\text{claim})B - \text{NSV}$. Further, applying $PC(.25)$ gives $.25 = q\,P(Y > d + G)$.

Example 4.9

The deductible insurance policy of Example 4.8 is investigated where the benefit takes the form of (4.21) and the loss follows an approximately normal distribution, or $Y \sim N(\mu, \sigma^2)$. The normal distribution is not typically utilized to model claim structures but is used here to demonstrate the effect of the deductible insurance structure. For deductible d and claim probability q, expectation (4.22) takes the form

$$\text{NSV} = q \left[(\mu - d)\Phi\left(\frac{\mu - d}{\sigma} \right) + \sigma\phi\left(\frac{\mu - d}{\sigma} \right) \right] \tag{4.23}$$

where Φ and ϕ are the df and pdf of a standard normal variable. Applying $PC(.25)$ produces

$$\text{NSV} = G = \mu - d + z_{(1-.25/q)}\,\sigma \tag{4.24}$$

In particular, in terms of $1,000$ units, let $\mu = 10$, $\sigma^2 = 9$, and $q = .05$. Using RC, the NSV with no deductible, from (4.20), is .5 units or $500. If a deductible amount of 2 units or $2,000 is imposed, then computing (4.23), NSV = .40017 or $400.17. The NSV is decreased by $100 with the addition of the deductible amount of 2 units when RC are employed. We remark that $PC(.25)$ is not applicable since $q = .05 < .25$.

This section concludes with an example of applying the RC to disability insurance where the payments are fixed and made at specified time periods. The duration or number of payments is stochastic and the force of interest is ignored.

Example 4.10

Disability insurance pays fixed benefits over consecutive time periods where the number of payments is stochastic. There may be an elimination period, a length of time after which benefit payments start, and there may be an upper limit on the number of payments denoted by m. In this example, the force of interest is ignored. The payment amounts in each time period are denoted b and are indexed by $J = j$, with associated pmf $f(j)$ for $j = 1, 2, \ldots$. If we let the probability of a claim be q, then the present value of the total claims is $X = b\, J\, I(\text{claim})$, where $J = 1, 2, \ldots$ and $I(\text{claim}) = 1$ if there is a claim and zero otherwise. For positive integer r, the rth moment corresponding to J is

$$E\{J^r\} = \sum_{j=1}^{m} j^r\, f(j) + m^r P(J > m) \qquad (4.25)$$

Applying moment formulas (1.44) and (1.45) the first and second moments of J are

$$E\{J^1\} = \sum_{j=1}^{m} P(J \geq j) \quad \text{and} \quad E\{J^2\} = \sum_{j=1}^{m} (2j-1)P(J \geq j) \qquad (4.26)$$

Based on the single-risk model and RC, the mean and variance of X are

$$\text{NSV} = q\, b\, E\{J\} \quad \text{and} \quad \text{Var}(X) = qb^2 E\{J^2\} - q^2 b^2 E\{J\}^2 \qquad (4.27)$$

In particular, suppose $3,000 is paid each month a person is disabled, starting after the second month and continuing for at most 20 months. Let $q = .01$ and number of payments have the geometric pdf given by $f(j) = .2\,(.8^{j-1})$ for $j = 1, 2, \ldots$. In this case, $P(J \geq j) = .8^{j-1}$ so that from (4.26) and (4.27) we compute $E\{J\} = 4.94235$ and $E\{J^2\} = 42.1753$ and

$$\text{NSV} = .01(3,000) \sum_{j=1}^{20} .8^{j-1} = 30\,\frac{1 - .8^{20}}{.2} = \$148.27$$

Thus, the estimated cost for this insurance is $148.27. For a collection of individual single-risk models, prediction intervals for the aggregate sum of claims can be approximated using the compound Poisson distribution (see Problem 4.7).

4.3.2 Investment Pricing

The theory and related applications of investment pricing have attracted much attention in modern financial literature. In this section, the topic of investment pricing is introduced and analysis is developed from a statistical point of view. An investment of amount F_0 is made in the hopes of its value increasing over time, and the return rates are considered to be random variables. The stochastic rate follows the normal distribution with constant mean and variance functions. Specifically, after time t the future value of the investment is

$$FV(t) = F_0 \exp(\delta_t) \quad \text{where} \quad \delta_t \sim N(t\mu, t\sigma^2) \tag{4.28}$$

for parameters μ and $\sigma > 0$. Using the expected future value formula in (3.43), which incorporates the normal mgf, the expectation of (4.28) is

$$E\{FV(t)\} = F_0 \exp(t(\mu + \sigma^2/2)) \tag{4.29}$$

The applicability of the normal distribution to model the financial rates can be assessed through statistical techniques, such as probability plotting, as discussed in Section 2.8.1.

In investment pricing, the growth of an investment, over a minimum guaranteed threshold, is estimated and utilized in decision making. In this context we let a risk-free, or guaranteed, rate over the time $(0, t]$ be denoted by r. The investment pricing formula represents the present value of the difference between the stochastic rate growth and the guaranteed rate growth corresponding to future time t, as measured by

$$PV(t) = F_0 \exp(-rt + \delta_t) \tag{4.30}$$

Under normality assumption (4.28) and RC, the expectation of (4.30) is the investment pricing net single value

$$\text{NSV} = F_0 \exp(t(-r + \mu + \sigma^2/2)) \tag{4.31}$$

In application, the mean associated with the rate may not be easily assessed. In a noninformative, in terms of mean rate approach, NSV is equated to F_0 in (4.31) to yield the rate mean

$$\mu = r - \frac{\sigma^2}{2} \tag{4.32}$$

The rate mean defined by (4.32) is very conservative in nature, especially when the risk-free rate is r, indicating a small growth in the investment. Relation (4.32) is utilized in option pricing in Section 4.3.3.

Percentile computations and associated investment pricing analysis can be
done based on the present value function in (4.30). Applying the normality
assumption (4.28), the survival or reliability function (see Section 1.5) associated with constant $c > 0$ is

$$P(PV(t) > c) = \Phi\left(\frac{\ln\left(\frac{F_0}{c}\right) + t(\mu - r)}{t^{1/2}\sigma}\right) \quad (4.33)$$

where Φ is the standard normal cdf. Under a conservative approach, the
mean rate takes forms (4.32) and (4.33) and becomes

$$P(PV(t) > c) = \Phi\left(\frac{\ln\left(\frac{F_0}{c}\right) - t\frac{\sigma^2}{2}}{t^{1/2}\sigma}\right) \quad (4.34)$$

for $c > 0$. Reliabilities (4.33) and (4.34) can be used to compute percentile measurements and prediction intervals, as demonstrated in Example 4.11.

Example 4.11

An investment of $5,000 is to be made where the guaranteed return rate
is $r = .02$. The investor is considering a more risky investment option
where the return rate is a normal random variable with mean and standard deviation, per year, given by $\mu = .12$ and $\sigma = .03$, respectively. After 2
years, investment pricing expectation (4.31) computes as NSV = $6,112.5.
The median associated with the investment pricing present value calculation, using (4.33), is

$$\$5,000 \exp(2(.12 - .02)) = \$6,107$$

A small increase in the investment pricing expectation over the median
implies a slight skewness to the right associated with the distribution of
the present value random variable.

4.3.3 Options Pricing

An option is a contract that gives the right, but not the obligation, to buy or
sell an asset for a fixed price on, or possibly before, a specified date called
an expiration date. The fixed price is called the strike price, and the option
to buy is referred to as a call option, while the option to sell is a put option.
If the holder may exercise the option only on the expiration date, the option

is a European call option. If the option may be exercised on a date prior to the expiration date, it is an American call option. Further, in this chapter we concentrate on options that do not involve a dividend.

In a European call option, the expiration is at future time t and the strike price is denoted K. In a call option, the exercise to buy will only occur if the future value exceeds the strike price and the option is "in the money." The present value of the option at future time t, based on an asset whose value at time 0 is S_0 and a guaranteed rate r, is

$$PV(t) = \exp(-rt)(S_0 \exp(\delta_t) - K) I\left(\delta_t > \ln\left(\frac{K}{S_0}\right)\right) \qquad (4.35)$$

where the indicator function $I(A) = 1$ if A holds and otherwise is zero. In a put option, the exercise to sell will occur if the future value is less than the strike price, and the present value takes the form

$$PV(t) = \exp(-rt)(K - S_0 \exp(\delta_t)) I\left(\delta_t < \ln\left(\frac{K}{S_0}\right)\right) \qquad (4.36)$$

In the balance of this section, we consider call options with present value (4.35) where the cumulative rate is a normal random variable.

The single-risk model approach is applied with the RC under the normal distribution assumption for the cumulative rate defined in (4.28). The investment pricing formula for call options is found by taking the expectation of (4.25) and reduces to

$$NSV = \exp(-rt)[S_0 \exp(t(\mu + \sigma^2/2)) \, \Phi(d_1) - K \, \Phi(d_2)] \qquad (4.37)$$

where

$$d_1 = \frac{\ln\left(\frac{S_0}{K}\right) + t(\mu + \sigma^2)}{t^{1/2}\sigma} \quad \text{and} \quad d_2 = \frac{\ln\left(\frac{S_0}{K}\right) + t\mu}{t^{1/2}\sigma} \qquad (4.38)$$

NSV defined by (4.37) is interpreted as the fair price of the option where the volatility is defined as σ. Statistical analysis of the American call option is more complex, and simulation methods, such as those in Section 9.2, can be applied to complete risk computations.

The celebrated Black–Scholes formula for the pricing of call options uses the conservative setting for investment growth where no mean rate is assumed. Putting (4.32) into (4.37) yields the investment pricing option formula

$$NSV = S_0 \, \Phi(d_1) - K \exp(-rt) \, \Phi(d_2) \qquad (4.39)$$

where the modified quantities are

$$d_1 = \frac{\ln\left(\frac{S_0}{K}\right) + t\left(r + \frac{\sigma^2}{2}\right)}{t^{1/2}\sigma} \quad \text{and} \quad d_2 = \frac{\ln\left(\frac{S_0}{K}\right) + t\left(r - \frac{\sigma^2}{2}\right)}{t^{1/2}\sigma} \tag{4.40}$$

Much work has been done in connection with the Black–Scholes option pricing formula, (4.38), introduced in Black and Scholes (1973).

European call option pricing is often used to approximate the pricing of an American call option. This approximation is applied under the rationale that the option to buy is likely to be exercised close to the expiration date. In this exploration of options pricing, we assume that no dividends are paid and no commissions are charged. The alternative to the RC method for the analysis of single-risk models, namely, the $PC(.25)$, can be applied to produce an options pricing formula.

The percentile approach to the analysis of single-risk option formulas utilizes a survival function associated with present value (4.35). Assuming $\delta_t \sim N(t\mu, t\sigma^2)$ for constant $c > 0$, the survival function reduces to

$$P(PV(t) > c) = 1 - \Phi\left(\frac{\ln\left(\frac{K + c\exp(rt)}{S_0}\right) - t\mu}{t^{1/2}\sigma}\right) \tag{4.41}$$

To apply $PC(.25)$, we set (4.41) equal to .25, and solving for c yields the desired NSV. Under the condition that $S_0 \exp(\alpha + z_{.75}\,\beta) > K$, we find the investment pricing formula.

$$NSV = \exp(-rt)\,[S_0 \exp(t\mu + t^{1/2}\,z_{.75}\,\sigma) - K] \tag{4.42}$$

If the conservative value of μ, given in (4.32), is applied in (4.42), the investment pricing formula becomes

$$NSV = S_0 \exp(-t\,\sigma^2/2 + t^{1/2}\,z_{.75}\,\sigma) - K\exp(-rt) \tag{4.43}$$

The two single-risk model evaluation approaches, namely, RC and $PC(.25)$, are demonstrated and calculations are compared in the option pricing example that follows.

Example 4.12

A European call option is examined under parameters $S_0 = 100$, $t = 1$, $r = .04$, and $\sigma = .02$. The future value based on rate r after 1 year is

$$FV(1) = 100 \exp(.04) = 104.08$$

If the strike price is $K = 105$, the investment pricing NSV under both the *RC* and *PC*(.25) methods is computed. The Black–Scholes option pricing formula (4.39) produces NSV = .436124. Thus, the option to buy 1,000 shares is valued at $435.12. The *PC*(.25) pricing method, utilizing (4.43), gives NSV = .45496, and the 1,000 shares of stock are valued at $454.96. This computational result is consistent with other comparisons in which the percentile approach yields a greater NSV over the mean or risk approach. Option pricing analysis for a specified mean rate is considered in Problem 4.10, and Excel computations are discussed in Problem 4.20.

The option pricing formulas and concepts presented in this section can be modified to accommodate the put option setting, the development of which is left to the reader. This section concludes with a discussion of option pricing diagnostics.

4.3.4 Option Pricing Diagnostics

The Greeks are a collection of option pricing diagnostics that measure the sensitivity of the options price NSV to changes in financial features, namely, price, interest rate, volatility, and time. The diagnostics are based on the European call option under *RC* with option price NSV defined in (4.39), with computational illustrations presented in Problem 4.24. The first two diagnostics examine the effect of changes in the stock price. The most significant Greek option diagnostic is delta, which measures the sensitivity of NSV to changes in the price S_0, and is

$$\Delta = \frac{dNSV}{dS_0} = \Phi(d_1) \tag{4.44}$$

Options with a greater delta will increase in value more than those with lower delta values, with the same positive movement of the initial stock price. The call option is "in the money," and the realized price exceeds the strike price as (4.44) increases but is of little value and "is out of the money" as (4.44) decreases. The sensitivity of delta (4.44) with respect to changes in price S_0 is gamma, defined as

$$\Gamma = \frac{d\Delta}{dS_0} = \frac{\phi(d_1)}{S_0 \, \sigma \, t^{1/2}} \tag{4.45}$$

Gamma measures the rate of change of delta with respect to changes in the underlying asset value.

Other diagnostics measure the effect on the price of the factors of volatility, expiration time, and risk-free interest rate. We consider three such measurements. First, vega measures the sensitivity of the NSV to changes in the volatility, as defined by σ, and is

$$\upsilon = S_0 \, \phi(d_1) \, t^{1/2} \tag{4.46}$$

where d_2 is defined in (4.40). Vega is largest when the market value is close to the strike price and decreases as the asset goes above or below the strike price. Second, theta measures the sensitivity of NSV to time change and utilizing (4.40) is

$$\theta = -\frac{dNSV}{dt} = -S_0 \, \phi(d_1) \frac{\sigma}{2t^{1/2}} - r \, K \exp(-rt) \, \Phi(d_2) \tag{4.47}$$

Theta measures how much the option price or NSV will diminish with increases in time.

Last, the effect of changes to the risk-free interest rate on the NSV is measured by rho given by

$$\rho = \frac{dNSV}{dr} = K \, t \exp(-rt) \, \Phi(d_2) \tag{4.48}$$

Rho values are often low, indicating a modest increase in interest rates does not greatly affect the stock option value.

Example 4.13

The European call option setting of Example 4.12 is revisited where $S_0 = 100$, $t = 1$, $r = .04$, $\sigma = .02$, and $K = 105$ to demonstrate the computation of option diagnostics. Computing (4.44) we find $\Delta = .3338$ indicating a moderate move in option value with a change in the underlying price. From (4.45) we find $\Gamma = .1669$. Further, utilizing (4.46), (4.47), and (4.48), we find $\upsilon = 33.3777$, $\theta = -1.03766$, and $\rho = 32.9415$. The interpretation of these diagnostics is left to sources in finance.

4.4 Collective Aggregate Models

In this section, the analysis of a collection of deterministic status models is considered where individual statuses contain stochastic components. The number of status models may either be fixed or random. Single-risk model analysis is applied to the aggregate of the present values associated with the individual statuses, where risk or reliability measures are utilized. Applications of collective aggregate or risk models in insurance include works by Butler et al. (1998), Butler and Worall (1991), and Cummins and Tennyson (1996).

4.4.1 Fixed Number of Variables

In this section, we explore the analysis of an aggregate model consisting of n independent random variables, such as accident or insurance claims, each denoted X_i, for $i = 1, 2, \ldots, n$. It is assumed that the number of random variables, n, is fixed and the aggregate is given by

$$S_n = \sum_{i=1}^{n} X_i \tag{4.49}$$

Applying a single-status model, the present value of the expenditure is $PVE = S_n$, and based on a model criterion, such as RC or $PC(.25)$, the net single value (NSV) can be computed. Computation of percentiles and prediction intervals rely on approximation techniques such as those of Section 2.4, or simulation methods as described in Section 9.2.

In general, there may be contracts of different types signified by different conditions. Let there be m unique types of contracts with associated random variables X_{ij} for $i = 1, 2, \ldots, n_j$ and $j = 1, 2, \ldots, m$, where $n_1 + \cdots + n_m = n$. Individual moments are denoted $E\{X_{ij}^r\} = \mu_j^r$ for $r \geq 1$, and moments of (4.49) are computed assuming statistical independence. In particular, the mean and variance of the aggregate are

$$NSV = \mu_s = \sum_{j=1}^{m} n_j \mu_j \quad \text{and} \quad \sigma_s^2 = \sum_{j=1}^{m} n_j Var\{X_{ij}\} \tag{4.50}$$

The approximation methods of Section 2.4, namely, central limit theorem (CLT), Haldane type A approximation (HAA), and saddlepoint approximation (SPA), can be applied to approximate the distribution of the aggregate sum. The CLT can be applied using (4.50) directly. To apply HAA, the first three moments must be computed, while the SPA requires the mgfs corresponding to the different types of contracts. In the independent and identically distributed (iid) case, the mgf of the aggregate is given in (2.41) and the SPA can be applied where (4.49) is treated as one random variable. In this scenario, the effective sample size is taken to be 1. Two illustrative examples follow.

Example 4.14

An aggregate of $n = 5$ fire-related house insurance claims is to be analyzed. The distribution of the claim amounts is skewed to the right and is modeled using an exponential distribution with mean, in thousands of dollars, of $\theta = 100$. For any claim, X_i, $E\{X_i^i\} = i\,\theta^i$, for $i = 1, 2, \ldots$, so that $Var\{X_i\} = \theta^2$. Using (4.50), we find $E\{S_5\} = 500$ and $Var\{S_5\} = 50,000$. For comparison purposes applying the CLT, the probability the sum of the claims is more than 1,000 is approximated to be

$$P(S_5 > 1,000) = 1 - \Phi\left(\frac{1,000 - 500}{50000^{1/2}}\right) = 1 - \Phi(2.236) = .01268$$

The SPA formulas in Example 2.16 are now applied with $S = 1,000$ and $t = (1,000 - 500)/(500) = 1.0$. From (2.77), $\beta = \frac{1}{2} = .5$, $c = (1 - 1/2)\exp(2/2) = 1.359141$, and $\alpha^2 = 4$, and from (2.76) the SPA is

$$P(S_5 > 1,000) = 1 - \Phi((10\ln(c))^{1/2}) + c^{-5}\frac{\left(1 - \dfrac{1}{(2\ln(c))^{1/2}}\right)}{(10\pi)^{1/2}} = .016173$$

If we compute the true probability of the sum exceeding 1,000, as in Example 2.16, utilizing the gamma distribution (Problem 2.3c), we observe that the SPA gives a superior approximation over CLT.

Example 4.15

In this example, there are m types of insurance contracts that, in the case of an accident, pay a benefit that is a random variable. For contract of type j the probability of a claim is q_j and the benefit is the random variable B_{ij}, for $i = 1, 2, \ldots, n_j$ and $j = 1, 2, \ldots, m$. In this case, (4.50) becomes

$$NSV = \sum_{j=1}^{m} n_j q_j E\{B_{ij}\} \quad \text{and} \quad Var\{S_n\} = \sum_{j=1}^{m} n_j \left[q_j Var\{B_{ij}\} + q_j(1 - q_j)(E\{B_{ij}\})^2\right]$$

$$(4.51)$$

If the benefit is fixed at $B_{ij} = b_j$, then (4.51) reduces to

$$NSV = \sum_{j=1}^{m} n_j q_j b_j \quad \text{and} \quad Var\{S_n\} = \sum_{j=1}^{m} n_j q_j(1 - q_j)b_j^2 \qquad (4.52)$$

In particular, let $m = 3$, where the required quantities are given in Table 4.1.

TABLE 4.1

Three Types of Insurance Policies

Policy Type	n_j	q_j	$E\{B_{1j}\}$	$\text{Var}\{B_{1j}\}$
1	100	.10	1.0	.09
2	150	.15	1.2	.16
3	90	.20	2.5	.25

Computing (4.51) yields NSV = 82.0 and $\text{Var}\{S_n\}$ = 135.54. Assuming an approximate normal distribution applies, a 95% prediction interval for the aggregate claim takes the form (2.16) or

$$\text{NSV} - z_{.975} \, (\text{Var}\{S_n\})^{1/2} \leq S_n \leq \text{NSV} + z_{.975} \, (\text{Var}\{S_n\})^{1/2} \qquad (4.53)$$

Computation yields the prediction interval (*PI*) for the aggregate claim as $59.19 \leq S_n \leq 104.82$. The reliability of *PI* (4.53) can be verified by the various simulation techniques discussed in Section 9.2.

4.4.2 Stochastic Number of Variables

Often a portfolio (collection) of financial or actuarial statuses such as stocks, bonds, and insurance policies may contain a stochastic number of elements. The number of status models is random variable N with associated pmf $P(N = n)$ for $n = 1, 2, \ldots$, where the mgf is assumed to exist. The aggregate takes the form of the compound aggregate variable discussed in Section 2.5 with mgf given by (2.81). The first moment, or the NSV, is found to be

$$\text{NSV} = E\{S_N\} = \sum_{n=0}^{\infty} P(N = n) \, E\{S_n \mid N = n\} \qquad (4.54)$$

Higher-order moments of the aggregate sum can be computed directly or by using the proper mgf. The cases of compound geometric and Poisson aggregate distributions were introduced in Examples 2.17 and 2.18.

For a compound aggregate variable, the distribution of the number is often taken to be Poisson with mean λ. The development of this stochastic structure can be found in Seal (1969, Chapter 2), and in Section 2.5.2 the limiting distribution was explored. The mgf of the compound Poisson aggregate variable is

$$E\{\exp(S_N t)\} = \sum_{n=0}^{\infty} \left(\exp(-\lambda) \frac{\lambda^n}{n!} \right) M_x(t)^n = \exp(\lambda(M_x(t) - 1)) \qquad (4.55)$$

Applying standard differentiation techniques through (1.40), the mean and variance take the reduced form

$$E\{S_N\} = \mu_1 \lambda \quad \text{and} \quad \text{Var}\{S_N\} = \mu_2 \lambda \qquad (4.56)$$

Other distributions have been used to model collective risk models. Dropkin (1959) and Simon (1960) applied the negative binomial distribution to model the stochastic claim size. An example demonstrating the compound Poisson aggregate variable follows.

Example 4.16

Consider m types of insurance policies with number of policies, claim probabilities, and amounts given by n_j, q_j, and b_j for $j = 1, 2, ..., m$. Hence, for $J = j$, the number of claims, N_j, is binomial with parameters n_j and q_j. If the claim probabilities are small, a Poisson approximation can be employed to estimate the binomial distribution. In this case, N_j is taken to be Poisson with mean $\lambda_j = n_j q_j$, and the total number of claims, $N = N_1 + N_2 + \cdots + N_m$, is considered Poisson with parameter λ. Hence, we take

$$\lambda = \sum_{j=1}^{m} \lambda_j = \sum_{j=1}^{m} n_j q_j \quad \text{and} \quad P(B = b_j) = \frac{n_j q_j}{\lambda} \quad \text{for} \quad j = 1, ..., m \quad (4.57)$$

Corresponding to the possible claim amounts, the moments μ_j are computed conditioning on the event of a claim, and using (4.57) take the form

$$\mu_r = \sum_{j=1}^{m} b_j^r P(B = b_j) = \sum_{j=1}^{m} \frac{b_j^r n_j q_j}{\lambda} \quad (4.58)$$

for $r = 1, 2,$ and 3. These formulas are demonstrated with quantities given in Table 4.2, where the probabilities follow (4.57), where $\lambda = 79$.

Computing (4.58) gives $\mu_1 = 2.24$ and $\mu_2 = 5.58$ and (4.56) yields $E\{S_N\} = 176.96$ and $\text{Vas}\{S_N\} = 440.82$. A normality-based approximate 95% *PI* for the aggregate variable is

$$\lambda \mu_1 - z_{.975} (\lambda \mu_2)^{1/2} \leq S_N \leq \lambda \mu_1 + z_{.975} (\lambda \mu_2)^{1/2} \quad (4.59)$$

or $135.808 \leq S_N \leq 218.112$. These prediction intervals are useful in assessing future liabilities connected with the aggregate variables.

TABLE 4.2

Three Fixed Benefits Policy Types

Type	b_j	q_j	n_j	$P(B = b_j)$
1	1	.1	150	.19
2	2	.15	100	.38
3	3	.2	170	.43

The above approximation methods used to construct prediction intervals for the aggregate variable are based on normal approximations stemming from the CLT (Section 2.4.1). There are other approximations to the distribution of aggregate variables. Borowiak (1999a), for example, applied the saddlepoint approximation method to the collective risk models in the case of heavy tailed claim distributions.

4.4.3 Aggregate Stop-Loss Reinsurance and Dividends

In a reinsurance arrangement for a collection of insurance policies, the insurer may desire to limit the possible benefits that are paid out. The aggregate claim variable is denoted by S, and the insurer absorbs aggregate claims up to amount d. For amounts S in excess of d, the insurer pays only d and takes out reinsurance with a separate insurer for the claim overage. In aggregate claim theory this is referred to as stop-loss coverage, while in the case of individual claims this is excess of loss coverage.

In stop-loss insurance, the insurer, referred to as the ceding company, retains aggregate claims below $d > 0$, S_d, which takes the form

$$S_d = \begin{cases} S & \text{if } S \le d \\ d & \text{if } S \ge d \end{cases} \tag{4.60}$$

The reinsurer pays to the ceding insurer the aggregate claim amount in excess of d, R_d, that is,

$$R_d = \begin{cases} 0 & \text{if } S \le d \\ S - d & \text{if } S \ge d \end{cases} \tag{4.61}$$

Thus, the aggregate claim is decomposed as

$$S = S_d + R_d \tag{4.62}$$

The expectations of (4.60) to (4.62), computed using the pdf of S, denoted $f(s)$, characterize the claim obligations of the insurer and the reinsurer, and are

$$E\{R_d\} = \int_d^\infty s\, f(s)\, ds - d\, P(S \ge d) \tag{4.63}$$

and

$$E\{S_d\} = E\{S\} - E\{R_d\} \tag{4.64}$$

In terms of a single-risk model based on loss function (4.1), the expenditures for the insurer are S_d plus the premium paid to the reinsurer, c_d, while the revenues are the aggregate premiums π_d. Applying RC and (4.64) yields

$$\pi_d = E\{S_d\} + c_d = c_d + E\{S\} - E\{R_d\} \qquad (4.65)$$

for $d > 0$. We note that in (4.65) no allowances are made for general expenses or contingencies. These expectations are treated as a function of the value d. Applications utilizing the normal and gamma distributions follow.

In the context of utility theory, optimal insurance for a single policy takes the form of a deductible model given by (4.61). As proved by Arrow (1963), this optimal holds if the expected claim amount is given by (4.64), and we refer to Bowers et al. (1997, Section 1.5) for details.

Example 4.17

Let aggregate claim S follow an approximate normal random variable with mean μ_s and standard deviation σ_s. In this case this distributional assumption may be unrealistic. For a fixed $d > 0$, the expectation (4.63) computes as

$$E\{R_d\} = (\mu_s - d)\left(1 - \Phi\left(\frac{d - \mu_s}{\sigma_s}\right)\right) + \sigma_s\phi\left(\frac{d - \mu_s}{\sigma_s}\right) \qquad (4.66)$$

where Φ and ϕ are the df and pdf of a standard normal random variable. For example, let $\mu_s = 10$ and $\sigma_s = 2$. Without any stop-loss structure on the aggregate claims, the insurer has an expected future obligation of NSV = 10. To reduce this obligation, the deductible amount, $d = 8$, is applied, and from (4.66) we find the reinsurer takes on an expected obligation of $E\{R_8\} = 2.167$. Using (4.64), the ceding insurer takes on the expected future obligation of $E\{S_8\} = 7.833$. If the insurer pays to the reinsurer 130% of the expected reinsurance claims, then $c_3 = 1.3(2.167) = 2.817$, and from (4.65) the aggregate premium is $\pi_3 = 7.833 + 2.817 = 10.65$.

Example 4.18

In this example the aggregate claim variable, S, is an approximate gamma random variable with positive parameters α and β where the cdf is given by $P(S \le s | \alpha, \beta) = G(s, \alpha, \beta)$. Using the gamma pdf the expectation of (4.60) computes as

$$E\{S_d\} = \alpha\beta\, G(d, \alpha + 1, \beta) + d\,(1 - G(d, \alpha, \beta)) \qquad (4.67)$$

In particular, if $\alpha = \beta = 2$, then $E\{S\} = \alpha\beta = 4$. If we take $d = 3$, applying (4.67) the insurer's expected liability reduces to

$$E\{S_3\} = 4\, G(3, 3, 2) + d\,(1 - G(3, 2, 2)) = 2.43809$$

Further, the expected obligation of the reinsurer, using (4.64), computes as $E\{R_3\} = 4 - 2.43809 = 1.56101$. If the reinsurer charges the insurer 120% of the expected reinsurance claims, then using (4.65) the insurer charges aggregate net premium $\pi_3 = E\{S_3\} + 1.2E\{R_3\} = 4.311302$.

The concepts and applications of stop-loss insurance and resulting premiums have been explored by many authors. References include work on expectation calculation by Bohman and Esscher (1964) and Bartlett (1965), and approximations and bounds by Bowers (1969), Taylor (1977), and Goovaerts and DeVylder (1980).

Similar to the setting of individual insurance in Section 4.3.1, a dividend can be associated with aggregate claims. In this context, the expected aggregate claim, denoted NSV, plays the part of the cost or premium payment. A gross premium for the aggregate insurance claims is greater than the NSV and is denoted by G. The gross premium takes into account other fixed and variable costs, including those related to the claims. For $0 < k < 1$, a dividend of D_{kG} is given where

$$
D_{kG} = \begin{cases} kG - S & \text{if } S \le kG \\ 0 & \text{if } S \ge kG \end{cases} \tag{4.68}
$$

If S is a continuous random variable with pdf $f(s)$, the expected dividend, from (4.68), takes the form

$$
E\{D_{kG}\} = kG\,P(S < kG) - \int_0^{kG} s\,f(s)\,ds \tag{4.69}
$$

Utilizing (4.63), with $d = kG$, we observe a relation between $E\{R_{kG}\}$ and $E\{D_{kG}\}$ exists as

$$
E\{D_{kG}\} = kG - E\{S\} + E\{R_{kG}\} \tag{4.70}
$$

Thus, we only need to compute one expectation in (4.70), and the other can be indirectly computed.

Example 4.19

The normally distributed aggregate claim variable discussed in Example 4.17 is now revisited in terms of the dividend of the form (4.68). Applying (4.66), we have

$$
E\{D_{kG}\} = (kG - \mu_s)\,\Phi\left(\frac{kG - \mu_s}{\sigma_s}\right) + \sigma_s\phi\left(\frac{kG - \mu_s}{\sigma_s}\right) \tag{4.71}
$$

For the aggregate claims with zero deductible NSV = $E\{S\}$ is computed as 10. A gross premium of $G = 15$ is considered and the dividend paid is 20% of G in excess of S as defined by (4.68). To compute (4.69) with $k G = .2(15) = 3$, first (4.66) yields $E\{R_3\} = 7.000117$. Applying (4.70) gives $E\{D_3\} = 3 - 10 + 7.000113 = .000113$. This is the expected value of the resulting dividend.

For further discussions of stop-loss insurance models in the context of reinsurance or deductible insurance we refer to Bowers et al. (1997, Sections 1.5 and 14.4). This chapter concludes with an introduction to a different type of stochastic status model where the analysis is based on probability assessments.

4.5 Stochastic Surplus Model

Our investigation into various deterministic status models concludes with an introduction to the topic of stochastic surplus modeling. The discussion is limited to the probability of ultimate ruin as a function of the initial surplus fund. Authors have explored the distribution of the ruin random variable (see Seal, 1978; Beekman and Bowers, 1972; Panjer and Willmot, 1992, Chapter 11), and much research exists on related topics. Specifically, the theory has been developed for the joint distribution of ruin time and the surplus amount after ruin, a review of which appears in Gerber and Shiu (1997). Simulation techniques discussed in Chapter 9 are demonstrated on stochastic surplus models in Section 9.4.

4.5.1 Discrete Surplus Model

In this section, a stochastic surplus model constructed on a purely discrete time period framework is introduced while an exploration of the continuous setting is left to the next section. Financial actions, such as claims and premium payments, are assumed to occur only at the discrete time periods denoted by t_j, for $j = 1, 2, \dots$. The aggregate claim variable corresponding to time period interval $E_j = (t_{j-1}, t_j]$ is denoted by S_j, for $j = 1, 2, \dots$. The payment of the claims is financed by an amount u at time 0, the initial surplus, and a series of level payments, each of value c, made at the beginning of each year. From the point of view of the insurer, the discrete surplus model is

$$U_j = u + jc - S_j \tag{4.72}$$

for $j \geq 1$. The aggregate claim is $S_j = X_1 + \cdots + X_j + \cdots$, where X_i are iid with mean $\mu = E\{X\} < c$. Letting $W_j = X_j - c$, the surplus model (4.72) becomes $U_j = u - (W_1 + \cdots + W_j)$

The discrete time period nature of the stop-loss model given in (4.72) is discussed in Bowers et al. (1997, Chapter 13).

There is a negative surplus at time $T = n$ if $U_n < 0$. The time of ruin is defined as the first time, if it exists, the surplus is negative and is denoted by $T_r = \min\{n : U_n < 0\}$. The economic soundness of the stochastic surplus model is manifested in the probability of future ruin, denoted

$$\psi(u) = P(\text{ruin given initial surplus } u) \tag{4.73}$$

In application, the exact calculation of (4.73) is difficult and mathematical bounds are often utilized. To construct bounds for the ruin probability, we first fix positive integer n and consider the mgf of U_n given by the expectation

$$E\{\exp(-rU_n)\} = \exp(-ru) E\left\{\exp\left(\sum_{i=1}^{n} rW_i\right)\right\} = \exp(-ru)[M_w(r)]^n \tag{4.74}$$

where $M_w(r)$ is the mgf of W_i. The adjustment coefficient is defined as the value r_a where $1 = M_w(r)$, so that

$$r_a c = \ln(M_x(r_a)) \tag{4.75}$$

It is interesting to note that theoretically there is exactly one positive real value r satisfying the mgf relation $M_w(r) = 1$ (Rohatgi, 1976, p. 624). From (4.74)

$$E\{\exp(-r_a U_n)\} = \exp(-r_a u) \tag{4.76}$$

The computation of (4.75), evaluated at r_a, is done by conditioning on the set $T_r = j \leq n$. Applying (4.76) for a fixed n, we find

$$E\{\exp(-r_a u)\} = E\{\exp(-r_a U_n) \mid T_r = j \leq n) \, P(T_r \leq n) + \alpha(n) \tag{4.77}$$

where

$$\alpha(n) = P(T_r > n) \, E\{\exp(-r_a U_n) \mid T_r > n\} \tag{4.78}$$

Now, in (4.78), for $j < n$, $U_n = U_j - (W_{j+1} + \cdots + W_n)$ and we rewrite (4.77) as

$$E\{\exp(-r_a u)\} = E\{\exp(-r_a U_j + r_a \Sigma \, W_i) \mid T_r = j \leq n\}P(T_r \leq n) + \alpha(n) \tag{4.79}$$

Since U_j and ΣW_i are independent, utilizing (4.77),

$$E\{\exp(-r_a u)\} = E\{\exp(-r_a U_j) \mid T_r = j \leq n\}P(T_r \leq n) + \alpha(n) \tag{4.80}$$

Now, $T_r = j \leq n$ implies $U_j < 0$, so $-r_a U_j > 0$ and

$$1 \leq E\{\exp(-r_a U_n) \mid T_r = j \leq n\} \tag{4.81}$$

Further, since $U_{j-1} > 0$, then $U_j > c - X_j$, and then

$$E\{\exp(-r_a U_n) \mid T_r = j \leq n\} \leq E\{\exp(-r_a (c - X_j)) \mid X_j > c\} \tag{4.82}$$

Combining (4.80), (4.81), and (4.82) gives

$$P(T_r \leq n) + \alpha(n) \leq \exp(-r_a u) \leq E\{\exp(-r_a (c - X_j)) \mid X_j > c\} P(T_r \leq n) + \alpha(n) \tag{4.83}$$

Bowers et al. (1997, p. 426) have shown that $\alpha(n)$ converges to zero as n approaches infinity, so that mathematical bounds for the probability of ruin are

$$LB(u) \leq \psi(u) = \lim_{n \to \infty} P(T_r \leq n) \leq UB(u) \tag{4.84}$$

where

$$LB(u) = \frac{\exp(-r_a(u - c))}{E\{\exp(r_a X_j) \mid X_j > c\}} \quad \text{and} \quad UB(u) = \exp(-r_a u) \tag{4.85}$$

In some cases, as discussed in the following examples, the adjustment coefficient and the upper and lower bounds can be solved explicitly.

Example 4.20

Let discrete surplus model (4.72) hold where we assume aggregate claim amounts $X_i \sim N(\mu, \sigma^2)$. Formula (4.74) gives $r_a c = \ln(\exp(\mu \, r_a + r_a^2 \, \sigma^2/2)$ and the adjustment coefficient reduces to

$$r_a = \frac{2(c - \mu)}{\sigma^2} \tag{4.86}$$

The upper probability of ruin bound in (4.85) is computed using (4.76). To derive a lower bound observe that for fixed c

$$\int_c^\infty \exp(rx) \exp(-(x-\mu)^2/(2\sigma^2))/(2\pi\sigma^2)^{1/2} \, dx = \exp(r\mu + r^2\sigma^2/2) \, \Phi\left(\frac{\mu + r\sigma^2 - c}{\sigma}\right)$$

$$\tag{4.87}$$

TABLE 4.3

Probability of Ruin Bounds for Normal Claims

u	0	.5	1	1.2	1.5	1.7	2
$LB(u)$	.526	.236	.106	.077	.048	.035	.021
$UB(u)$	1.0	.449	.202	.147	.091	.066	.041

so that

$$E\{\exp(-r_a(c-x))|x>c\} = \exp(r_a(\mu-c)+r_a^2\sigma^2/2)\frac{\Phi\left(\dfrac{\mu+\dfrac{r_a\sigma^2}{2}-c}{\sigma}\right)}{\Phi\left(\dfrac{\mu-c}{\sigma}\right)} \quad (4.88)$$

Applying (4.86) in (4.88) achieves the formula for the lower bound. Hence, bounds (4.85) become

$$LB(u) = \exp(-r_a u)\frac{\Phi\left(\dfrac{\mu-c}{\sigma}\right)}{\Phi\left(\dfrac{c-\mu}{\sigma}\right)} \quad \text{and} \quad UB(u) = \exp(-r_a u) \quad (4.89)$$

In particular, let $\mu = 1$, $\sigma = .5$, and $c = 1.2$ so that (4.86) gives $r_a = 1.6$. The bounds for $R(u)$ given in (4.89) are computed for various values of u and are listed in Table 4.3. From Table 4.3 we observe that as the initial amount u increases the bounds converge together.

Example 4.21

In the case where the aggregate claims follow an exponential distribution with mean θ the adjustment coefficient r_a, from (4.75), solves $r_a c = -\ln(1 - \theta r_a)$. Thus, $\theta r_a = 1 - \exp(-r_a c)$ and the upper bound on the ruin is $\psi(u) \le \exp(-r_a u)$. For example, if $\theta = 1$ and $c = 2$, then solving we find $r_a = .797$ and $\psi(u) \le \exp(-.797 u)$. For various initial amounts u the upper bound on the ruin probability is computed. For example, $\psi(1) \le .451$, $\psi(2) \le .203$, and $\psi(3) \le .092$.

A graph of the upper bound on the probability of ruin as a function of initial amount u is given in Figure 4.1. As expected, as the initial amount u increases the ruin probability upper bound decreases at an increasingly slower rate. To find the lower bound

$$\int_c^\infty \exp(rx)\,\theta^{-1}\exp(-x/\theta)\,dx = (1-r\theta)^{-1}\exp(-c(\theta^{-1}-r)) \quad (4.90)$$

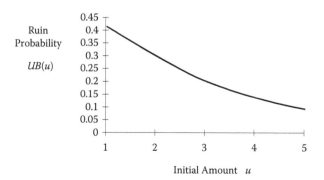

FIGURE 4.1
Exponential surplus model.

Since $(1 - r_a\theta)^{-1} = \exp(r_a c)$ we have

$$E\{\exp(-r_a(c - x))|x > c\} = \frac{\exp(r_a c - c/\theta)}{\exp(-c/\theta)} = \exp(r_a c) \qquad (4.91)$$

Using (4.91) the lower bound in (4.85) is

$$LB(u) = \exp(-r_a(u + c)) \qquad (4.92)$$

This example is explored further in the context of a fixed number of intervals in Section 8.4 utilizing simulation analysis.

4.5.2 Continuous Surplus Model

Stochastic surplus models based on a continuous-time structure exist where the financial actions, such as claims and premium payments, occur at future random times. As in the discrete time period model (4.72), an initial amount and premium payments finance aggregate claim payments. The continuous surplus model defines the surplus at future time t

$$U(t) = u + c(t) - S(t) \qquad (4.93)$$

where u is the initial fund, $c(t)$ is aggregate premiums through time t, and $S(t)$ is the aggregate claims through time t for $t > 0$. Analogous to the discrete model ruin is the first time that the surplus is negative and is denoted by $T_r = \min\{t : U(t) < 0\}$, and the probability of future ruin is $R(u)$ analogous to (4.73).

As developed in Section 4.5.1 for the discrete time period model a similar approach yields an upper bound on the probability of ruin. In continuous surplus model (4.93) we utilize the compound Poisson aggregate variable as discussed in Section 4.4.2 and assume

$$c(t) = c\,t \quad \text{and} \quad S(t) = S_t \text{ is compound Poisson with parameter } \lambda t \quad (4.94)$$

for $t > 0$. Analogous to continuous-time settings (4.93) and (4.94) for $t > 0$

$$E\{\exp(-r\,U(t))\} = \exp(-ru)\exp(-crt)\,M_{S(t)}(r)) \quad (4.95)$$

where $M_{S(t)}$ is the mgf of $S(t)$. For the continuous-time model the adjustment coefficient, r_a, is defined so that (4.95) equals $\exp(-r_a u)$. Applying the mgf of a compound Poisson aggregate variable (4.55), the adjustment coefficient solves

$$cr_a = \lambda[M_x(r_a) - 1] \quad (4.96)$$

where $M_x(r)$ is the mgf for the individual claims. Modifying the development (4.76) to (4.81) the continuous-time surplus model yields the upper bound

$$\psi(u) \le \exp(-r_a u) \quad (4.97)$$

A lower bound similar to the bound for the discrete time model in (4.85) exists. The continuous-time surplus model analysis utilizing the compound Poisson distribution is now applied to insurance and reinsurance settings in a series of examples.

Example 4.22

In the continuous surplus model we consider a compound Poisson aggregate claim variable for parameter λ where the first two moments of the individual claims are specified as $\mu_1 = E\{X\}$ and $\mu_2 = E\{X^2\}$. The three-term Taylor series approximation $M_x(r) = 1 + \mu_1 r + \mu_2 r^2/2$ reduces (4.96) to $cr_a = \lambda(r_a\mu_1 + r_a^2\,\mu_2/2)$, leading to adjustment coefficient:

$$r_a = \frac{2(c - \lambda\mu_1)}{\lambda\mu_2} \quad (4.98)$$

In particular, if $\lambda = 50$, $c = 52$, $\mu_1 = 1$, and $\mu_2 = 4.0$, then from (4.98) we find $r_a = 1.0$. Thus, from (4.97) we have $R(u) \le \exp(-u)$.

Example 4.23

Ruin probabilities are explored for the collection of insurance policies modeled by the compound Poisson aggregate claim variable described in Example 4.16, where the initial amount required so that the probability of ruin is at most 5% is desired. The distribution of the aggregate claim variable is approximated by a compound Poisson aggregate random variable with parameter and distribution given in (4.57). From Table 4.2 we compute $\lambda = .1(150) + .15(100) + .2(170) = 79$. Further, the first two moments following (4.58) are

$$\mu_1 = \Sigma b_j\, P(B = b_j) = 2.24 \quad \text{and} \quad \mu_2 = \Sigma b_j^2\, P(B = b_j) = 5.58 \qquad (4.99)$$

The mean and variance for the aggregate using (4.56) are approximated as

$$E\{S\} = \lambda\,\mu_1 = 176.96 \quad \text{and} \quad \text{Var}\{S\} = \lambda\,\mu_2 = 440.82 \qquad (4.100)$$

Based on the quantities in Table 4.2, the distribution given in (4.57), the mgf of the aggregate claim variable is

$$M_B(r) = \Sigma P(B = b_j)\exp(r\,b_j) = .19\,e^r + .38\,e^{2r} + .43\,e^{3r} \qquad (4.101)$$

Applying (4.96) the adjustment coefficient r_a is the positive root of

$$h(r) = \lambda\,M_B(r) - \lambda - cr \qquad (4.102)$$

For premium $c = 200$ setting (4.102) equal to zero results in the positive root or adjustment coefficient $r_a = .18609$. Thus, $R(u) \leq \exp(-.18609\,u)$, and setting $.05 = \exp(-.18609)$ gives the required initial amount $u = -\ln(.05)/.18609 = 16.098$.

Example 4.24

Continuous-time surplus model analysis is applied to a reinsurance setting. For compound Poisson aggregate claims proportional reinsurance is taken out to mitigate possible losses. In proportional reinsurance the reinsurer rebates a function of the aggregate claims of the form $h(S) = \alpha\,S$, for $0 < \alpha < 1$, to the insurer for a cost of c_r. Thus in (4.93), $c(t) = t(c - c_r)$, $S(t) = (1 - \alpha)\,S_t$, and utilizing (4.96) the adjustment coefficient r_a solves

$$(c - c_r) = \lambda\,M_x(r_a(1 - \alpha)) \qquad (4.103)$$

In particular, let the individual claims follow an exponential distribution with mean 1. Thus, $M_x(a) = (1 - a)^{-1}$ and (4.103) give

$$r_a = \frac{(c - c_r) - \lambda(1 - \alpha)}{(1 - \alpha)(c - c_r)} \qquad (4.104)$$

Further, assume the insurer charges an aggregate premium of 135% of the expected aggregate claims and the reinsurer premium is 150% of the expected reinsurance rebate. Thus,

$$c = 1.35\,\lambda \quad \mu_1 = 1.35\,\lambda \quad \text{and} \quad c_r = 1.5\,\alpha\,\lambda \quad \mu_1 = 1.5\,\alpha\,\lambda \qquad (4.105)$$

and (4.104) yields

$$r_a = \frac{(.35 - .5\alpha)}{(1 - \alpha)(1.35 - 1.5\alpha)} \tag{4.106}$$

If $\alpha = .1$ from (4.106), we find $r_a = .27778$, and the upper bound on the probability of ruin is $R(u) \leq \exp(-.27778\ u)$. Is there a better choice for α?

Problems

4.1. A sum of $1,000 is required in 2 years and the effective annual interest rate is $i = .05$. How much money is needed to be deposited if the interest is compounded (a) quarterly and (b) continuously?

4.2. A couple wishes to buy a $150,000 house where their initial down payment is $20,000 and fixed payments are made at the end of each month. Find the total interest paid for (a) a 20-year mortgage at 11.5% annual interest and (b) a 30-year mortgage at 9.5% annual interest.

4.3. A factory building is sold for $550,000. What are the monthly payments on a 30-year loan where the annual interest rate is 8.6%?

4.4. A financial advisor invests $30,000 for 5 years where the financial rate is an approximate normal random variable with mean .08 and standard deviation .02.

   a. Find the expectation for the future value or NSV using RC and EP.

   b. Apply (4.15) to use $PC(.25)$.

   c. If the guaranteed interest rate over 5 years is .03, give investment pricing computations using both RC and $PC(.25)$.

4.5. An insurance policy over a 1-year time period pays a benefit of B in the advent of an accident occurring with probability q. Let $q = .05$ and $B \sim N(2,000, 200^2)$. Compute the NSV (a) using RC and (b) using and $PC(.01)$.

4.6. Disability insurance is considered where a fixed benefit of b is paid for up to m months where the probability structure for the number of months of claims, J, is discrete uniform, where $P(J = j) = 1/m$ for $1 \leq j \leq m$.

   a. Find $E\{J\}$.

   b. Find $Var\{J\}$.

4.7. We consider a collection of n independent disability insurance policies as discussed in Example 4.10 where the probability of a disability claim is q. The benefit is b and the duration is J. Let the aggregate claim variable S be approximated as a compound Poisson aggregate variable with parameter $\lambda = nq$ and individual variables $X_j = bJ$.

  a. Give formulas for $E\{S\}$ and $Var\{S\}$.

  b. Using the quantities of Example 4.10 for $n = 100$ policies, construct an approximate 95% upper bound for S.

4.8. Consider 5,000 disability insurance policies under the stochastic structure of Problem 4.6. Let the probability of a claim be .5% where the benefit per period is $b = 2$ for at most 10 periods. Applying the results of Problem 4.7, construct an approximate 95% upper bound on the aggregate claim.

4.9. An investment of $1,000 is made where the guaranteed financial rate is $r = .03$. The investor considers an investment where the rate is distributed approximately normal with parameters $\mu = .08$ and $\sigma = .03$. After 18 months compute (a) $E\{FV(t)\}$ using (4.29), (b) the investment pricing NSV, (c) the probability the investment pricing present value exceeds 1,000, and (d) the probability the conservative investment pricing present value, applying (4.32), exceeds 1,000.

4.10. Consider the call option of Example 4.12. Instead of applying the Black–Scholes formula (4.39) utilize the NSV formula of (4.37) with growth rate $\mu = .045$. How does this compare to the Black–Scholes computation?

4.11. A portfolio contains 20 investments where each investment is valued between $10,000 and $20,000. Without assuming any special knowledge of the distribution of values, estimate the mean and variance for the aggregate value of the portfolio. Give an approximate 95% prediction interval on the value of the investments.

4.12. A collection of 50 insurance policies is considered over a short time period where for each the probability of a claim is .03. If there is a claim the claim amount, B, follows a distribution with survival function $S(b) = (1 + .01b)^{-2.5}$.

  a. Find $E\{B\}$ and $Var\{B\}$.

  b. For $m = 1$ find the mean and variance of the aggregate sum of the policies.

  c. Use (4.53) to approximate a 95% prediction interval for the aggregate claim.

4.13. For a collection of four types of insurance policies the probability of a claim over a short time period is given by q and the claim amounts are denoted by B. The following data hold:

Type	n	q	$E\{B\}$	$Var\{B\}$
1	25	.10	1.2	.31
2	30	.05	1.5	.35
3	35	.08	2.1	.45
4	40	.12	2.3	.55

a. Compute the NSV or the expectation for the aggregate.

b. Find the variance of the aggregate.

c. Find an approximate 95% prediction interval for the aggregate.

4.14. In a portfolio of insurance policies the number of claims over a month follows a Poisson distribution with mean equal to 25. Claim amounts follow an exponential distribution with a mean of 1.25 units.

a. Find the mean and variance of the aggregate claim variable.

b. Using the CLT approximate a 95% prediction interval for the aggregate claim.

4.15. We have four types of insurance policies with relevant quantities following Example 4.15:

Type	n_j	q_j	b_j
1	125	.05	1.0
2	130	.03	1.5
3	95	.08	2.0
4	80	.10	2.5

a. Following (4.57) give the pdf of B.

b. Compute $E\{B\}$, $E\{B^2\}$, and $E\{B^3\}$.

c. Using a compound Poisson aggregate variable give an approximate 95% prediction interval for the aggregate.

4.16. Aggregate claims follow a Pareto distribution as described in Section 1.6.1, with pdf and df given in (1.46) where parameter $\alpha > 1$. For deductible d compute (a) $E\{R_d\}$ and (b) $E\{S_d\}$. (c) If d is selected so that the ceding insurer retains obligation $E\{S_d\}$ equal to ½ of $E\{S\}$, find the obligation of the reinsurer as defined by $E\{R_d\}$.

4.17. Let the aggregate claim variable follow a gamma distribution with pdf given in Problem 2.3c. For constant $d > 0$ show $E\{S_d\}$ is computed as (4.67).

4.18. Consider a put option under assumption (4.28) with present value (4.36).

  a. Show the Black–Scholes NSV takes the form $NSV = K \exp(-rt)$ $\Phi(-d_2) - S_0\Phi(-d_1)$ where (4.40) holds.

  b. Using (4.44) show $\Delta = -\Phi(d_1)$.

  c. Using (4.45) show $\Gamma = \phi(d_1)/(S_0 \sigma t^{1/2})$.

  d. Using (4.46) $\nu = S_0 \phi(d_1) t^{1/2}$.

  e. Using (4.47) show $\theta = S_0 \sigma\phi(d_1)/(2 t^{1/2}) + r K \exp(-rt)\Phi(-d_2)$.

  f. Using (4.48) $\rho = -K t \exp(-rt) \Phi(-d_2)$.

Excel Problems

4.19. Let the insurance conditions of Problem 4.5 hold. Deductible insurance of the form (4.21) is considered.

  a. If $d = 200$ use (4.23) to compute the NSV.

  b. Use Solver to find deductible d so that NSV = 80% of the NSV without a deductible. Excel: NORMDIST, Data, Solver, setting target cell equal to the value of 72.

4.20. We consider a call option analysis. A European call option has an expiry of 2 years and the initial value is $50. The guaranteed interest rate is 5%, $\mu = .06$, $\sigma = .06$, and the strike price is 55. Compute the NSV (a) using the Black–Scholes formula, (b) using the estimated growth, (c) using the PC(.25) method with (4.43), and (d) using the PC(.25) method with (4.42).

  Excel: Basic operations for (4.38), (4.40), NORMSDIST, and NORMSINV.

4.21. We consider the continuous surplus model based on the compound Poisson aggregate distribution introduced in Problem 4.15. Here $\lambda = 25.75$, $c = 50 > E\{S_N\} = 47.299$, and $r = .06$. What is the initial fund so that the probability of ruin is at most 10%?

  Excel: Data, Solver, setting the target cell equal to value of 0.

4.22. Aggregate claims follow a gamma distribution with parameters $\alpha = 1.5$ and $\beta = 2$. For deductible d the NSV is $E\{S_d\}$.

  a. Find $E\{S_1\}$.

  b. Find d so that $E\{S_d\} = 2$.

  Excel: GAMMADIST and Data, Solver, setting the target equal to the value 2.

4.23. Insurance claims follow an approximate normal distribution with mean μ and variance σ^2. The loaded premium is G and the dividend follows (4.68). Compute the expected dividend for $\mu = 10$, $\sigma = 3.5$, $G = 20$, and $k = .25$.

Excel: NORMDIST.

4.24. We compute the call option diagnostics presented in Section 4.3.4 for Problem 4.20.

   a. Using (4.44) compute δ.

   b. Using (4.45) compute Γ.

   c. Using (4.46) compute υ.

   d. Using (4.47) compute θ.

   e. Using (4.48) compute ρ.

Excel Problem 4.20 extension: NORMSDIST and NORMDIST.

Solutions

4.1. Use (4.2).

   a. $PVE(2) = 1,000(1 + .05/4)^{-8} = PVR(2) = F_0$ gives $F_0 = 905.4$.

   b. $\delta = \ln(1.05) = .04879$, $PVE(2) = 1,000\exp(-2(.04879)) = PVR(2) = F_0$ gives $F_0 = 907.03$.

4.2. $PVE(20) = 130,000$ and (3.22) holds.

   a. $PVR(20) = \pi \ (1 - (1 + .115/12)^{-240})/(.115/12)$ gives $\pi = 1,386.36$, so interest paid is $1,386.36(240) - 130,000 = 202,726.05$.

   b. $PVR(30) = \pi \ (1 - (1 + .095/12)^{-360})/(.095/12)$ gives $\pi = 1,093.11$, so interest paid is $1,093.11(360) - 130,000 = 263,519.77$.

4.3. Using (3.22) and (4.2), $PVE(30) = 550,000 = PVR(30) = \pi \ (1 - (1 + .086/12)^{-360})/(.086/12)$, so $\pi = 4268.06$.

4.4. a. (4.29) gives $NSV = 30,000\exp(5(.08 + .02^2/2)) = 44,799.52$.

   b. $NSV = 30,000\exp(5(.08) - .6745(5^{1/2})(.02)) = 43,424.89$.

   c. Using (4.31) for RC, $NSV = \exp(-.03(5))(44,799.52 = 38,559.30$, and for $PC(.25)$, $NSV = \exp(-.03(5)) \ 43,424.89 = 37,376.15$.

4.5. a. (4.20) gives $NSV = .05(2,000) = 100$.

   b. Applying (4.24), $z_{1-.01/.05} = z_{.8} = .84162$ and $NSV = 2,000 + .84,162(200) = 2,168.32$.

4.6. $P(J \geq j) = (m - j + 1)/m$ for $1 \leq j \leq m$.

   a. (4.26) gives $E\{J\} = (m + 1)/2$.

   b. (4.27) gives $E\{J^2\} = (m + 1)(2m + 1)/6$.

4.7. a. Using (4.27) $E\{S_N\} = nqbE\{J\}$ and $Var\{S_N\} = nqb^2E\{J^2\}$.

b. $E\{S_N\} = 100(.01)(3,000)(4.94235) = 14,827.05$, $Var\{S_N\} = 100(.01)$ $(3,000^2)(42.17534)$, and $(Var\{S_N\})^{1/2} = 19,482.76$, so $S_N \leq 14,827.05 + 1.645(19,482.76) = 46,876.20$.

4.8. $E\{S_N\} = 5,000(.005)(2)(11/2) = 275$, $Var\{S_N\} = 5,000(.005)(4)(11(21)/6) = 3,850$, and $S_N \leq 275 + 1.645(62.0484) = 377.07$.

4.9. a. Using (4.29), $SNV = 1,000\exp(1.5(.08 + .03^2/2)) = 1,128.26$.

b. Using (4.31) gives $NSV = 1,000\exp(1.5(-.03+ .08+ .03^2/2)) = 1,078.61$.

c. Using (4.33) $P(PV(1.5) > 1,000) = \Phi(2.0412) = .9794$.

d. Using (4.34) $P(PV(1.5) > 1,000) = \Phi(-.01837) = .4927$.

4.10. NSV = .661865.

4.11. Individual values $X_i \sim U(10,000, 20,000)$, so $E\{X\} = 15,000$ and $Var\{X\} = 10,000^2/12$. Thus, $E\{S\} = 300,000$ and $[Var(S)]^{1/2} = 12,909.94$, leading to $274,696.51 \leq S \leq 325,303.49$.

4.12. a. (1.44) and (1.45) give $E\{B\} = 1/.15 = 6.6666$ and $E\{B^2\} = 26,666,667$, so $Var\{B\} = 26,622.222$.

b. $E\{S_N\} = 50(.03)(6.666) = 10$ and $Var\{S_N\} = 50[.03(26,622.22) + .03(.97)(6.6666^2)] = 39997.967$.

c. $0 \leq S_N \leq 402$.

4.13. a. (4.51) gives NSV = 22.17.

b. (4.51) gives $Var\{S_n\} = 45.35137$.

c. Using (4.53), $8.9707 \leq S_N \leq 35.3693$.

4.14. a. $E\{X\} = 1.25$, $E\{X^2\} = 2(1.25^2) = 3.125$, so (4.56) gives $E\{S_N\} = 31.25$ and $Var\{S_N\} = 78.125$.

b. $13.9258 \leq S_N \leq 48.5741$.

4.15. a. $\lambda = 25.75$, $P(B = 1) = .2427$, $P(B = 1.5) = .1515$, $P(B = 2.0) = .2951$ and $P(B = 2.5) = .3107$.

b. $E\{B\} = 1.8369$, $E\{B^2\} = 3.7058$, $E\{B^3\} = 7.9642$.

c. Using (4.56) $E\{S_N\} = 47.3$ $Var\{S_N\} = 95.425$.

4.16. a. (4.63) gives $E\{R_d\} = \beta^\alpha d^{1-\alpha}/(\alpha - 1)$.

b. (4.64) gives $E\{S_d\} = (\alpha\beta - \beta^\alpha d^{1-\alpha})/(\alpha - 1)$.

c. $E\{S_d\} = \alpha\beta/(2(\alpha - 1))$ gives $\alpha\beta/(2(\alpha - 1))$.

4.17. Put (4.63) into (4.64) and reduce and use the form of the gamma pdf.

4.19. a.   90.

    b.   Set NSV = 80, find $d = 400$.

4.20. a.   (3.51) NSV = 1.80765.

    b.   (4.37) NSV = 2.516304.

    c.   NSV = 1.219376.

    d.   NSV = 4.248595.

4.21. Using Problem 4.15a, (4.101) and (4.102) give $r_a = .054397$ and $u = -\ln(10)/.054397$.

4.22. a. $d = 1$ in (4.67) gives $E\{S_1\} = 1.5588$.

    b. $E\{S_d\} = 2$ and Solver gives $d = 1.2815$.

4.23. From (4.71) $E\{D_5\} = .120474$.

4.24. a. $\Delta = .53891$.

    b. $\Gamma = .09358$.

    c. $v = 28.20715$.

    d. $\theta = -1.62803$.

    e. $\rho = 50.27603$.

5

Future Lifetime Random Variables and Life Tables

Stochastic status models are characterized by economic actions at future times triggered by specific events. Often these models are defined in terms of a lifetime variable referred to as the future lifetime random variable. The future lifetime random variable can be either discrete, denoted by the curtate random variable K, continuous, denoted T, or a combination of these with the general structure following Sections 1.2.1, 1.2.2, and 1.2.3. For example, a life insurance policy is written for a person age 30 associated with the future lifetime random variable T, initially set to zero. The benefit is paid at future lifetime T corresponding to age $30 + T$. Associated with the future time random variable are the statistical concepts and functions, such as the cumulative distribution function (cdf), probability density function (pdf), and survival function, introduced in Chapter 1. Through this stochastic structure, statistical inference procedures and techniques, such as expectations, percentiles, and prediction intervals, can be applied in the analysis of stochastic status models.

This chapter starts with the theory and formulas associated with the future lifetime random variable that are central to the study of financial and actuarial science. In particular, distributional and computational formulas for continuous and discrete structures are presented in Sections 5.1 and 5.2. In the continuous case, the general concept of the force of mortality is discussed in Section 5.3 and associations with distributional formulas are outlined. The concepts and formulas associated with the future lifetime random variable are extended to describe the techniques associated with fractional age modeling and select future lifetimes in Sections 5.5 and 5.4. The discussion moves to the construction of life tables using the techniques of survivorship groups, introduced in Section 5.6, for single-status models. Life models and table construction and applications are discussed in Section 5.7. The life tables are used to construct prediction intervals for actuarial functions in Section 5.8 based on the techniques of Section 2.1. Further, life table adjustments, such as fractional age adjustments and parameter computation, are discussed in Section 5.9. Select and ultimate life tables are discussed in Section 5.10.

5.1 Continuous Future Lifetime

Assume a stochastic status model is based on a defined function of an individual lifetime. Let continuous random variable X denote the underlying lifetime of (x). Associated with X are the pdf and cdf, denoted, respectively, $f_X(x)$ and $F_X(x)$, and the survival function $S_X(x)$ for $x \geq 0$. If the status model is initiated at x, the future lifetime random variable associated with the stochastic status model is defined by $T(x) = X - x$ and depicted in Figure 5.1.

For the stochastic status model associated with (x), the future lifetime random variable is often denoted by T and follows a distribution formed by conditioning on surviving to the fixed age x, introduced in Section 1.7. The pdf of T is found by utilizing conditioning pdf (1.64) and takes the form $f(t) = f_X(X - x \mid X > x)$. Similarly, using a conditioning argument the survival function of T is dependent on x and is

$$S(t) = P(X - x > t \mid X > x) = \frac{S_X(t + x)}{S_X(x)} \tag{5.1}$$

for $t > 0$. The support of $f(t)$ and the form of (5.1) dictate the structure of the failure probabilities, called mortality rates, for individual ages. The concepts and formulas associated with the survival function, such as those presented in Section 1.5, hold for (5.1).

Actuarial science has standard notations associated with probabilities associated with a status initiated at time x based on future lifetime random variable T. Following Bowers et al, (1997, Chapter 3) standard notations are now described and presented. For $t > 0$, the cdf of T, representing the probability of status failure within time t, is

$$_t q_x = P(T \leq t) \tag{5.2}$$

and the associated survival function to future time t is

$$_t p_x = 1 - {}_t q_x = P(T > t) \tag{5.3}$$

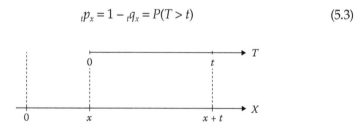

FIGURE 5.1
Future lifetime random variable.

Here $1 = {}_tp_x + {}_tq_x$, and these quantities are often referred to as mortality and survival rates or probabilities associated with the future lifetime random variable. In terms of conditioning arguments, (5.2) represents the probability that the status, upon surviving x years, will fail within t years or $P(X \le x + t \mid X > x)$, while (5.3) gives the probability that (x) will attain age $x + t$ or $P(X > x + t \mid X > x)$. For simplicity, in rare cases, if $x = 0$, the notation is streamlined by suppressing x

$$ {}_tp_0 = {}_tp \quad \text{and} \quad {}_tq_0 = {}_tq \tag{5.4} $$

Commonly, if $t = 1$ in both (5.2) and (5.3), the 1 is suppressed and we have notations

$$ {}_1q_x = q_x \quad \text{and} \quad {}_1p_x = p_x \tag{5.5} $$

Other notations exist; for example, the probability that (x) will survive t years and fail within the subsequent u years is

$$ {}_{t|u}q_x = P\,(t < T \le t + u) = {}_{t+u}q_x - {}_tq_x \tag{5.6} $$

If $u = 1$ in (5.6), then the $u = 1$ is deleted and the notation is simply ${}_{t|1}q_x = {}_{t|}q_x$.

Conditioning on an underlying failure or mortality structure, probabilities associated with the future lifetime random variable can be computed using (5.1). The survival function associated with the future lifetime random variable T, given in (5.3), can be computed as

$$ {}_tp_x = \frac{{}_{x+t}p_0}{{}_xp_0} = \frac{S_X(x+t)}{S_X(x)} \tag{5.7} $$

Partitioning future time and applying conditional arguments can be instructive in these computations. For example, the probability that (x) survives t additional years but then fails within $t + u$ additional years, (5.6), can be written as

$$ {}_{t|u}q_x = {}_tp_x\,{}_uq_{x+t} \tag{5.8} $$

These formulas often have practical intuitive interpretations: in (5.8), the first factor is the probability of survival of the status past $x + t$, and the second factor measures the probability of the status failing in u years past $x + t$. A series of illustrative examples demonstrating these concepts and formulas is now presented. Further, applications of these concepts are presented in Problem 5.1.

Example 5.1

Let the individual lifetime random variable, X, have an exponential distribution introduced in Example 1.11 with mean θ and pdf given by (1.16) where the survival function is

$$ S_X(x) = {}_xp_0 = \exp(-x/\theta) $$

Conditioning on age x, the survival function associated with T, following (5.7), becomes

$$_tp_x = \frac{\exp(-(t+x)/\theta)}{\exp(-x/\theta)} = \exp(-t/\theta)$$

The survival function of T demonstrates the lack of memory property of the exponential distribution, depending only on the future lifetime value t and not on the initial age x.

Example 5.2

Let the lifetime of a stochastic status model have underlying survival function $S_x(x) = (100 - x)^{1/2}/10$, for $0 \le x \le 100$. For (x), applying (5.7), the survival and cdf corresponding to the future lifetime random variable T are

$$_tp_x = \left[\frac{100 - x - t}{100 - x}\right]^{1/2} \quad \text{and} \quad _tq_x = 1 - {_tp_x}$$

Following (5.8), the probability of failure between future lifetimes t and $t + u$ is

$$_{t|u}q_x = \left[\frac{100 - x - t}{100 - x}\right]^{1/2}\left[1 - \left(\frac{100 - x - t - u}{100 - x - t}\right)^{1/2}\right]$$

For illustration, an individual age 40, designated as (40), buys a life insurance policy where the status is the survival of the insured person. Applying (5.7), the probability (40) lives an additional 30 years is

$$_{30}p_{40} = \left(\frac{30}{60}\right)^{1/2} = .7071$$

Also, the probability (40) lives 20 years but dies in the next 5 years is computed as

$$_{20|5}q_{40} = {_{20}p_{40}}\,{_5q_{60}} = \left(\frac{40}{60}\right)^{1/2}\left(1 - \left(\frac{35}{40}\right)^{1/2}\right) = .052733$$

Many other conditional reliability and mortality computations are possible.

This section is concluded with an example of a stochastic status model from a financial investment setting. The example is simplistic in nature but introduces a basic structure that is utilized in future financial assessments.

Example 5.3

A stock investment of initial value S_0 is made where it is assumed that the future value is given by (3.4) and the growth rate, per year, is fixed at δ. The stock is to be held for at most t years. If the future time of sale, denoted by T, is uniform over $(0, t)$, discussed in Example 1.10, what is the expected future value at the time of sale? Using the moment generating function (mgf) associated with uniform pdf (1.14), the expected future value takes the form

$$E\{FV(T)\} = E\{S_0 \exp(T\delta)\} = S_0\, M_T(\delta) = S_0\, \frac{\exp(t\delta)-1}{t\delta} \qquad (5.9)$$

Thus, if $S_0 = \$1{,}000$, $\delta = .1$, and $t = 5$, then the expected sale price is $\$1{,}297.44$. In this example, a combination of statistics and economics concepts and formulas that are central in financial and actuarial model analysis are utilized.

5.2 Discrete Future Lifetime

In many applications of interest, computations are based on the number of whole years an individual survives. In this case, a discrete random variable is appropriate to model the future lifetime, and it is given special attention and nomenclature. For (x) the curtate future lifetime is the number of full years a status holds and is

$$K(x) = \text{Number of whole years a status completes past } x \qquad (5.10)$$

Mathematically, $K(x)$ is the greatest integer function of the future lifetime variable T. Hence, $K(x) = K$ is a discrete random variable with corresponding support S_K and pdf:

$$P(K(x) = k) = P(k \le T < k + 1) = {}_kp_x - {}_{k+1}p_x \qquad (5.11)$$

Analogous to the case of the continuous future lifetime random variable, individual probabilities (5.11) can be written as the product of the probabilities of survival to age $x + k$ followed by a failure probability during the year $k + 1$. For (x) the probability mass function (pmf) associated with K can be written as

$$f(k) = P(K(x) = k) = {}_kp_x\, q_{x+k} = {}_{k|}q_x \qquad (5.12)$$

The form of the discrete pmf is general and accommodates many applications. In particular, form (5.12) is quite useful when the probabilities are listed in a tabular form, as is done in the life tables introduced in Section 5.7.

As with all probability density functions, there is an associated cumulative distribution function. If the discrete random variable $K(x) = K$ has pdf (5.12) defined on support $S = \{0, 1, ...\}$, then the cdf takes the form, for $k = 0, 1, ...,$

$$F(k) = \sum_{j=0}^{k-1} {}_j p_x \, q_{x+j} \tag{5.13}$$

The survival function associated with the curtate random variable K, evaluated at k in S, is

$$S(k) = P(K > k) = 1 - F(k)$$

An application of these formulas based on the geometric random variable introduced in Example 1.20 is given in the next example, where the conditioning structure in the continuous random variable setting is extended to the discrete setting.

Example 5.4

The underlying probability structure for a population is based on U, the number of whole years lived. The geometric distribution of Example 1.20 is applied with specific pmf:

$$P(U = u) = .1(.9^u) \quad \text{for} \quad u = 0, 1, ... \tag{5.14}$$

Here $P(U \geq m) = .9^m$, for $m = 0, 1, ...$ For (x), let the curtate future lifetime be K, where conditioning of survival to age x gives the survival probability past integer age $x + k$:

$$_k p_x = P(K \geq k) = \frac{P(U \geq x + k)}{P(U \geq x)} = \frac{.9^{x+k}}{.9^x} = .9^k \tag{5.15}$$

for $k = 0, 1, ...$ Applying (5.15) the pmf associated with K can be written as

$$f(k) = P(K = k) = {}_k p_x - {}_{k+1} p_x = .9^k - .9^{k+1} = .1(.9)^k$$

for $k = 0, 1, ...$ Hence, the probability the status fails within 1 year is $f(0) = .1$, and for positive integer k the probability of survival for at least k years is given by (5.15). In particular, for (60) the probability of surviving at least 5 years is ${}_5 p_{60} = .9^5 = .59049$, and the probability of survival of 10 years and then failure within an additional 5 years is ${}_{10|5} q_{60} = {}_{10} p_{60} - {}_{15} p_{60} = .9^{10} - .9^{15}$ $= .142787$.

As before, in the continuous exponential random variable case, the lack of memory property for the geometric random variable implies the survival rate or probability depends only on the number of future integer years lived and not on the conditioning age.

5.3 Force of Mortality

A standard concept in the field of reliability and actuarial life modeling is the instantaneous failure or death rate associated with the lifetime of equipment or individuals, respectively. In engineering reliability modeling this is referred to as the hazard function or rate, while in the life and health-related disciplines of actuarial science this is called the force of mortality. As in the previous discussions, let the underlying population lifetime be the continuous random variable X with pdf $f_X(x)$, cdf $F_X(x)$, and survival function $S_X(x)$. The instantaneous failure or mortality rate at age x is called the force of mortality and is defined as

$$\mu_x = \frac{f_X(x)}{S_X(x)} = -\frac{d}{dx}\ln(S_X(x)) \tag{5.16}$$

Given the event that the status has survived to age x, the quantity (5.16) is interpreted as the failure rate in the next instant. The shape of the hazard function, or force of mortality, can be used to model mortality distributions (see Nelson, 2005, Chapter 4). The graphs of three types of force of mortality curves are shown in Figure 5.2.

In the case of infant mortality, the graph of the force of mortality is decreasing, signifying a decrease in the mortality rate with time. If the force of mortality is increasing with time, then wear-out mortality is present. A combination of both infant and wear-out mortality exists for many populations, such as human beings, and is characterized by the bathtub-shaped force of mortality curve.

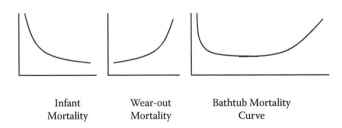

| Infant | Wear-out | Bathtub Mortality |
| Mortality | Mortality | Curve |

FIGURE 5.2
Force of mortality.

The force of mortality function is unique and determines the distribution associated with the future lifetime random variable. It is interesting to note that the pdf and cdf of the future lifetime random variable can be written as a function of the force of mortality. First, from (5.16), with $n > 0$,

$$-\int_{x}^{x+n} \mu_y \, dy = \ln(S(x+n)) - \ln(S(x)) = \ln({}_n p_x) \tag{5.17}$$

Applying (5.17) for (x) the conditional survival probability can be written in terms of the force of mortality as

$$_n p_x = \exp\left(-\int_{x}^{x+n} \mu_y \, dy\right) = \exp\left(-\int_{0}^{n} \mu_{x+s} \, ds\right) \tag{5.18}$$

Based on survival probabilities (5.18), the force of mortality needs to satisfy certain conditions, namely,

$$\mu_y \geq 0 \quad \text{and} \quad \int_{0}^{\infty} \mu_{x+s} \, ds = \infty$$

Survival probability (5.18) plays a central role in both the theory and applications in actuarial science.

For a specified population the distribution of the lifetimes is a function of the force of mortality. If we set the initial age to 0 and take the future lifetime to be $T = X$, then from (5.18)

$$_x p_0 = S_X(x) = \exp\left(-\int_{0}^{x} \mu_s \, ds\right) \tag{5.19}$$

and the cdf is

$$F_X(x) = 1 - \exp\left(-\int_{0}^{x} \mu_s \, ds\right) \tag{5.20}$$

Since X is a continuous random variable, as discussed in Section 1.2.2, taking the derivative of (5.19) yields the pdf of X:

$$f_X(x) = \exp\left(-\int_{0}^{x} \mu_s \, ds\right) \frac{d}{dx}\left[-\int_{0}^{x} \mu_s \, ds\right] \tag{5.21}$$

Since the derivative in (5.21) is equal to μ_x, the pdf of X can be written as

$$f_X(x) = {}_x p_o \, \mu_x \qquad (5.22)$$

Every continuous type pdf allows for the decomposition in (5.22) in that it can be written as a product of two functions. The first is the probability of survival to age x, while the second denotes the instantaneous failure rate associated with age x. In the following example, the structure of the force of mortality is used to model the mortality distribution of a specified population.

Example 5.5

For a population, the individual lifetime random variable X has a constant force of mortality given by μ. From (5.19), the survival function evaluated at future age x takes the form

$$S_X(x) = P(X \geq x) = \exp\left(-\int_0^x \mu \, dy\right) = \exp(-\mu x) \qquad (5.23)$$

and matches the survival function of the exponential random variable given in Example 1.26. Hence, if the force of mortality is constant, indicating a level instantaneous failure rate over time, then the corresponding survival function corresponds to the exponential random variable.

The previous population mortality and survival formulas can be applied for stochastic status models. For a stochastic status age x, consider the future lifetime random variable $T = T(x)$ with pdf and cdf denoted by, respectively, $f(t)$ and $F(t)$. Applying standard conditioning (5.1) at future time t, the cdf is

$$F(t) = 1 - P(X > x + t \mid X > x) = 1 - \frac{S_X(x+t)}{S_X(x)} \qquad (5.24)$$

so that the survival function is

$$S(t) = \frac{S_X(x+t)}{S_X(x)} \qquad (5.25)$$

Typical actuarial and life science notation equates (5.24) with ${}_t q_x$ as defined by (5.2). Taking the derivative of (5.24) with respect to t and using the fact that

$$\mu_{x+t} = -\frac{S_X^{(1)}(x+t)}{S_X(x+t)}$$

the pdf of the future lifetime random variable T can be expressed as

$$f(t) = -\frac{S_X^{(1)}(x+t)}{S_X(x)} = {}_tp_x\,\mu_{x+t} \qquad (5.26)$$

where the derivative of (5.25) with respect to t is $S_X^{(1)}(x+t)$. Analogous to (5.22) the pdf of the future lifetime random variable T given by (5.26) is a product of the survival function to time t and the force of mortality. An example demonstrating distributional formulas associated with a stochastic status follows. Further work relating the basic concepts and formulas is given in Problems 5.2 and 5.3.

Example 5.6

The mortality structure for the population discussed in Example 5.2 is explored where the survival function for future age x takes the form

$$S_X(x) = \frac{(100-x)^{1/2}}{10} \quad \text{for} \quad 0 \le x < 100$$

For (x) where $x < 100$, utilizing (5.16) and (5.26), the force of mortality and pdf associated with T are desired. For future lifetime $T = t, 0 \le t \le 100 - x$, we find

$$\mu_{x+t} = \frac{1}{2(100-x-t)} \quad \text{and} \quad f(t) = \frac{1}{2}[(100-x-t)(100-x)]^{-1/2} \quad (5.27)$$

Further, the survival function takes the form

$$_tp_x = \left[\frac{100-x-t}{100-x}\right]^{1/2} \qquad (5.28)$$

for $0 \le t \le 100 - x$. These formulas can be used to compute various mortality and survival probabilities as well as descriptive measures associated with the probability distributions.

Many applications of the force of mortality exist, and we now investigate two useful formulas revealing the structure of the force of mortality. For (x), the probability of failure within 1 year, using notation (5.5), can be written as a weighted integral involving the survival probabilities.

$$q_x = \int_0^1 {}_tp_x\,\mu_{x+t}\,dt \qquad (5.29)$$

Further, the force of mortality evaluated at age $x + t$ is related to the conditional survival probability by

$$-\frac{d}{dt}\ln(_tp_x) = \mu_{x+t} \tag{5.30}$$

These formulas demonstrate basic relations and are utilized in theoretical and application developments. A classical distribution defined in terms of the force of mortality is now presented.

Example 5.7

In this example we explore the Gompertz and Makeham laws of mortality used to model adult mortality. Gompertz (1825, p. 514) observed that "law of geometrical progression pervades, in an approximate degree, large portions of different tables of mortality." He hypothesized (p. 518) that in such instances "the average exhaustions of a man's power to avoid death were such that at the end of equal infinitely small intervals of time, he lost equal portions of his remaining power to oppose destruction which he had at the conclusion of those intervals." From this he concluded that the force of mortality, at age $x \geq 0$, should be

$$\mu_x = Bc^x \tag{5.31}$$

for constants $B > 0$ and $c > 1$. To see why this is so, we follow Jordan (1967, p. 21) and note that under the conditions of Gompertz's hypothesis, man's power to resist death decreases at a rate proportional to itself. Since μ_x is a measure of man's susceptibility to death, the reciprocal, $1/\mu_x$, is a measure of man's resistance to death. Thus, for constant of proportionality

$$\frac{d}{dx}(1/\mu_x) = -h/\mu_x \quad \text{and} \quad \text{integrating } \ln(1/\mu_x) = -hx - \ln(B)$$

where B is the constant of integration. Setting $e^h = c$ gives the celebrated Gompertz law of mortality (5.31).

In his article, Gompertz (1825, p. 517) stated, "It is possible that death may be a consequence of two generally co-existing causes; the one, chance, without previous disposition to death or deterioration; the other, a deterioration, or an increase inability to withstand destruction." Nonetheless, in his derivation, he incorporated only the second of the causes. It remained for Makeham (1860) to combine these two causes. Hence, what would become the Makeham law of mortality may be written, in terms of the force of mortality, as $\mu_x = A + Bc^x$ for positive constants A, B, and c at age x.

For (x), the probability of surviving an additional t years (5.18) applying force of mortality (5.31) is

$$_tp_x = \exp\left(-B\,c^x \int_0^t c^s\,ds\right) = \exp(-Bc^x(c^t - 1)/\ln(c)) \tag{5.32}$$

The pdf of T is found by taking the derivative of (5.32) with respect to t and is

$$f(t) = (B\ c^{x+t})\exp(-B\ c^x(c^t - 1)/\ln(c)) \tag{5.33}$$

A specified form of the Gompertz distribution utilized in statistical reliability is discussed in Problem 5.7.

In application the force of mortality may need to be defined differently over disjoint intervals of time. For example, as represented in Figure 5.2, the force of mortality may take the shape of a "bathtub curve" decreasing at the start, leveling off, and then increasing as the future time increases. The decreasing structure is referred to as infant mortality, and the rising part is called wear-out mortality (Nelson, 1982, p. 26). Two examples follow where the first demonstrates the multiruled force of mortality situation and the second involves a moment computation defined in terms of the survival function.

Example 5.8

For a population, the individual lifetimes have an associated force of mortality defined differently in two disjoint regions by the multirule

$$\mu_x = \frac{b}{x^{1/2}} \quad \text{for} \quad 0 \le x \le 1,$$

$$= b\exp(a(x-1)) \quad \text{for} \quad x > 1$$

for positive constants a and b. The force of mortality μ_x is decreasing on $0 < x < 1$ and is increasing on $x > 1$, and using (5.19) the survival function is defined in two pieces. For $0 < x < 1$,

$$S_X(x) = \exp\left(-\int_0^x b\ s^{-1/2}\ ds\right) = \exp(-2bx^{1/2})$$

and for $x > 1$,

$$S_X(x) = \exp\left(-\int_0^1 b\ s^{-1/2}\ ds - \int_1^x b\exp(a(s-1))\ ds\right) = \exp(-2b + (b/a)\exp(a(x-1)))$$

For (x), mortality probabilities and reliabilities are computed using the condition structure. For example, using (5.1), the probability of survival past future time t, $1 < x$, is computed as

$$_tp_x = \frac{S_X(x+t)}{S_X(x)} = \exp(-(b/a)\exp(a(x-1))(\exp(at)-1))$$

If $x > 1$, then the conditional survival rate would be a combination of the two rules for the survival rates.

Example 5.9

The lifetime, X, for a population of individuals follows survival function $S(x) = 1 - x^{1/2}/10$, for $0 < x < 100$. For (x) the survival probability associated with future lifetime $t < 100 - x$ is

$$_t p_x = \frac{10 - (t+x)^{1/2}}{10 - x^{1/2}}$$

For example, given $x = 20$, the probability of failure before age 50 is $_{30}q_{20} = 1 - {_{30}}p_{20} = .470150$. Standard statistical computation functions of survival and probability distributions for a stochastic status can be made. For example, for (x) the expected number of future survival years is computed integrating the survival function as in (1.44):

$$E\{T\} = \int_0^{100-x} {_t}p_x \, dt = \int_0^{100-x} \frac{10 - (t+x)^{1/2}}{10 - x^{1/2}} \, dt = \frac{\frac{1000}{3} - 10x + \frac{2}{3}x^{3/2}}{10 - x^{1/2}}$$

Specifically, for newborns, $x = 0$ and $E\{X\} = 100/3$, and when $x = 36$, then $E\{T\} = 11.333$. Other moments, such as the second moment in (1.45) leading to the variance, are computed using the basic statistical rules.

5.4 Fractional Ages

In actuarial analysis the future lifetime associated with a stochastic status model is often modeled by a mixture of discrete and continuous random variables. For a specified population actuarial life tables list discrete failure and survival rates for disjoint intervals of time such as the consecutive years. To estimate mortality and survival rates within future time periods a continuous stochastic structure is required. The distribution of continuous future lifetimes between boundaries is often approximated. Three possible interpolation techniques have been given by Bowers et al. (1997, Section 3.6): linear, exponential, and harmonic interpolation. In our discussions, we explore only linear interpolation.

For (x) let the future lifetime random variable T be continuous with curtate future lifetime K. A natural decomposition defines $T = K + S$ where S, $0 \le S < 1$, is the fractional part of the year lived. If information on the failure structure within years is lacking, then a noninformative distribution such as the uniform distribution is often applied as a default option. For modeling mortalities and survival rates for fractional ages, we utilize a uniform distribution (see Example 1.10) of fractional ages.

Assuming the conditional failure rates within any year follow a uniform distribution is referred to as the uniform distribution of death (UDD) assumption. For any age x we have $S_X(x) \geq S_X(x + s) \geq S_X(x + 1)$ for $0 \leq s \leq 1$. Under UDD model mortalities at fractional ages, we take $S_x(x + s)$ to be a linear function in terms of s as

$$S_X(x + s) = (1 - s)S_X(x) + s\, S_X(x + 1) \tag{5.34}$$

For (x) the conditional mortality corresponding to fractional age $0 < s < 1$ utilizes (5.7) in (5.34) and becomes

$$_s q_x = 1 - \frac{S_X(x+s)}{S_X(x)} = \frac{s[S_X(x) - S_X(x+1)]}{S_X(x)} = s q_x \tag{5.35}$$

The probability structure for the continuous future lifetime random variable T follows

$$P(k < T \leq k + s) = {}_k p_x \, {}_s q_{x+k} \tag{5.36}$$

Applying the linearity approximation (5.35) we write (5.36) as

$$P(K = k, S \leq s) = s\, [{}_k p_x\, q_{x+k}] \tag{5.37}$$

Taking the derivative of (5.37) with respect to s gives the joint pdf of K and S, $f(k, s) = {}_k p_x\, q_{x+k}$, where S is uniform over $[0, 1]$. Further, this construction demonstrates that random variables K and S are statistically independent (Section 1.8, formula (1.74). The relationship between independence and the fractional age assumption is well known; for a discussion see Willmot (1998). An example of fractional age modeling in terms of an assumed integer age mathematical structure follows, and another is discussed in Problem 5.5.

Example 5.10

For (x) the distribution of the future lifetime random variable is unknown, but probabilities over yearly intervals are approximated by the geometric pdf given in Example 5.4. The curtate future lifetime K has pdf

$$P(K = k) = .1(.9^k) \quad \text{for} \quad k = 0, 1, \ldots$$

The distribution of future lifetimes between integer years is assumed to be uniform between 0 and 1 so that the UDD assumption holds. The probability that the status survives an additional 5½ years, using (5.35) and (5.15), is

$$_{5+1/2} p_x = {}_5 p_x\, {}_{1/2} p_{x+5} = {}_5 p_x\, (1 - .5\, q_{x+5}) = .9^5\, (1 - .5(.1)) = .56096$$

A decrease in the survival probability for the additional ½ year is noted as calculated in Example 5.4, where $_5p_{60} = .59041$.

The structure for fractional ages (5.37) plays an important role in actuarial science theory and applications. For (x) the future lifetime is decomposed as $T = K + S$, where K is the curtate future lifetime and K and S are independent. In this decomposition, the cumulative distribution associated with T can be replaced by the joint cdf for (K, S):

$$P(K \leq k, S \leq s) = P(K \leq k-1) + P(K = k) P(S \leq s) \tag{5.38}$$

If we further assume S is uniform on $[0, 1]$, them the joint mixed probability function (mpf) associated with (5.38) is of a mixed type (see Section 1.2.3) given by

$$f(k, s) = {_k}p_x\, q_{x+k} \tag{5.39}$$

for $k = 0, 1, \ldots$ and $0 \leq s < 1$. Alternate distributions to the uniform distribution can be used to model S assuming (5.38) holds. One such distributional alternative is presented in the next example.

Example 5.11

For a population let the force of mortality be constant so that from Example 5.5 the survival function takes the form $S_X(x) = \exp(-\mu x)$, $_kp_x = \exp(-\mu k)$, and $_sq_{x+k} = 1 - \exp(-\mu s)$. Under (5.38) decomposition $T = K + S$ yields

$$P(K = k, S \leq s) = {_k}p_x\, {_s}q_{x+k} = \exp(-\mu k)(1 - \exp(-\mu s)) \tag{5.40}$$

Taking the derivative of (5.40) with respect to s yields the joint mpf of (SK). We remark that the modeling of the fractional age mortality under a constant force of mortality differs from that of the uniform approach of (5.37).

5.5 Select Future Lifetimes

The survival function associated with a stochastic status model may be influenced by the time of related events, such as the construction of a financial contract or life table. In such cases the structure of the force of mortality corresponding to the entire class of individuals differs based on related conditions. This concept is reflected in the increases in survival probabilities and mean expected future lifetimes for newborns in advancing subsequent years. In life insurance there is thought to be lower mortality rates associated with lifetimes closer to the time the policy was issued. At policy issue time, the class of individuals tends to be healthier than at times away from the initial insuring. Probabilities, such as mortality rates and survival probabilities, based on the selection of a construction point or select age, are referred to as select probabilities.

For individual (x) there is a corresponding fixed age referred to as the select age and denoted $[x]$. At this age (x) is deemed to have a lower mortality rate than the general insured population. Possibly the individual is examined at age $[x]$ when the insurance policy is written. The underlying survival probability for an individual to age $[x] + u$ is denoted $S_x([x] + u)$. For a stochastic status model age $[x] + u$ the probability of survival of t additional years or failing within t additional years applying (5.7) is

$$_t p_{[x]+u} = \frac{S_X([x]+u+t)}{S_X([x]+u)} \quad \text{and} \quad _t q_{[x]+u} = 1 - {_t p_{[x]+u}} \tag{5.41}$$

Similar to notations (5.5), if $t = 1$, the t is suppressed in (5.41), leaving the notations $p_{[x]+u}$ and $q_{[x]+u}$. Since the survival function decreases with age, the effect of the age of selection wears off as u increases in (5.41). For conceptual clarity, a hypothetical mathematical modeling example is now presented where the survival function demonstrates the relevant concepts and formulas.

Example 5.12

The lifetime of an individual for whom life insurance is purchased takes values between 0 and 100. The age at which the life insurance policy is issued is select age $[x]$, and the survival function evaluated at age $[x] + u$ is

$$S_X([x]+u) = \frac{(100 - 1.2[x] - .8u)^{1/2}}{10} \tag{5.42}$$

where $0 \le 100 - 1.2[x] - .8u$. For a stochastic status model age $[x] + u$ from (5.41) the survival probability corresponding to t additional years is

$$_t p_{[x]+u} = \frac{(100 - 1.2[x] - .8(u+t))^{1/2}}{(100 - 1.2[x] - .8u)^{1/2}} \tag{5.43}$$

For example, a life insurance policy is issued to a person age 50, so that the select age is $[x] = [50]$. After 5 years, the conditional probability the individual survives an additional 3 years is

$$_3 p_{[50]+5} = \frac{(100 - 1.2(50) - .8(8))^{1/2}}{(100 - 1.2(50) - .8(5))^{1/2}} = .96609 \tag{5.44}$$

Stochastic status probabilities are dependent on select age $[x]$. Consider a person age 55 where the policy was issued at select age 52. The likelihood of survival 3 additional years using (5.43), $_3p_{[52]+3} = .96531$, is different from (5.44). Both survival probabilities correspond to individual age 55 and survival of an additional 3 years.

The concept of select future lifetimes is useful in the construction of accurate mortality and survival rates. For instance, the structure can also be used to model a time trend in the survival function based on increasing select ages. The construction of select and ultimate life tables is discussed in Section 5.10.

5.6 Survivorship Groups

The notation used in the development of survivorship group theory is standard in the life sciences and is found in Bowers et al. (1997, Section 3.3). A population is identified where individuals possess similar mortality characteristics. Statistically, a survivorship group is a collection of individuals with associated lifetimes, X, assumed to be continuous random variables, each associated with survivor function $S_X(x)$. The group is considered closed in that no new individuals may enter the group at a later time. The number of initial individuals is denoted l_0, and the number of survivors past age x is the random variable $\mathcal{L}_x$. A limiting age is defined as an age ω such that $S_X(x) > 0$ for $0 \le x < \omega$ and $S_X(x) = 0$ for $x \ge \omega$.

The individuals in the survivorship group have independent lifetimes with identical mortality structure. For the jth individual the indicator function for survival past x is defined as

$I_j(x) = 1$ if survival is at least to age x
$\quad\quad\quad 0$ if the life fails before age x

For a survivorship population consisting of l_0 individuals, the number of survivors to age x can be written as

$$\mathcal{L}_x = \sum_{j=1}^{n} I_j(x)$$

As introduced in Example 1.8 the random variable $\mathcal{L}_x$ is binomial with parameters $P(X \ge x) = S_X(x)$ and $l_0 = n$. From the survivorship group, the expected number of survivors to age x, $E\{\mathcal{L}_x\}$, is

$$l_x = l_0 \, S_x(x) \tag{5.45}$$

Further, using the fact that $\mathcal{L}_x$ is binomial, the variance of $\mathcal{L}_x$ is given by

$$\text{Var}\{\mathcal{L}_x\} = l_0 \, S_X(x) \, [1 - S_X(x)] \tag{5.46}$$

Statistical measurements for the survivorship group, such as the mean future lifetime for various ages, are functions of the expected quantities (5.45).

Decrement in terms of the survivorship group refers to individuals leaving the group. Let the random variable $_n\mathcal{D}_x$ denote the number of individuals that leave the group in the time interval between x and $x + n$. Utilizing (5.45) the expected number of individual status failures between age x and $x + n$ is

$$_n d_x = E\{_n\mathcal{D}_x\} = l_0 \, [S_x(x) - S_x(x + n)] = l_x - l_{x+n} \tag{5.47}$$

Similar to the notational convention (5.5), if $n = 1$, then the 1 is suppressed and we have the mortality notations for the random variable $\mathcal{D}_x$ and the expectation d_x. Further, using basic probability laws,

$$l_x = l_{x-1}(1 - q_{x-1}) = l_0 - \sum_{y=0}^{x-1} d_y \tag{5.48}$$

Life tables consist of survival and decrement expectations (5.45) and (5.47) over subsequent years ending at a limiting age.

Survivorship group characteristics and parameters can be written in terms of the force of mortality as defined in Section 5.3. The force of mortality associated with an individual age x from (5.16) is $\mu_x = -(dS_x(x)/dx)/S_x(x)$. Since $(d/dx)l_x = l_0 \, (dS_x(x)/dx)$, the instantaneous change in the size of the population at age x, as measured by the derivative, is

$$\tfrac{d}{dx} l_x = -l_0 \mu_x S_x(x) = -l_x \mu_x \tag{5.49}$$

From (5.49), we observe that l_x is a decreasing function of x indicating a wear-out type of mortality. Applying the decomposition of the pdf (5.22), a relation defining the expected number of decrements at age x is

$$l_x \mu_x = l_0 \, _x p_0 \, \mu_x = l_0 \, f_X(x) \tag{5.50}$$

Further, using the survival probability and the force of mortality relation (5.19) the expected number of survivors from the group at age x takes the form

$$l_x = l_0 \exp\left(-\int_0^x \mu_s \, ds\right) \tag{5.51}$$

Utilizing (5.51), the expected number of decrements in the survivorship group between ages x and $x + n$ can be written as

$$l_x - l_{x+n} = l_0 \exp\left(-\int_0^x \mu_s \, ds\right)\left[1 - \exp\left(-\int_x^{x+n} \mu_s \, ds\right)\right] = l_0 \, S_X(x) \,_n q_x \qquad (5.52)$$

These formulas are applicable in population modeling where either the force of mortality is assumed known or estimated. In insurance pricing theory adjusting the force of mortality can be utilized to model high-risk populations or strata.

Group survivorship expectations can be used to compute the survival and failure probabilities introduced in Chapter 5. For (x), the mortality and the survivor probabilities associated with n additional years are written as

$$_n q_x = \frac{S_x(x) - S_x(x+n)}{S_x(x)} = \frac{l_x - l_{x+n}}{l_x} = \frac{_n d_x}{l_x} \qquad (5.53)$$

and

$$_n p_x = 1 - \,_n q_x = \frac{S_x(x+n)}{S_x(x)} = \frac{l_{x+n}}{l_x} \qquad (5.54)$$

Utilizing group survivorship notation, life tables consisting of survival expectations (5.45) and decrement expectations (5.47) for various ages x are constructed for target populations. Thus, for (x) applying (5.53) and (5.54), decrement and survival probabilities can be computed. In Chapter 6, expectations for various forms of life insurance and life annuities are computed using life table expectations.

In surveys, observed or empirical data yield statistics based on $\hat{l}_x$ and $\hat{d}_x$ for various ages x. Actuarial statistics are defined as a function of these random variables, and their development and application are explored in Section 10.3. Analysis of actuarial statistics allows for a reliability assessment for actuarial measurements. In particular, various forms of prediction intervals for common actuarial computations are discussed in Sections 10.3.1, 10.3.2, and 10.3.3.

In standard life tables, adjustments to the survivorship group expectations, such as of l_x or d_x, are often made. Empirical values of q_x are often adjusted to "smooth" the life table to improve their fit and reflect logical underlying mortality patterns. The smoothing of life tables is called graduation of the table. The graduation process is often ad hoc in nature and produces mortality rates such that

$$0 < q_x < 1 \quad \text{and} \quad q_x \leq q_{x+1} \qquad (5.55)$$

for all $x \geq 0$. Further, the difference between the smoothed mortalities and the observed mortalities is to be minimized. The choice of the magnitude of graduation is a selection between smoothness and fit to the observed data.

5.7 Life Models and Life Tables

In this section, the construction and utilization of life tables based on a group survivorship structure are explored. In application, the underlying mortality structure is implicitly defined through expectations l_x and d_x defined by (5.45) and (5.47) for integer ages $x \geq 0$. For a discrete stochastic status model involving the curtate future lifetime introduced in Section 5.2, statistical quantities, such as survival and mortality probabilities, as well as single net values associated with risk models, as discussed in Section 4.3, are computed using life table expectations. For continuous-time stochastic status models, probabilities associated with a continuous lifetime random variable can be estimated using the UDD assumption discussed in Section 5.4. An example follows that demonstrates these concepts and computations.

Example 5.13

The mortality structure of a population is modeled by a survivorship group with $l_0 = 100,000$ initial newborns and the limiting age $\omega = 100$. The underlying population survival function is taken to be

$$S_X(x) = \frac{(100-x)^{1/2}}{10} \quad \text{for} \quad 0 \leq x \leq 100 \tag{5.56}$$

A life table consisting of the expected number of survivors, l_x, and using (5.47), the number of decrements, d_x, at yearly intervals is constructed and the quantities for certain years are listed in Table 5.1. In practical applications, the form of $S_X(x)$ is indeterminate and the

TABLE 5.1

Life Table Listing Expected Numbers of Survivors and Deaths

x	l_x	d_x	x	l_x	d_x	x	l_x	d_x
0	100,000	501	50	70,711	711	70	54,772	920
1	99,499	504	51	70,000	718	71	53,852	936
2	98,995	506	52	69,282	725	72	52,915	953
3	98,489	509	53	68,557	734	73	51,962	972
4	97,980	512	54	67,823	741	74	50,990	990
5	97,468	514	55	67,082	750	75	50,000	1,010

mortality structure is implicitly defined through the age-based life table quantities.

Utilizing Table 5.1, typical probability and statistical questions can be answered. For example:

1. The expected number of deaths in the first year is $d_0 = l_0 - l_1 = 100,000 - 99,499 = 501$.
2. The expected number of survivors to age 50 is $l_{50} = 70,711$.
3. The probability an individual age 70 survives another 5 years is

$$_5p_{70} = \frac{l_{75}}{l_{70}} = \frac{50,000}{54,772} = .912875$$

4. The probability an individual age 50 lives 2 years and dies in the following year is

$$_{2|1}q_{50} = (_2p_{50} - _3p_{50}) = \frac{l_{52} - l_{53}}{l_{50}} = \frac{d_{52}}{l_{50}} = .01025$$

Probabilities based on life table computations can be used to compute single net values associated with single-risk discrete stochastic status. For example, an increasing endowment insurance for a newborn pays $k + 1$ units, $k = 0, 1, 2, 3$, at the end of the year of death if death occurs prior to age 4, and 4 at age 4, if death occurs after that age. Ignoring the force of interest the net single value is computed as the expectation

$$\text{NSV} = \sum_{k=0}^{3}(1+k)\frac{d_k}{l_0} + 4\frac{l_4}{l_0} = 3.95$$

Other statistical quantities, such as variances, can be computed using the life table expectations.

An alternate form of a life table to the one described in Example 5.13 lists mortality rates for individual years. For each age x, the life table lists the yearly failure rates, q_x in (5.53), with $n = 1$. For general survival, failure probabilities such as those discussed in Section 5.1 can be computed utilizing the yearly rates. For example, the probability (x) survives to age $x + n$, given in (5.54), can be computed as

$$_np_x = \prod_{k=0}^{n-1} p_{x+k} = \frac{l_{x+n}}{l_x} \tag{5.57}$$

Similarly, for (x) the probability of surviving n years and failure in the next year, defined in (5.8) with $u = 1$, can be computed utilizing (5.57) as

$$_{n|}q_x = \prod_{k=0}^{n-1} p_{x+k}\, q_{x+n} = \frac{d_{x+n}}{l_x} \tag{5.58}$$

The probability of failure of (x) within n additional years can be computed by combining subsequent years calculations as

$$_n q_x = q_x + \sum_{k=1}^{n-1} q_{x+k} \prod_{s=0}^{k-1} p_{x+s} = \sum_{0}^{n-1} \frac{d_{x+j}}{l_x} \tag{5.59}$$

Computations utilizing quantities in the alternate form of the life table listing yearly mortality rates are now demonstrated. Excel computations utilizing life tables are investigated in Problem 5.17.

Example 5.14

For the population given in Example 5.13 following underlying population mortality structure (5.56), a life table listing the yearly mortality rates is constructed. For selected years, the mortality rates and survival rates, given in formula (5.53) with n, are listed in Table 5.2. As in Example 5.13, probability calculations conditioning an age x can be made utilizing table quantities. For example, since $p_x = 1 - q_x$:

1. The probability an individual age 70 survives to age 74, using (5.57), is

$$_4 p_{70} = (1 - q_{70})(1 - q_{71})(1 - q_{72})(1 - q_{73}) = (.98320)(.98260)(.98199)(.98129)$$
$$= .93093$$

2. The probability of decrement for an individual age 70 in the third year, using (5.58), is

$$_{2|1} q_{70} = (1 - q_{70})(1 - q_{71})q_{72} = (.98320)(.98260)(.01801) = .01740$$

The two forms of the life tables, namely, Tables 5.1 and 5.2, are based on the same underlying mortality structure and yield identical information on the lifetimes of prospective discrete stochastic statuses. The quantities of these tables can be combined to yield a more complete life table, the column headings often being x, q_x, l_x, and d_x, in that order.

TABLE 5.2

Life Table Listing Mortality Rates

x	q_x	x	q_x	x	q_x
0	.00501	50	.01005	70	.01680
1	.00506	51	.01026	71	.01740
2	.00511	52	.01046	72	.01801
3	.00517	53	.01071	73	.01871
4	.00523	54	.01092	74	.01942

Stochastic status models involving high-risk strata may have lower survival rates through time than counterparts from the general population. Life tables based on the general population may be adjusted, reflecting changes to the force of mortality, to model individuals from high-risk strata. Many types of risk-based adjustments to mortalities exist, such as the one demonstrated in Example 5.15. Stress model techniques originating in reliability modeling are adapted to the actuarial life table setting in Section 10.1.

Example 5.15

An individual age 50 is subject to the mortality rates listed in Table 5.2, except between ages 50 and 51. At this age the individual is subject to an additional stress manifested in the form of a linear addition to the force of mortality, μ_x, that starts at .05 and decreases to zero. Hence, during the 51st year of age the total force of mortality is $\mu'_{50+t} = \mu_{50+t} + .05(1 - t)$ for $0 \le t \le 1$. Applying formula (5.18) the adjusted survival rate for the individual at age 50 is

$$p'_{50} = \exp\left(-\int_0^1 (\mu_{50+t} + .05(1-t))\,dt\right) = p_{50}\exp\left(-.05\int_0^1 (1-t)\,dt\right) = .97521$$

The adjusted survival rate at age 50 is less than the implied table value of .98995. In this way, mortality rates given in life tables may be adjusted for individuals in higher-risk categories who are affected by an additional stress or hazard.

5.8 Life Table Confidence Sets and Prediction Intervals

There are many research opportunities in the general area of estimation and statistical inference in the context of tables. Background in the area of actuarial life tables and measurements includes the textbooks of Neill (1977) and Jordan (1967). Many approaches to life table analysis, particularly to biostatistics, exist, and for a review of these we refer to Chiang (1968) and Elandt-Johnson and Johnson (1980). Further, London (1997) discusses estimation topics in life tables. The concept of prediction sets and associated statistical inference, introduced in Section 2.1.4, was adapted to interval data in Section 2.8.3. In this section, techniques associated with interval data are applied to the group survivorship structure through life table decrements. For a statistic associated with an aggregate collection of identical discrete stochastic statuses, approximate prediction intervals are constructed.

For a collection of identical discrete stochastic statuses age x, we assume the mortality structure follows life table decrements where the limiting age is ω. For individuals, the curtate future lifetime K takes on probabilities $_{k|}q_x = d_{x+k}/l_x$ for $k = 0, 1, \ldots, \omega - x$. Applying group survivorship concepts, the observed number of survivors to age $x + k$ is $\mathcal{L}_{x+k}$, and $\mathcal{D}_{x+k}$ denotes the number of failures at age $x + k$ so that $\mathcal{D}_{x+k} = \mathcal{L}_{x+k} - \mathcal{L}_{x+k+1}$. The interval data discussion of Section 2.8.3 is adapted by relating $X(k)/n = {}_{k|}\mathcal{Q}_x = \mathcal{D}_{x+k}/l_x$ and $p_k = {}_{k|}q_x$ in (2.138) and observing that the random vector $(\mathcal{D}_{x+1}, \mathcal{D}_{x+2}, \ldots, \mathcal{D}_{x+m})$ is multinomial with parameters l_x and $_{k|}q_x$ for $1 \le k \le \omega - x$. The generalized least squares measure (GLSM) for the observed decrement totals, relative to the life table parameters from forms (2.141) and (2.142), is

$$W_x = l_x \sum_{k=0}^{\omega-x-1} \frac{({}_{k|}\mathcal{Q}_x - {}_{k|}q_x)^2}{{}_{k|}q_x} \tag{5.60}$$

where $_{0|}\mathcal{Q}_x = \mathcal{D}_x/l_x$ and $_{0|}q_x = q_x = d_x/l_x$. Using life table notations (5.60) reduces to

$$W_x = \sum_{k=0}^{\omega-x-1} (\mathcal{D}_{x+k} - l_x\, {}_{k|}q_x)^2 / (l_x\, {}_{k|}q_x) \tag{5.61}$$

Following asymptotic statistical theory (5.61) is an approximate chi square random variable with degrees of freedom $\omega - x$ for suitable large $l_x\, {}_{k|}q_x$ for $1 \le k \le \omega - x$. Using the notation of (2.20), an approximate $(1 - \alpha)100\%$ prediction set for $\mathcal{D}_x^t = (\mathcal{D}_x, \mathcal{D}_{x+1}, \ldots, \mathcal{D}_{x+m})$ takes the form

$$PS_{1-\alpha} = \{\mathcal{D}_x^t : \mathcal{D}_{x+k} \ge 0, \textstyle\sum \mathcal{D}_{x+k} = l_x, W_x \le \chi^2_{1-\alpha, df=\omega-x}\} \tag{5.62}$$

Prediction intervals for functions of $\mathcal{D}_x^t$ can be formed by optimization within the prediction set (5.62) as demonstrated in Example 2.5. A life table example dealing with prediction set techniques now follows.

Example 5.16

A discrete stochastic status at age x is associated with curtate future lifetime K where the probabilities are defined as $_{k|}q_x = .2$ for $k = 0, 1, 2, 3$, and 4. The expectation of K is

$$E\{K\} = \sum_{k=0}^{4} k\, {}_{k|}q_x = 2.00 \tag{5.63}$$

For a collection of $l_x = 100$ independent statuses with the above probability structure the random vector $\mathcal{D}_x^t = (\mathcal{D}_x, \mathcal{D}_{x+1}, \mathcal{D}_{x+2}, \mathcal{D}_{x+3}, \mathcal{D}_{x+4})$ is observed. An asymptotic 95% prediction set, PS_{95}, for the random vector $\mathcal{D}_x^t$ defined by (5.62) is formed where the degrees of freedom are 4. An

TABLE 5.3

Life Table Prediction Interval

	$\mathcal{D}_x$	$\mathcal{D}_{x+1}$	$\mathcal{D}_{x+2}$	$\mathcal{D}_{x+3}$	$\mathcal{D}_{x+4}$	$S(K)$
$UB(S(K))$	11.288	15.644	20.000	24.356	28.712	243.561
$LB(S(K))$	28.712	24.357	20.000	15.643	11.288	156.439

interval estimate for the aggregate the total number of whole survival years survived

$$S(K) = \sum_{k=0}^{4} k \, \mathcal{D}_{x+k} \qquad (5.64)$$

is desired. Applying the approach of Section 2.1.4, an approximate prediction interval for (5.64) is

$$LB(S(K)) \le S(K) \le UB(S(K)) \qquad (5.65)$$

where

$$LB(S(K)) = \min\{S(K)\} \quad \text{and} \quad UB(S(K)) = \max\{S(K)\} \qquad (5.66)$$

where the minimum and maximum are taken over $\mathcal{D}_x^t$ in $PS_{.95}$. An approximate 95% prediction interval for (5.64) given by (5.65) and (5.66) is found using modern computing techniques and the results given in Table 5.3. As expected, the prediction interval, $156.439 \le S(K) \le 243,561$, contains the expected aggregate total (5.63).

Prediction set-based prediction intervals tend to be conservative, resulting in wide intervals. The construction of prediction sets (5.62) and associated prediction intervals requires the utilization of modern computational methods such as Solver in Excel, as outlined in Problem 5.18. These methods are useful in the context of actuarial statistics introduced later. We now turn our attention to statistical measures applied to life tables.

5.9 Life Models and Life Table Parameters

For a stochastic status model, time-sensitive conditions are modeled by a future lifetime random variable that can be either discrete or continuous in nature. In the continuous case, the pdf of the future lifetime random variable follows (5.26). In the discrete case, probabilities associated with the curtate future lifetime, discussed in Section 5.2, are given in life table form as presented in Section 5.8. Statistical measurements of the mortality structure that include the mean, median, and percentiles are now considered for these two cases. Also, measurements for aggregate models based on a collection of individual statuses and fractional age adjustments are discussed.

5.9.1 Population Parameters

First, we consider (x) with continuous future lifetime random variable T. The most basic characteristics of the distribution of the future lifetime random variable are center and variability. Two common measures of center are the expectation or mean (1.24) and the 50th percentile or median (2.13). For a continuous stochastic status age x the median future lifetime is the value $m(x)$ such that

$$P(T \geq m(x)) = \frac{1}{2} \tag{5.67}$$

Utilizing (5.7), Equation (5.67) becomes

$$\frac{1}{2} = \frac{S_X(x + m(x))}{S_X(x)} \tag{5.68}$$

and to find the median we solve for the constant $m(x)$. The expected value of the future lifetime random variable T is called the complete expectation of life, and can be calculated using either the pdf as in (1.22) or the survivor function following (1.44), explicitly

$$\overset{\circ}{e}_x = E\{T\} = \int_0^{\omega - x} t \, {}_t p_x \, \mu_{x+t} \, dt = \int_0^{\omega - x} {}_t p_x \, dt \tag{5.69}$$

Other moments of T can be computed, leading to the computation of additional measurements. To be specific, using (1.45), the second moment of T can be computed as

$$E\{T^2\} = 2 \int_0^{\omega - x} t \, {}_t p_x \, dt \tag{5.70}$$

A major measurement of the variability of the future lifetime random variable is the variance, computed as

$$Var\{T\} = 2 \int_0^{\omega - x} t \, {}_t p_x - [E\{T\}]^2 \tag{5.71}$$

Other distributional measurements associated with the continuous random variable are the skewness and kurtosis as defined in (1.26) and can be computed using moment computations (see Problem 1.8). Two continuous lifetime examples are now presented.

Example 5.17

For a continuous stochastic status model the force of mortality follows a variation of the Gompertz law (5.31) defined by $\mu_x = \exp(x)$, for age $x > 0$. For (x) the survival probability to future time t, applying (5.18), is

$$_t p_x = \exp\left(-\int_0^t \exp(x+s)\,ds\right) = \exp(-(\exp(x+t) - \exp(x))) \quad (5.72)$$

For (x) the median future lifetime, defined by (5.67), is found by setting (5.72) equal to ½ and computes to be

$$m(x) = -x + \ln(\ln(2) + \exp(x))$$

Utilizing the survivor function (5.72), the mean of the future lifetime random variable associated with (x) is

$$E\{T\} = \int_0^{\omega-x} \exp(-(\exp(x+t) - \exp(x)))\,dt \quad (5.73)$$

This computation is either approximated or completed using numerical methods.

Example 5.18

The underlying mortality structure for a population of individuals follows survival function (5.56), namely, $S(x) = (100 - x)^{1/2}/10$, for $0 \le x \le 100$. For an individual age x the survivor function to future lifetime t, from (5.54), is

$$_t p_x = \left[\frac{100 - x - t}{100 - x}\right]^{1/2} \quad (5.74)$$

Setting (5.74) equal to 1/2 and solving for t, the median of the future lifetime random variable is found to be $m(x) = (3/4)(100 - x)$. The mean of the future lifetime random variable for (x), applying (5.69), is

$$E\{T\} = \int_0^{100-x} \left[\frac{100 - x - t}{100 - x}\right]^{1/2} dt = \frac{2(100 - x)}{3} \quad (5.75)$$

Thus, in this example, the median is greater than the mean for the future lifetime random variable, indicating the distribution of the future lifetimes is skewed to the left.

Distributional measurements for a discrete stochastic status model are now explored. For (x), the distribution of future lifetime is the curtate random variable K. As in the continuous case, the expectation or mean and the variance characterize the center and variability. Based on age x, the mean curtate future lifetime, found using either (1.21) or (1.44), is

$$E\{K\} = \sum_{k=0}^{\omega-x-1} {}_{k+1}p_x = \sum_{k=0}^{\omega-x-1} k\,{}_k p_x\, q_{x+k} \tag{5.76}$$

Applying (1.45) the second moment takes the form

$$E\{K^2\} = \sum_{k=0}^{\omega-x-1} (2k+1)\,{}_{k+1}p_x \tag{5.77}$$

and the variance of K is

$$\mathrm{Var}\{K\} = \sum_{k=0}^{\omega-x-1} (2k+1)\,{}_{k+1}p_x - [E\{K\}]^2 \tag{5.78}$$

The curtate future lifetime is a discrete random variable so that of the median is not uniquely defined. Two examples demonstrating discrete future lifetime computations end this section.

Example 5.19

For a discrete stochastic status model the associated curtate future lifetime random variable has the discrete geometric distribution as discussed in Example 1.19. The survival function is given in (5.15) by ${}_k p_x = .9^k$ for $k = 0, 1, \ldots$. The mean future curtate lifetime applying (5.76) is

$$E\{K\} = \sum_{k=1}^{\infty} .9^k = \frac{.9}{.1} = 9.0$$

For the geometric distribution the second moment, leading to the computation of the variance of the curtate future lifetime, is most easily found using the mgf as demonstrated in Section 1.4.

TABLE 5.4

Life Table Moment Example

x	0	1	2	3	4	5	6
l_x	1,000	975	850	725	300	250	0

Example 5.20

For (x) the survival probabilities associated with a curtate random variable, from (5.54), are $_k p_x = l_{x+k}/l_x$ for $k = 0, 1, \dots$. The first and second moments associated with the distribution of K are computed using life table quantities and following (1.44) and (1.45) are

$$E\{K\} = \sum_{k=0}^{\omega-x-1} {}_{k+1}p_x = \sum_{k=0}^{\omega-x-1} \frac{l_{x+k+1}}{l_x} \quad \text{and} \quad E\{K^2\} = \sum_{k=0}^{\omega-x-1} (2k+1)\frac{l_{x+k+1}}{l_x}$$

where the upper bound on the summation is $\omega - x - 1$. The mean of K is referred to as the curtate expectation of life and is denoted by an example demonstration. Let the survival expectations of Table 5.4 exist. For a discrete-status age $x = 2$, we compute $E\{K\} = 1.5$, $E\{K^2\} = 3.3823$, and $\text{Var}\{K\} = 3.3823 - 1.5^2 = 1.1323$. These measurements can be used in connection to a collection of identical status models to form asymptotically normal prediction intervals.

Additional computational examples in the continuous future lifetime setting are outlined in Problems 5.10, 5.11, and 5.12. Probability and parameter computations in discrete setting are demonstrated through Problems 5.13 and 5.14.

5.9.2 Aggregate Parameters

In many applications, measurements of a collection of stochastic status models are desired. For example, to estimate the future costs associated with a retirement system, the number of years lived by the entire survivorship group may need to be determined. For a collection of discrete stochastic status models aggregate measurements, such as the expected total number of years lived by the group over specified ages, are straightforward calculations using life table quantities. In our discussion, we investigate aggregate measurements or parameters associated with a collection of continuous stochastic statuses.

Group survivorship concepts and notations are applied to a collection of continuous stochastic statuses. We define $T(x, n)$ as the future lifetime random variable between ages x and $x + n$. The total number of years lived between ages x and $x + 1$ is $l_x T(x, 1)$. The expected value of $l_x T(x, 1)$ is the expected number of survivors to age $x + 1$. For the group the expected number of years lived between ages x and $x + 1$ is

$$L_x = l_x\, E\{T(x,1)\} = l_{x+1} + \int_0^1 t\, l_{x+t}\, \mu_{x+t}\, dt \qquad (5.79)$$

Applying integration by parts an alternate expectation for (5.79) is

$$L_x = l_x\, E\{T(x,1)\} = \int_0^1 l_{x+t}\, dt \qquad (5.80)$$

Other measurements of the mortality structure of the survivorship group utilizing the expected number of survivors exist. For example, the expected total number of years lived by the group beyond age x is denoted by $l_x\, E\{T(x, \omega)\}$ and is computed as

$$T_x = l_x\, E\{T(x,\omega)\} = \int_0^\infty t\, l_{x+t}\, \mu_{x+t}\, dt = \int_0^\infty l_{x+t}\, dt \qquad (5.81)$$

Measurements (5.79), (5.80), and (5.81) give a good idea of the shrinkage structure of the population as time increases.

Expectations based on a collection of stochastic statuses can also be considered. Using group survivorship concepts, the mean number of future lifetime years of the l_x survivors to age x is computed as $E\{T(x)\}$, that is,

$$\overset{o}{e}_x = \int_0^\infty t\, {}_t p_x \mu_{x+t}\, dt \qquad (5.82)$$

Also, the mean number of years lived between ages x and $x + 1$ by the members of the group whose status fails between x and $x + 1$ is

$$\frac{\int_0^1 t\, l_{x+t}\, \mu_{x+t}\, dt}{\int_0^1 l_{x+t}\, \mu_{x+t}\, dt} \qquad (5.83)$$

These group parameters are directly applicable to theoretical population models. Fractional age techniques, such as described in Section 5.4, can be used to approximate continuous stochastic status aggregate measurements in the presence of discrete life table expectations. We end this section with a computational example.

Example 5.21

Initially, a survivorship group consists of 100,000 individuals and, using underlying survivorship function $S_X(x)$, the number of survivors to age x is $l_x = 100{,}000\, S_X(x)$. In particular, as in (5.56), let $S_X(x) = (100 - x)^{1/2}/10$ for $0 \le x \le 100$, so that

$$l_x = 10{,}000(100 - x)^{1/2} \quad \text{for} \quad 0 \le x \le 100$$

Applying (5.81), the total number of years lived by the group beyond age x is

$$T_x = 10{,}000 \int_0^{100-x} (100 - x - t)^{1/2}\, dt = 10{,}000 \left(\frac{2}{3} \right)(100 - x)^{3/2}$$

The mean number of years lived by the group past age x (5.82) is found to be

$$\overset{\circ}{e}_x = \left(\frac{2}{3} \right)(100 - x)^1 \quad \text{for} \quad 0 \le x \le 100$$

In many applications, the underlying continuous mortality structure may be unknown with only life table expectations available. In this case fractional age assumptions and adjustments to life table-based aggregate parameters are required.

5.9.3 Fractional Age Adjustments

Parameter quantities for individual and aggregate collections associated with continuous future lifetime random variables need to be partly estimated when the only survival and mortality expectations exist in the form of a life table. The mortality structure for the continuous future lifetime random variable T, within periods, such as years, must be assumed. Following the development of Section 5.4, we assume UDD holds, implying $T = K + S$, where K is the curtate lifetime, S is uniform over $[0, 1]$, and K and S are independent. Survival and failure probabilities containing fractional ages are computed with the help of (5.34). For example, assuming UDD, the probability (x) survives an additional $n + s$ years, for $0 < s < 1$, is

$$_{n+s}p_x = {}_np_x\, {}_sp_{n+x} = {}_np_x\, (1 - s\, q_{n+x}) \tag{5.84}$$

Expectations for fractional ages can also be computed under the UDD assumption. The uniform distribution of S implies that the expected future lifetime takes the form

$$\overset{\circ}{e}_x = E\{T\} = E\{K+S\} = E\{K\} + \frac{1}{2} = e_x + 1/2 \qquad (5.85)$$

Further, the variance of T is computed applying the independence of K and S and the uniform pdf given in (1.14) with $b = 1$ and $a = 0$

$$\text{Var}\{T\} = \text{Var}\{K\} + \text{Var}\{S\} = \text{Var}(K) + \frac{1}{12} \qquad (5.86)$$

Other assumptions, alternate to UDD, may be applied to model the distribution of future ages within periods. A prime example is discussed in Example 5.11 where a constant force of mortality assumption leads to the probability adjustment given by (5.40).

5.10 Select and Ultimate Life Tables

To model the mortality structure and analyze actuarial contracts, such as life insurance and life annuities, select future lifetime random variables, as introduced in Section 5.5, have been widely utilized. The life tables based on select ages and limiting ages can be very accurate and are referred to as select and ultimate life tables. It is thought that at the select age, or the age when the life insurance policy is issued, mortality probabilities should be at a low point due, in part, to the approval process of the life insurance policy. Utilizing the structure of the select future lifetime random variable presented in Section 5.5 and modifying group survivorship notations of Section 5.6, we now extend group survivorship concepts and notations to construction of select and ultimate life tables.

Group survivorship theory introduced in Section 5.6 is extended to model the mortality structure of a population dependent on a fixed select age. The number of individuals alive at select age $[x]$ is denoted by the random variable $\mathcal{L}_{[x]}$, and following notation (5.45), the associated expectation is $l_{[x]}$. After t additional years the number of survivors from the group of $\mathcal{L}_{[x]}$ individuals is denoted $\mathcal{L}_{[x]+t}$, where the expectation can be computed as

$$l_{[x]+t} = E\{\mathcal{L}_{[x]+t}\} = l_{[x]} \frac{S_X([x]+t)}{S_X([x])} \qquad (5.87)$$

for $t = 0, 1, 2, \ldots$ Similarly, decrement notations are modified. Out of the collection of select age $[x]$ individuals the number of decrements between ages $[x] + t$ and $[x] + t + n$ is denoted ${}_nD_{[x]+t}$ with corresponding expectation

$$_n d_{[x]+t} = E\{_n D_{[x]+t}\} = l_{[x]+t} - l_{[x]+t+n} \tag{5.88}$$

for nonnegative integers t and n. Utilizing quantities (5.87) and (5.88), mortality and survivor rates based on the select age structure can be found. For an individual age $[x] + t$ the mortality and survival probability associated with n additional years are

$$_n q_{[x]+t} = \frac{l_{[x]+t} - l_{[x]+t+n}}{l_{[x]+t}} \tag{5.89}$$

and

$$_n p_{[x]+t} = 1 - {}_n q_{[x]+t} = \frac{l_{[x]+t+n}}{l_{[x]+t}} \tag{5.90}$$

Select life tables follow the life table structure of Section 5.5, based on the select age, and show yearly survival totals or mortality rates, $l_{[x]+t}$ and $q_{[x]+t}$, for positive integers of $[x]$ and t. An illustrative example follows.

Example 5.22

For the population in Example 5.12, the underlying survival function evaluated at age $[x] + t$ follows (5.42) given by

$$S_X([x]+t) = \frac{(100 - 1.2[x] - .8t)^{1/2}}{10} \tag{5.91}$$

for $0 < 100 - 1.2[x] - .8t < 1$. A select life table based on a survivorship group consisting of 100,000 newborns includes the select life totals

$$l_{[x]} = 100,000 \frac{(100 - 1.2x)^{1/2}}{10}$$

for $0 \le x \le 8$. Included in the select life table for chosen values of $[x]$ and t are mortality rates defined by (5.91) for $t = 1$. For this example, the select life table is given in Table 5.5.

From Table 5.5, probabilities such as the following can be evaluated:

1. The probability a person [50] survives to age 53 is

$$_3 p_{[50]} = p_{[50]} \, p_{[50]+1} \, p_{[50]+2} = .98994(.98974)(.98952) = .99515$$

2. The probability a person [51] dies in the third year is computed as

$$_{2|1} q_{[51]} = p_{[51]} \, p_{[51]+1} q_{[51]+2} = .98963(.98942)(.01081) = .010585$$

TABLE 5.5

Select Life Table

$[x]$	$l_{[x]}$	$l_{[x]+1}$	$l_{[x]+2}$	$q_{[x]}$	$q_{[x]+1}$	$q_{[x]+2}$
50	63,246	62,610	61,987	.01006	.01026	.01048
51	62,290	61,644	60,992	.01047	.01058	.01081
52	61,319	60,663	60,000	.01070	.01093	.01117
53	60,332	59,666	58,992	.01105	.01130	.01157
54	59,330	58,652	57,966	.01144	.01171	.01198
55	58,310	57,619	56,921	.01184	.01211	.01242

As demonstrated in Section 5.7, a wide variety of probability computations are possible utilizing quantities given in a select life table.

In select age actuarial applications the mortality rates are often assumed to converge in a uniform manner as a function of duration since selection. To be precise, for positive integer r and small positive constant $\alpha > 0$, assume

$$|q_{[x]+r} - q_{[x-t]+r+t}| < \alpha \qquad (5.92)$$

for all ages of selection $[x]$ and $t \geq 1$. Thus, for all $[x]$ the future failure rates associated with subsequent years converge to ultimate failure or mortality probabilities as the future time increases. If r is the smallest positive integer such that (5.92) holds, then the select tables can be truncated by taking

$$q_{[x]+t} = q_{x+t} \qquad (5.93)$$

for $t = r, r + 1, \dots$. The first r years are called the select period and the convergent rates q_{x+t} make up the ultimate table. The combination of select and convergent mortality rates makes up a select-ultimate life table. An example of a select-ultimate table now follows.

Example 5.23

Part of a select-ultimate table with a 3-year select period is given in Table 5.6. The failure rates are hypothetical, but the listed values converge by criteria (5.92) with $\alpha = .001$. In a full table other columns corresponding to q_{x+t} would be listed for $t \geq 4$. The ultimate table values are in the q_{x+t} columns for $t = 3, 4, \dots$.

From Table 5.6, we see various decrement probabilities listed for individuals age 43, namely, $q_{[43]}, q_{[42]+1}, q_{[41]+2}$, and q_{40+3}. Also, for future lifetimes greater than 2 years past $[x]$, these mortality rates are functions of the future age and not $[x]$.

Mortality rates given in select and ultimate life tables are often smoothed or graduated to create a table that reflects a reasonable mortality pattern. Guidelines for the smoothing or graduation of mortality tables have been discussed by the Society of Actuaries (see SOA, 2000, 2001). Life tables based on empirical data are often adjusted to achieve common mortality modeling goals. For example, select life tables are often adjusted so that, except for extreme circumstances such as the very young, mortality rates should increase with age. Thus, we desire life tables where

$$q_{[x]} \leq q_{[x]+1} \leq q_{[x]+2} \leq \dots \text{ and } q_{[x]+s} \leq q_{[x+1]+s} \leq q_{[x+2]+s} \dots \qquad (5.94)$$

Also, the mortality at an attained age should increase with the length of time, or duration, since issue, or

$$q_{[x]} \leq q_{[x-1]+1} \leq q_{[x-2]+2} \leq \dots \qquad (5.95)$$

Using the constraints (5.94) and (5.95), select and ultimate tables are formed using expert opinion and ad hoc methods, which is now demonstrated.

Example 5.24

The select and ultimate life table given in Table 5.6 contains mortality rates that increase within the rows and columns as the age increases so that (5.94) holds. However, condition (5.95) does not hold and the mortality rates require adjustment. In this case, mortality rates are not a minimum around initial ages. Moving along diagonal entries, according to (5.95), we adjust the entries about their mean to form an adjusted table given in Table 5.7.

TABLE 5.6

Select and Ultimate Life Table

$[x]$	$q_{[x]}$	$q_{[x]+1}$	$q_{[x]+2}$	q_{x+3}
40	.010	.013	.015	.017
41	.014	.016	.018	.019
42	.018	.019	.020	.023
43	.021	.022	.024	.027
44	.024	.026	.028	.030

TABLE 5.7

Graduated Select and Ultimate Life Table

$[x]$	$q_{[x]}$	$q_{[x]+1}$	$q_{[x]+2}$	q_{x+3}
40	.010	.013	.018	.020
41	.013	.017	.019	.023
42	.016	.018	.022	.026
43	.018	.021	.024	.027
44	.021	.024	.027	.030

Many adjustments to mortality tables exist that are done so as to minimize the change in the select and ultimate mortality rates. These mortality tables are considered to have a better underlying mortality pattern where inconsistent mortality computations do not arise.

Problems

5.1. Let the lifetime in a status model, X, have survival function $S_X(x)$ $= 1 - (x/100)^2$ for $0 \le x \le 100$. Find (a) μ_x, (b) $F_X(x)$, (c) $f_X(x)$, and (d) $P(10 \le X \le 40)$. (e) For initial time x let $T(x) = T$ be the corresponding future lifetime random variable. Give formulas for $f(t)$ and $F(t)$. (f) Given the status lasts 30 years, what is the probability it will last an additional 5 years? (g) Compute $_{5|8}q_{20}$ and interpret this value.

5.2. Given $\mu_x = \tan(x)$ for $0 < x \le \pi/2$ find simplified formulas for (a) $S_X(x)$, (b) $F_X(x)$, and (c) $f_X(x)$.

5.3. Given $S_X(x) = 1/(1 + x)$ for $x \ge 0$, find (a) $F_X(x)$, (b) $f_X(x)$, and (c) μ_x. (d) For $T(x) = T$ what is the form of $f(t)$ and $F(t)$?

5.4. For the following distributions find the force of mortality and for the future lifetime random variable find $F(t)$ and $f(t)$:

   a. Pareto of Section 1.6.1.

   b. Weibull of Section 1.6.3.

5.5. For a status model let the integer lifetime X with pmf $f_X(x) = 1/(1 + m)$ for $S = \{0, 1, ..., m\}$ where $m > 0$. At integer age x let the curtate future lifetime be K.

   a. Find the pmf of K.

   b. Assume UDD holds and compute the probability the status survives 1.5 years.

5.6. Let the pmf of an integer-valued random variable U have the form $f(u) = \exp(-5) \, 5^u/u!$ for $S = \{0, 1, ...\}$. Let $K = K(x)$ be the curtate future lifetime random variable conditioned on age x. Find the pmf of $K(1)$.

5.7. Consider mortality rates following the Gompertz law given in Example 5.7, where $x = 0$ and T represents the lifetime.

   a. Compute the survival function where $B = 1$ and $c = e$. The resulting random variable is one of the extreme-valued random variables (Nelson, 1982, p. 39).

   b. Find the pdf.

5.8. For a status the future lifetime random variable is T, where the notations of this chapter hold. For each of the following give the notations for an individual status age 30:

   a. Survives 40 additional years.

   b. Fails between the ages of 60 and 65.

   c. Fails before age 75.

5.9. Let the lifetime of a status have a survival function given by $S_X(x) = 10x^{-2}$ for $0 \leq x \leq 100$.

   a. Give formulas for $f_X(x)$ and μ_x.

   b. For the future lifetime random variable $T = T(x)$ compute the survival function.

   c. Compute the probabilities defined in Problem 5.8 using this lifetime distribution.

5.10. Let future lifetime T have $G(t) = t/(100 - x)$ for $0 \leq t < 100 - x, = 1$ for $t \geq 100 - x$. Find (a) $E\{T\}$, (b) $\text{Var}\{T\}$, and (c) median of T.

5.11. Let the force of mortality be defined by $\mu_x = x$ for $x > 0$. Find (a) $_tp_x$, (b) pdf $g(t)$, and (c) $E\{T\}$.

5.12. Let X have survival function $S_x(x) = (1 - x/w)^a$ for $0 \leq x \leq w$ and $a > 0$. Find (a) $_tp_x$, (b) $E\{T\}$, and (c) $\text{Var}\{T\}$.

5.13. For a status over a limited time period a life table is given by

x	l_x	d_x	q_x	x	l_x	d_x	q_x
0	500			3	280		
1	495			4	100		
2	365			5	0		

   a. Compute d_x in the table.

   b. Compute q_x in the table.

   c. Find the probability a person age 1 lives 2 years and dies in the next year.

   d. Approximate the probability a person age 2 survives at least 1.5 years.

5.14. Let the life table in Problem 5.13 hold. For (x) let K be the curtate future lifetime. Find (a) the pdf of K, (b) $E\{K\}$, and (c) $\text{Var}\{K\}$.

5.15. Consider the Pareto distribution discussed in Problem 5.4a for parameter $\alpha > 2$. Find (a) $E\{T\}$, (b) $\text{Var}\{T\}$, and (c) median $m(T)$.

5.16. Consider the graduated select-ultimate life table given in Table 5.7. Find the probability a person with initial age 40 (a) is now age 41 and survives 2 additional years, and (b) is now 41, survives 1 year, and then dies within 1 year.

Excel Problems

5.17. For a closed survivorship group we construct a life table for the quantities listed below:

x	l_x	x	l_x	x	l_x	x	l_x
0	1,000	3	860	6	495	9	110
1	995	4	750	7	350	10	51
2	931	5	635	8	225	11	0

Construct a mortality table listing d_x, q_x, and p_x for each age. For (2) compute (a) the probability of failure within 2 years, (b) the probability of survival for 2 years and the failure within 3 additional years, (c) the probability of survival for 2½ years (assume UDD), and (d) the mean of the curtate future lifetime. Excel: Basic operations.

5.18. Let the life table in Table 5.4 hold. For 20 independent statuses all age $x = 2$ let the observed aggregate whole years lived be $S(K)$ given by (5.64). An approximate 95% prediction interval for $S(K)$ is constructed by applying the prediction set approach (5.61) and (5.62). Excel: CNIINV, and Data, Solver, setting the target cell equal to min or max.

Solutions

5.1. a. Following (5.16), $\mu_x = 2x/(100^2 - x^2)$.
   b. (1.42) gives $F_X(x) = (x/100)^2$.
   c. (1.43) gives $f_X(x) = x/5{,}000$.
   d. .15.
   e. (5.24) gives $F(t) = 1 - (100^2 - (x+t)^2)/(100^2 - x^2)$, $f(t) = 2(x+t)/(100^2 - x^2)$.
   f. $_5p_{30} = (100^2 - 35^2)/(100^2 - 30^2) = .964286$
   g. $_{5|8}q_{20} = {_5p_{20}}\, _8q_{25} = .04833$.
5.2. a. (5.19) gives $S_x(x) = \cos(x)$.
   b. $F_x(x) = 1 - \cos(x)$.
   c. $f_x(x) = \sin(x)$.
5.3. a. $F_X(x) = x/(1+x)$.
   b. $f_X(x) = (1+x)^{-2}$.

c. (5.31) gives $\mu_x = (1 + x)^{-1}$.

d. (5.25) gives $_tp_x = (1 + x)/(1 + x + t)$ so $F(t) = t/(1 + x + t)$ and $f(t) = (1 + x)/(1 + x + t)^2$.

5.4. a. From (1.46) $\mu_x = \alpha/x$.

b. From (1.54) and (1.55) $\mu_x = \alpha\beta^{-\alpha}x^{\alpha-1}$.

5.5. a. $P(K = k) = 1/(m - x + 1)$ for $k = 0, 1, \dots, m - x$.

b. (5.37) gives $_{1.5}q_x = {}_1p_x (1 - .5\ q_{x+1}) = ((m - x)/(m - x + 1))(1 - .5(m - x - 1)/(m - x))$.

5.6. (5.15) gives $_kp_x = P(U \geq k + x)/P(U \geq x)$ so $P(K(1) = k) = 5^{k+1}/((k + 1)!(\exp(5) - 1))$.

5.7. a. (5.32) gives $S_x(x) = \exp(-\exp(x) + 1)$.

b. (5.33) gives $f_x(x) = \exp(x)\exp(-\exp(x) + 1)$.

5.8. a. $_{40}p_{30}$.

b. $_{30|5}q_{30}$.

c. $_{45}q_{30}$.

5.9. a. $S_X(x) = 10x^{-2}$ and $\mu_x = 2/x$.

b. $_tp_x = S_X(x + t)/S_X(x) = (x/(x + t))^2$.

c. $_{40}p_{30} = (30/70)^2 = .18367$, $_{30}p_{30}\ _5q_{60} = (30/60)^2[1 - (60/65)^2] = .036982$, $_{45}q_{30} = 1 - (30/75)^2 = .84$.

5.10. a. (5.69) gives $E\{T\} = (100 - x)/2$.

b. (5.70) gives $E\{T^2\} = (100 - t)^2/3$, and from (5.71), $\text{Var}\{T\} = (100 - t)^2/12$.

5.11. a. (5.18) gives $_tp_x = \exp(-t^2/2 - xt)$.

b. $g(t) = (t + x)\exp(-t^2/2 - xt)$.

c. (5.69) gives $E\{T\} = (\pi/2)^{1/2}\exp(x^2/2)\Phi(-x)$.

5.12. a. (5.7) gives $_tp_x = [(w - x - t)/(w - x)]^a$.

b. (5.69) gives $E\{T\} = (w - x)/(1 + a)$.

c. (5.70) and integration by parts give $E\{T^2\} = 2(w - x)^2/[(1 + a)(2 + a)]$, and (5.71) gives $\text{Var}\{T\} = a(w - x)^2/[(1 + a)^2(2 + a)]$.

5.13. a. Using (5.47) $d_x = l_x - l_{x+1}$.

b. Using (5.53) $q_x = d_x/l_x$.

c. $_{2|1}q_1 = p_1 p_2 q_3 = .3636$.

d. $_{1.5}q_2 = {}_1p_2 (1 - .5\ q_3) = .52055$.

5.14. a. $f(k) = {}_{k|}q_1 = d_{1+k}/l_1$.

b. (5.54) gives $E\{K\} = \Sigma\ l_{2+k}/l_1 = 1.505$.

c. (5.55) gives $E\{K\} = \Sigma(2k + 1)l_{2+k}/l_1 = 3.444$ and (5.56) gives $\text{Var}\{K\} = 1.1794$.

5.15. $_tp_x = x^\alpha (x + t)^{-\alpha}$.

a. (5.69) gives $E\{T\} = x/(\alpha - 1)$.

   b. (5.70) gives $E\{T^2\} = 2x^2/[(\alpha - 1)(\alpha - 2)]$ and (5.71) gives $\mathrm{Var}\{T\} = \alpha x^2/[(\alpha - 1)^2(\alpha - 2)]$.

   c. $x(2^{1/\alpha} - 1)$.

5.16. a. $_2p_{[40]+1} = (.987)(.982) = .96923$.

   b. $_{1|2}q_{[40]+1} = p_{[40]+1}\,(q_{[40]+2} + p_{[40]+2}\,q_{43}) = .043935$.

5.17. a. $_2q_2 = .194415$.

   b. $_{2|3}q_2 = .429646$.

   c. $_{2+1/2}p_2 = .743824$.

   d. $E\{K\} = 3.73362$.

5.18. $LB(S(K)) = 1133.595$ at $D_2 = 125.045$, $D_3 = 420.319$, $D_4 = 41.947$, $D_5 = 269.294$, and $UB(S(K)) = 1361.728$ at $D_2 = 108.105$, $D_3 = 405.853$, $D_4 = 52.251$, $D_5 = 283.791$.

6

Stochastic Status Models

Stochastic status models are used to describe many financial and actuarial processes. As outlined in the introduction of Chapter 3, stochastic status models are characterized by one or more future economic actions initiated at future times. The structure of the future time conditions is stochastic in nature, leading to two types of stochastic status models: discrete and continuous. In a continuous-status model the time structure is modeled by the future lifetime random variable introduced in Section 5.1. In a discrete stochastic status model the lifetime structure is integer valued and is represented by the curtate future lifetime random variable of Section 5.2. This chapter introduces and analyzes various stochastic status models, specifically life insurance and life annuities that are central to actuarial models.

Stochastic present value functions associated with the stochastic status actions are introduced in Section 6.1. The loss model and analysis approach of Chapter 4 is extended to analyze stochastic status models. Of the two criteria used to analyze loss models, the risk approach based on the expectation of the loss function is predominant. Risk evaluations for continuous, discrete, and mixed stochastic structures are introduced in Sections 6.2.1, 6.2.2, and 6.2.3, respectively. The risk criteria applied to the single-risk models of Section 4.3 yield the net single value or net single premium (NSP). The percentile criterion applied to a stochastic loss function is briefly discussed in Section 6.3.

Various types of life insurance and life annuities, including whole life insurance and annuities, term life insurance, temporary life annuities, endowment life insurance, and deferred life annuities, are analyzed using the risk criteria in Sections 6.4 and 6.5. Expectation formulas, in the form of net single premium formulas, for these models along with standard actuarial notations are presented in Sections 6.4.1 and 6.5.1. Both continuous and discrete stochastic status life insurance and life annuity models are discussed, with relations among these expectations explored in Section 6.6. In Section 6.7 life tables introduced in Section 5.7 are utilized to compute net single premiums of typical life insurances and life annuities.

The topic of premium evaluation for life insurance and life annuities is explored through the stochastic loss model approach in conjunction with the risk criteria. Premium evaluation for common actuarial settings, an example being when life insurance is financed by premiums payable for the lifetime of the insured, is explored in Section 6.8 through the risk analysis of stochastic loss functions. Standard premium notations are presented in Section 6.8.1, and variance computations are considered in Section 6.8.2. The topic of reserves associated with these models is discussed in Section 6.9. Specifically, unit benefit reserves, relations among reserve computations, and a survivorship table approach to reserve calculations are discussed in Sections 6.9.1, 6.9.2, and Section 6.9.3. The chapter ends with a discussion of general time period models in Section 6.10 and the exploration of the effect of addition of expenses, both fixed and variable, on these stochastic status computations in Section 6.11.

6.1 Stochastic Present Value Functions

Stochastic present value functions are central in the analysis of stochastic status models. A continuous stochastic status model is comprised of one or more future economic actions occurring at future times defined by a continuous future lifetime random variable, T, discussed in Section 5.1. The future value of an action at time t is denoted by F_t and is evaluated at initial time 0 using a present value function, as discussed in Section 3.1.1. For fixed continuous rate δ, the present value (3.9) is applied. Utilizing notation common in actuarial science, the discount function $v = (1 + i)^{-1} = \exp(-\delta)$ is applied to the value at time t to give the stochastic present value function

$$PV(T) = F_T\, v^T \tag{6.1}$$

The support of the present value, denoted S_{PV}, consists of future lifetimes corresponding to positive (6.1) values. The distribution of T is modeled by either a theoretical statistical distribution or empirical-driven life table expectations as discussed in Chapter 5.

The stochastic structure of the future lifetime random variable can be partitioned into an integer and a continuous fractional part as presented in Section 5.4. For (x) the general decomposition of the future lifetime random variable T is $T = K + S$, where K is the curtate future lifetime and S takes values in $[0, 1)$. The distribution of S models the mortality structure within years, and the present value function (6.1) takes the form

$$PV(K + S) = F_{K+S}\, v^{K+S} \tag{6.2}$$

The probability density function (pdf) of S takes support inside the unit interval, and for most applications we assume S and K are independent. In some settings the distribution of S is continuous, and in some cases the distribution may be discrete.

In the analysis of actuarial stochastic status models, present value functions are combined to form stochastic loss functions similar to the loss models presented in Section 4.1. Analysis of a stochastic loss model depends on the criteria employed. In actuarial science modeling, the risk criteria based on expectation computations have been the main technique, but other methods, such as utility theory (see Bowers et al., 1997), have been explored. In this chapter, the risk criteria and percentile criteria introduced in Sections 4.2.1 and 4.2.2, respectively, are the main criteria employed.

6.2 Risk Evaluations

In this section, the risk criteria in conjunction with the equivalence principle, denoted *RC* and *EP* in Section 4.2.1, are used to evaluate stochastic loss models. For single-status models discussed in Section 4.3, the *RC* and *EP* approach leads to the expectation of a stochastic present value function referred to as the net single value (NSV). The form of the expectation depends on the structure and conditions of the stochastic status model. In some models, the financial action takes place only at the endpoints of time intervals. For example, insurance premium payments may be due at the start of each 6-month time period. In others, such as life insurance where a benefit payment is immediate or paid at the time of death, the financial action may occur within time periods. In the next three subsections expectations for stochastic loss functions are examined for continuous, discrete, and mixed time structure models.

6.2.1 Continuous-Risk Calculations

In this subsection the conditions associated with a stochastic status model involve a continuous random variable. For (x) a continuous stochastic status model defines future lifetime random variable T where the pdf is given by (5.26). For a single-status model based on stochastic present value (6.1), applying *RC* and *EP* yields the expectation or NSV

$$\text{NSV} = \int_{S_{PV}} F_t \, v^t \, {}_t p_x \, \mu_{x+t} \, dt \tag{6.3}$$

The integrand in (6.3) is a combination of three components: the future value quantity, the discount operator v, and the pdf of T. The NSV (6.3) is general in nature, applicable to a variety of situations, and in many specific cases can be simplified to yield standard actuarial computations. This subsection concludes with an example that demonstrates this approach in the area of investment modeling.

Example 6.1

An investment is to be sold at a future date for $1,000 where the annual rate is assumed to be $\delta = .1$. Due to other considerations, the investment will be sold at any time at or before 4 years. The future lifetime random variable, T, is the time to sale and may follow any continuous probability distribution. The expected present value of the sale is computed using (6.3) with $F_t = \$1,000$ as

$$\text{NSV} = \$1000 \ E\{\exp(-\delta T)\} = \$1000 \ M_T(-\delta)$$

where $M_T(a)$ is the moment generating function (mgf) of T as defined in Section 1.4. Specifically, we assume the distribution of T is uniform, with pdf (1.14), on (0, 4). Using the uniform mgf evaluated at $-.1$, the expectation is

$$\text{NSV} = \$1000 \frac{1 - \exp(-.4)}{(.4)} = \$824.20$$

Hence, the value of the investment considered as one lump sum evaluated at the initial time is $824.20.

6.2.2 Discrete Risk Calculations

In a discrete stochastic status model the future time structure is modeled by a discrete random variable, the curtate future lifetime random variable K. The support of K, denoted by S_K, is countable and defines a collection of disjoint intervals of time. In these models, economic actions occur at only one point in a time interval defined by $K + s$ for $0 \le s \le 1$, and the stochastic present value (6.2) holds where $S = s$. For example, if the financial action occurs at the end of the period, then $s = 1$, while action occurring at the start of the period is indicated by $s = 0$. The expectation of (6.2) using the pdf of K given by (5.12) is the net single value and is

$$\text{NSV} = \sum_{S_{PV}} F_{K+S} \ v^{k+s} \ {}_k p_x \ q_{x+k} \tag{6.4}$$

Computation (6.4) is completed utilizing a specifying discrete pdf or life table expectations as described in Chapter 5. This is demonstrated in Example 6.2.

Example 6.2

A stock investment is to be sold at the end of the first month its future value reaches a fixed price denoted by the quantity fv. Over the next 5 years the probability of sale is equal to 1/60 for each month and the yearly rate of return, computed continuously, is fixed at $\delta = .1$. Here, $P(K = k) = {}_k p_x \ q_{x+k} = 1/60$ for k in $S_{PV} = \{0, 1, 2, ..., 59\}$. Then net single value (6.4) is evaluated at $s = 1$ and the future value is the constant $F_{K+S} = fv$. Applying summation (1.31) simplifies to

$$\text{NSV} = fv \frac{1}{60} \sum_{k=0}^{59} v^{k+1} = \left(\frac{fv}{60} \right) \left(\frac{v - v^{61}}{1 - v} \right)$$

where v denotes the monthly discount value. Substituting $v = \exp(-.1/12)$, the net single value becomes NSV = .783636 fv. Thus, using the *RC* and *EP* approach the present value of the investment is estimated at 78.36% of the sales price.

The computational examples given so far have demonstrated the versatility in general stochastic status models. These models are applied as parts of overall structures to analyze stochastic economic events such as financial strategies and insurance and annuity models. As an additional modeling structure, we consider mixed-type future lifetime random variables and their associated computations, as these models form the theoretical basis of some important relations and actuarial measurements.

6.2.3 Mixed Risk Calculations

Mixed-type random variables play an important role in the modeling and analysis of stochastic status models. For example, the time structure of a financial action may be continuous in nature, but the only reliable associated probabilities are discrete. This is the case when life table data mortality rates, as described in Chapter 5, are utilized to analyze a continuous stochastic status model. For a stochastic status age x the decomposition of the future lifetime random variable T introduced in Section 5.4 is applied, namely, $T = K + S$, where K is the curtate future lifetime and $0 \leq S < 1$. The uniform distribution of death (UDD) assumption is assumed to hold, implying that S is a uniform random variable on $[0, 1)$ and independent of K. Expectation (6.4), written in terms of this decomposition and utilizing the mixed distribution (5.37), takes the form

$$\text{NSV} = \sum_{S_{PV}} \int_0^1 F_{k+s} \, v^{k+s} \, ds \, {}_kp_x \, q_{x+k} \tag{6.5}$$

Expectation (6.5) is general and can be applied to many actuarial settings. One common actuarial usage of (6.5) is the derivation of formulas relating discrete- and continuous-type risk-based expectations in life insurance and life annuities, as explored in Section 6.6.

In many applications, the future value quantity in the stochastic present value function (6.1) is fixed, which we denote as $F_{k+s} = fv$, for all $k \geq 0$ and $0 \leq s$

< 1. If the economic action associated with fv can take place only at the end of a time interval, then the event occurs at $k + 1$. To simplify expectation (6.5), we use $v^{k+1+s-1}$ and

$$\int_0^1 v^{s-1}\, ds = \int_0^1 \exp(-(\delta(s-1)))\, ds = \frac{\exp(\delta)-1}{\delta} = \frac{i}{\delta}$$

General expectation (6.5) reduces to

$$\text{NSV} = fv\left(\frac{i}{\delta}\right)\underbrace{\sum v^{k+1}\,{}_k p_x\, q_{x+k}}_{S_{PV}} \tag{6.6}$$

In (6.6) the relation between discrete and continuous stochastic status expectations, useful in actuarial science, is plainly demonstrated. A computational example follows.

Example 6.3

For the stock discussed in Example 6.2 the sale for future value fv occurs at the end of the first year, the price reaches fv, and the number of full years the stock is held is assumed to follow a Poisson distribution, as given in Example 1.9, with mean λ. Utilizing the Poisson probability mass function (pmf) (1.11), expectation (6.6) simplifies as

$$\text{NSV} = fv\left(\frac{i}{\delta}\right) v \sum_{k=0}^{\infty} \exp(-\lambda)\frac{(v\lambda)^k}{k!} = fv\left(\frac{i}{\delta}\right) v \exp(\lambda(v-1))$$

In particular, if $\delta = .1$ and $\lambda = 1$, the NSP $= .8652\, fv$, indicating the expectation of the present values is 86.52% of the sales price.

Examples of these techniques are outlined in Problems 6.1, 6.2, and 6.3. As mentioned, the RC and EP approach is only one method to evaluate stochastic status models. In the next section the percentile approach introduced in Section 4.2.2, namely, PC(.25), is used to analyze stochastic status models.

6.3 Percentile Evaluations

A general percentile criteria or approach to the analysis of continuous loss models sets a percentile for the loss function equal to a fixed value and solves for any unknown constants. In the 25th percentile criteria introduced in

Section 4.2.2, denoted $PC(.25)$, the probability of a positive loss is set equal to .25. In actuarial life insurance or life annuity settings involving single-status models, this approach yields a net single value referred to as a net single premium.

Consider a stochastic status model represented by the present value function (6.1) where the future value at time t is a constant, $F_t = fv$, and is explored. A stochastic loss function is constructed to reflect certain goals. If an upper bound on the present value function is desired, a single-status loss function is constructed from (4.1) where $PVE(T) = fv \exp(-T\delta)$ and $PVR(T) = NSV$. To apply $PC(.25)$, we first set the probability of a positive loss to 25. Utilizing the 25th percentile for the future lifetime random variable, denoted $t_{.25}$, and solving for the net single value produces

$$NSV = fv \exp(-\delta t_{.25}) \tag{6.7}$$

An alternate interpretation is to view (6.7) as the 75th percentile associated with the stochastic present value function (6.1) with $F_t = fv$. This technique is demonstrated in Example 6.4 and outlined in Problem 6.5.

Example 6.4

The $1,000 investment in Example 6.1 is evaluated using $PC(.25)$. Since the sale made at any time in the next 4 years follows a continuous uniform distribution, the 25th percentile for T is $t_{.25} = 1.0$. Applying (6.7) with rate $\delta = .1$ the $PC(.25)$ method yields

$$NSV = \$1,000 \exp(-.1) = \$904.84$$

The NSV of $904.84 represents an upper bound at the 75th percentile level on the stochastic present value of the sale and exceeds the RC approach where $NSP = \$824.20$.

In the situation where the stochastic status is discrete in nature, a percentile approach can be defined by interpolating the mortality function between defined time periods. The decomposition of the future lifetime random variable T and the UDD assumption of Section 5.4 are applied along with a definition of the discrete percentile measurement. For $0 < \alpha < 1$, the α100th percentile for K, denoted k_α, is the positive integer, where

$$P(K \le k_\alpha) \le \alpha \quad \text{and} \quad P(K \le k_\alpha + 1) > \alpha \tag{6.8}$$

Under these assumptions, $t_\alpha = k_\alpha + s_\alpha$, where the joint distribution (5.37) implies $\alpha = P(K \le k_\alpha) + s_\alpha [P(K = k_\alpha + 1) - P(K \le k_\alpha)]$. Solving for s_α we find

$$s_\alpha = \frac{\alpha - P(K \le k_\alpha)}{P(K = k_\alpha + 1)} \tag{6.9}$$

The $PC(.25)$ approach in conjunction with a single-status loss function yields

$$NSV = fv \exp(-\delta(k_{.25} + s_{.25})) \tag{6.10}$$

Percentile measurement (6.10) is based on a continuous future lifetime random variable but can be computed using life table data in conjunction with assumptions, such as UDD. A computational example of the discrete case percentile NSV follows.

Example 6.5

The mixed random variable case is considered where the pmf of the curtate future lifetime is given by the geometric distribution presented in Example 1.19 with pmf $f(k) = p\, q^k$ for $S_K = \{0, 1, ...\}$. The survival function given by (1.66) yields

$$P(K \geq k) = q^k \quad \text{for} \quad k = 0, 1, ...$$

To apply $PC(.25)$ and compute (6.10) we solve for $k_{.25}$, the smallest k such that $q^k \leq .75$. Hence, $k_{.25} = \text{gil}(\ln(.75)/\ln(p) + 1)$, where $\text{gil}(a)$ is the greatest integer less than a, and (6.10) can be computed. For example, if $\delta = .1$ and $p = .9$, then $k_{.25} = \text{gil}(3.73045) = 3$ and computing (6.10) the net single value is

$$NSV = fv \exp(-.325) = .722527\, fv$$

or 72.25% of future value (fv). Analogous to the continuous setting in the $PC(.25)$ approach, (6.10) represents an upper bound on the present value function at the 75th percentile level.

6.4 Life Insurance

People take out life insurance to mitigate the negative effects of possible future economic consequences associated with individual deaths. In the simplest form of life insurance, a policy is issued on (x) that after death pays a benefit to a beneficiary. The insurance policy is financed by one or more premium payments that are dependent, among other things, on the amount of the benefit and the risk category of the individual being insured. In more complex settings, the benefit may be a function of several factors, such as a reliance on multiple death combinations or the exact cause of death. The general nomenclature used in connection with an insurance policy is that of a general stochastic status model where the benefit is paid upon change of the status due to a specified decrement. The future time of the status change may be either continuous, modeled by the future lifetime random variable, or discrete, based on the curtate future lifetime. In this section, the interest rate is fixed at δ for the life of the contract. For a detailed analysis of life insurance and relevant derivations we refer to Bowers et al. (1997, Chapter 4).

Insurance policies vary in nature to meet desired needs, and there exist two stochastic structures analogous to the two types of random variables, continuous and discrete. In the first, the benefit is payable at the time of the status change, subject to satisfying qualification provisions. The second is based on discrete time periods where the benefit often is paid at the end of the period in which the status changes. In this section, different types of insurance are presented and analyzed. The analysis consists of constructing a single-status loss model and applying RC and EP as described in Section 4.2. Example 6.6 gives an introduction to some of the conditions that may be included in life insurance.

Example 6.6

An individual age $x = 50$ takes out a discrete life insurance policy where applying group survivorship theory the yearly expected survival and decrement totals are given in Table 5.1. The insurance benefit is paid at the end of the year of death and is defined as 3 units if death is in the first year, 2 units if death occurs in the second through fourth years, and 1 unit if death occurs during or after age 54. The yearly effective interest rate is taken to be $i = .05$, so that $v = (1 + .05)^{-1} = .95238$. For this insurance, present value is a function of the curtate future lifetime K and is

$$PV(K) = 3v^1 \quad \text{for} \quad K = 0 \tag{6.11}$$
$$2v^{K+1} \quad \text{for} \quad K = 1, 2, 3$$
$$v^4 \quad \text{for} \quad K \geq 4$$

where the support of the present value function is $S_{PV} = \{0, 1, \ldots\}$. To estimate the future cost of this insurance a single-status loss function is constructed using (6.11) and RC with EP applied. The expectation of (6.11) is computed using the life table expectations

$$E\{PV(K)\} = 3v^1 \frac{d_{50}}{l_{50}} + 2\sum_{k=1}^{3} v^{k+1} \frac{d_{50+k}}{l_{50}} + 1v^4 \frac{l_{54}}{l_{50}} = .833467$$

The insurance company estimates the expected cost of this policy, without expenses, to be 83.34% of the unit value. Further, following basic theory in Section 1.3 the second moment is computed as

$$E\{PV(K)^2\} = 9v^2 \frac{d_{50}}{l_{50}} + 4\sum_{k=1}^{3} v^{2(k+1)} \frac{d_{50+k}}{l_{50}} + 1v^8 \frac{l_{54}}{l_{50}} = .763043$$

and the variance of the present value function is

$$\text{Var}\{PV(K)\} = E\{PV(K)^2\} - [E\{PV(K)\}]^2 = .763043 - .833467^2 = .068376$$

Approximate prediction intervals for the future lifetime can be computed. For example, a two standard deviation prediction interval for $PV(K)$ is given by

$$.31049 = .833467 - 2(.068376^{1/2}) \le PV(K) \le .833467 + 2(.068376^{1/2}) = 1.356443$$

In the context of the insurance policy, if each unit is valued at $1,000, the estimated NSP of the policy is between $310.49 and $1,356.44. The accuracy or reliability of the prediction interval depends on distributional properties and can be assessed using simulation methods similar to those discussed in Section 9.2 of the life table quantities.

The previous example is general in nature and demonstrates computations based on life table quantities utilized in the analysis of life insurance contracts. Standard types of life insurance common in actuarial science are now investigated in terms of present value functions and related expectations.

6.4.1 Types of Unit Benefit Life Insurance

This section represents an introduction to standard life insurance models. Actuarial nomenclature and the associated present value functions for various life insurance models based on (x) are described. The *RC* and *EP* approach applied to single-status loss models constructed from these unit benefit present value functions yields the NSP expectation. Standard notations for the expectation of the present value function for standard life insurance policies paying a unit benefit where the interest rate is fixed have a long history (see Actuarial Society of America, 1947). For standard unit benefit life insurance models, the present value functions with support along with standard actuarial notation for the expectation, assuming a constant interest rate, are now listed.

1. Continuous-status whole life insurance: A unit benefit is payable at the moment of death. The present value function and expectation are

$$PV(T) = v^T \quad \text{for} \quad S_{PV} = [0, \infty) \quad \text{and} \quad \bar{A}_x = \int_0^\infty v^t \, {}_t p_x \, \mu_{x+t} \, dt \qquad (6.12)$$

2. Discrete-status whole life insurance: A unit benefit is payable at the end of the year of death. The present value function and the expectation are

$$PV(K) = v^{K+1} \quad \text{for} \quad S_{PV} = \{0, 1, ...\} \quad \text{and} \quad A_x = \sum_{k=0}^\infty v^{k+1} \, {}_k p_x \, q_{x+k} \qquad (6.13)$$

3. Continuous-status n-year term life insurance: A unit benefit is payable at the time of decrement if the decrement occurs before n years elapse. If the status survives n years no benefit is paid. The present value function and expectation are

$$PV(T) = v^T \quad \text{for} \quad S_{PV} = [0, n) \quad \text{and} \quad \bar{A}^1_{x:\overline{n}|} = \int_0^n v^t \, {}_t p_x \, \mu_{x+t} \, dt \qquad (6.14)$$

4. Discrete-status n-year term life insurance: A unit benefit is payable at the end of the year of decrement if the decrement occurs before n years elapse. If decrement occurs after n future years no benefit is paid. The present value and expectation are

$$PV(K) = v^{K+1} \quad \text{for} \quad S_{PV} = \{0, 1, \dots, n-1\} \quad \text{and} \quad A^1_{x:\overline{n}|} = \sum_{k=0}^{n-1} v^{k+1} \, {}_k p_x \, q_{x+k}$$

$$(6.15)$$

5. n-year pure endowment: A unit benefit is payable upon survival to future age $x + n$. The present value function and expectation are

$$PV(T) = v^n \quad \text{for} \quad S_{PV} = [n, \infty) \quad \text{and} \quad {}_n E_x = v^n {}_n p_x \qquad (6.16)$$

6. Continuous-status n-year endowment insurance: A unit benefit is payable at the time of decrement or upon survival for n years, whichever occurs first. The present value function and expectation are

$$PV(T) = v^T \quad \text{for} \quad S_{PV1} = [0, n) \qquad (6.17)$$
$$\text{for} \quad S_{PV2} = [n, \infty)$$

where the support of the present value is $S_{PV1} \cup S_{PV2}$. The NSP for unit benefit is

$$\bar{A}_{x:\overline{n}|} = \int_0^n v^t \, {}_t p_x \, \mu_{x+t} \, dt + v^n \, P(T > n)$$

$$(6.18)$$

$$= \bar{A}^1_{x:\overline{n}|} + {}_n E_x$$

7. Continuous-status n-year deferred insurance: A unit benefit is payable at the moment of decrement if decrement occurs after n future years. The future value function and expectation are

$$PV(T) = v^T \quad \text{for} \quad S_{PV} = [n, \infty) \quad \text{and} \quad {}_{n|}\bar{A}_x = \int_{t=n}^{\infty} v^t \, {}_t p_x \, \mu_{x+t} \, dt \qquad (6.19)$$

If the life insurance has a benefit valued at b for these policies, the present value formulas are multiplied by b and the net single premium (NSP) is found by multiplying the actuarial expectation by b. A series of examples dealing with the various types of life insurance is now presented.

Example 6.7

(50) is issued a discrete 4-year term life insurance policy where the life table quantities given in Table 5.1 apply. The benefit paid at the end of the year of death is \$100,000 and the annual effective interest rate is taken to be $i = .05$. The unit benefit expectation (6.15), where $v = (1.05)^{-1} = .9524$, is computed as

$$A^1_{50:\overline{4}|} = \sum_{k=0}^{3} v^{k+1} \frac{d_{50+k}}{l_{50}} = .036183$$

Since the benefit is valued at \$100,000, the net single premium is

$$NSP = \$100,000(.036183) = \$3,618.30$$

At the time the policy is written, the expected future cost of the life insurance policy is estimated to be \$3,618.30. In practice, the cost of the policy can be financed by one lump payment or a series of annuity payments.

Example 6.8

For an individual age x the future lifetime random variable is exponential with mean θ where the pdf is given by (1.16) or $f(t) = (1/\theta)\exp(-t/\theta)$ for $t > 0$. Continuous-status life insurance is considered where the benefit payment is at the moment of death. The force of interest is fixed at $\delta > 0$ and the discount operator is of the form $= \exp(-\delta)$. For a unit benefit whole life insurance, the expectation, using (6.12), is

$$\bar{A}_x = \int_0^\infty \exp(-\delta t)(1/\theta)\exp(-t/\theta)\,dt = (1/(1+\theta\delta))\exp(-t(\delta+1/\theta))\,|_0^\infty = \frac{1}{1+\theta\delta}$$

Also, for an n-year endowment policy with a unit benefit expectation, (6.18) is computed in two parts as

$$\bar{A}_{x:\overline{n}|} = \int_0^n \frac{1}{\theta}\exp(-t(\delta+1/\theta))\,dt + \exp(-\delta n)\,P(T \geq n)$$

$$= \frac{1}{1+\theta\delta}(1-\exp(-n(\delta+1/\theta))) + \exp(-n(\delta+1/\theta))$$

In contrast to the *RC* and *EP* approach, the *PC*(.25) can be applied to life insurance analysis. Following this approach for whole life insurance, the future cost of the life insurance is estimated by (6.7). From the exponential random variable the 25th percentile is $t_{.25} = -\ln(.75)/\theta$ and the computed NSP value is

$$\text{NSP} = \exp((\delta/\theta)\ln(.75)) = .75^{\delta/\theta}$$

Differing criteria applied to the loss function associated with single-status models result in different estimates. For example, if $\delta = .05$ and $\theta = 10$, then *RC* gives NSP = .66667, while *PC*(.25) results in NSP = .99857. This result is consistent with the findings of Example 6.4, where *PC*(.25) yields larger estimates than *RC*.

Computational examples concerning other types of life insurance policies are outlined in the problems at the end of the chapter. Discussions of moment and variance computations for life insurance present values are found in Section 6.8.2.

6.5 Life Annuities

In a life annuity, the general payment pattern for deterministic status annuities, introduced in Section 3.2, is followed, but a stochastic component is added. In a stochastic status life annuity the time frame for the series of payments is dependent on future changes to the status. For example, premium payments that finance life insurance may be paid until a certain age is reached or death occurs. As in deterministic status annuities, there are two basic types, defined as discrete and continuous lifetime annuities. In the discrete life annuities payments are made periodically during the life of an annuitant, while in the continuous annuity setting payments are made in a continuous manner. In this section, we assume the force of interest is fixed at δ over the life of the status and construct present value functions for stochastic status life annuities. Applying the *RC* and *EP* to relevant single-status models based on the life annuity present value functions yields NSP. Information on the development of annuity models can be found in Bowers et al. (1997, Chapter 5).

The simplest form of a discrete life annuity occurs when the payments are level (all equal), denoted by π, and are made at the start or end of each time period. The length of the time periods is taken to be 1 year, and the effective annual interest rate is i. If payments are made in advance, the accumulated value of the first $k + 1$ payments follows formula (3.24) and is

$$\ddot{s}_{\overline{k}|} = \sum_{r=1}^{k}(1+i)^r = \frac{(1+i)^k - 1}{d} \tag{6.20}$$

where $d = i/(1 + i)$. Applying the discount operator $v^{k+1} = (1 + i)^{-(k+1)}$ to (6.20) produces the present value function (3.25)

$$\ddot{a}_{\overline{k|}} = \frac{1-(1+i)^{-k}}{d}$$

For a discrete life annuity with level payments of π made at the start of each year during the lifetime of (x), the present value function depends on the curtate future lifetime K and simplifies to

$$PV(K) = \pi \sum_{r=0}^{K} (1+i)^{-r} = \pi \frac{1-v^{K+1}}{d} \tag{6.21}$$

Applying expectation formula (6.4) with $s = 1$, the expected value of (6.21) is the net single premium

$$NSP = \sum_{S_{PV}} \frac{\pi}{d}(1 - v^{k+1})\,_k p_x\, q_{x+k} = \pi E\left\{\frac{1-v^{K+1}}{d}\right\} = \pi \frac{1-E\{v^{K+1}\}}{d} \tag{6.22}$$

where S_{PV} is the support of the present value function. For a whole life annuity-due, payments continue until the conditions of status change, signified by individual decrement, and the support of the present value function is $S_{PV} = \{0, 1, 2, \ldots, \omega - k - 1\}$. Further, for a discrete whole life annuity an alternate form of (6.22) is found by applying expectation (1.30) and takes the form

$$NSP = \pi \ddot{a}_x = \pi \sum_{k=0}^{\infty} v^k\,_k p_x \tag{6.23}$$

Expectation (6.23) is a function of the yearly survival rates that are readily available in life tables for target populations. An example of a discrete annuity is now given.

Example 6.9

An individual age 50 pays 1 unit into an account at the start of each year, while the individual survives up to, but not including, age 54. For this individual, life table expectations given in Table 5.1 hold where the yearly effective rate is assumed to be $i = .05$. Thus, $d = .05/1.05 = .047619$, and applying (6.21) the present value of this annuity is

$$PV(K) = \frac{1-v^{K+1}}{d} \quad \text{for} \quad K = 0,1,2,3,$$

$$= \frac{1-v^4}{d} \quad \text{for} \quad K \geq 4$$

Applying expectation formula (6.22), the expected present value of the annuity is

$$\text{NSP} = \frac{1}{d}\sum_{k=0}^{3}(1-v^{k+1})\frac{d_{50+k}}{l_{50}} + \frac{1}{d}(1-v^4)\frac{l_{54}}{l_{50}} = 3.66903$$

If each unit payment is valued at \$1,000, the NSP of the annuity estimates the present value of the future payments to be worth \$3,669.03.

This section concludes by considering a continuous life annuity with level yearly payments denoted by π. The general form of the future value of the annuity is given in (3.27), where $\pi_t = \pi$ for $t > 0$. Applying the discount function (3.15) under the constant force of interest δ, the present value of the annuity associated with support S_{PV} is

$$PV(T) = \pi \int_0^T v^r \, dr = \frac{\pi}{\delta}(1-v^T) \tag{6.24}$$

For a continuous life annuity status, age x, expectation (6.24) is a function of the future lifetime random variable T, and it reduces to the net single premium

$$\text{NSP} = \pi \int_{S_{PV}} \int_0^t v^r \, dr \; {}_tp_x \, \mu_{x+t} dt = \frac{\pi}{\delta}E\{1-v^T\} = \frac{\pi}{\delta}(1-E\{v^T\}) \tag{6.25}$$

In the case of a continuous whole life annuity, payments continue until status failure so that the support is $S_{PV} = [0, \infty)$. Using integration by parts, (6.25) can be rewritten

$$\text{NSP} = \pi \int_{t=0}^{\infty} v^t \; {}_tp_x \, dt \tag{6.26}$$

Computing the expectation (6.26) requires the form of the survivor function associated with the future lifetime random variable.

As in the case of life insurance, a collection of typical life annuities plays a prominent role in actuarial modeling. These typical stochastic status annuities are utilized in the computation of life insurance premiums and pension plan evaluations.

6.5.1 Types of Unit Payment Life Annuities

Similar to life insurance, standard types of life annuities exist with associated nomenclature and formulas. In this section, standard life annuity models for individuals age x are discussed where the interest rate is fixed and annuity payments are level, or all equal, and valued at $\pi = 1$ per year. Based on the structure of the annuity status, the associated present value function is either discrete, as in (6.21), or continuous, following (6.24). A single-status loss model is constructed using the present value function. Applying *RC* and *EP* the expectation or NSP is computed. There is standard actuarial notation for the expectation of the present value function for these standard life annuities. We present the present value function with support and standard actuarial notation for the associated expectation for these standard life annuities.

1. Continuous whole life annuity: A unit payment per annum is paid continuously until a change in status. The present value function and the expected present value are

$$PV(T) = \frac{1 - v^T}{\delta} \quad \text{for} \quad S_{PV} = [0, \infty) \quad \text{and} \quad \bar{a}_x = \int_0^\infty v^t \, {}_t p_x \, dt \qquad (6.27)$$

2. Discrete whole life annuity-due: A unit payment is made at the start of each year (x) survives. The present value function and expectation are

$$PV(K) = \frac{1 - v^{K+1}}{d} \quad \text{for} \quad S_{PV} = \{0, 1, \ldots\} \quad \text{and} \quad \ddot{a}_x = \sum_{k=0}^\infty v^k \, {}_k p_x \qquad (6.28)$$

3. Continuous n-year temporary life annuity: Unit payments are paid continuously until status failure or survival to n future years, whichever occurs first. The present value function is

$$PV(T) = \frac{1 - v^T}{\delta} \quad \text{for} \quad S_{PV1} = [0, n),$$

$$= \frac{1 - v^n}{\delta} \quad \text{for} \quad S_{PV2} = [n, \infty) \qquad (6.29)$$

The expected present value is computed following either formula in

$$\bar{a}_{x:\overline{n}|} = \int_0^n \frac{1}{\delta}(1 - v^t) \, {}_t p_x \, \mu_{x+t} \, dt + \frac{1}{\delta}(1 - v^n) P(T \geq n)$$

$$= \int_0^n v^t \, {}_t p_x \, dt \qquad (6.30)$$

4. Discrete n-year temporary life annuity-due: A unit payment is made at the start of each year (x) survives, up to n years. The present value function is

$$PV(K) = \frac{1}{d}(1 - v^{K+1}) \quad \text{for} \quad S_{PV1} = \{0, \ldots, n-1\},$$

$$= \frac{1}{d}(1 - v^n) \quad \text{for} \quad S_{PV2} = \{n, n+1, \ldots\} \tag{6.31}$$

Two forms of the expectation of (6.31) are

$$\ddot{a}_{x:\overline{n}|} = \sum_{k=0}^{n-1} \frac{1}{d}(1 - v^{k+1})\,_kp_x\,q_{x+k} + \frac{1}{d}(1 - v^n)P(K \geq n) = \sum_{k=0}^{n-1} v^k\,_kp_x \tag{6.32}$$

5. Discrete n-year deferred whole life annuity-due: In this discrete life annuity there are no payments for the first n years, corresponding to curtate future lifetimes $K = 0, 1, \ldots, n - 1$. Unit payments are made at the beginning of each year, starting with year $n + 1$, and continuing for the lifetime of the status. The present value is

$$PV(K) = v^n\frac{1 - v^{K+1-n}}{d} \quad \text{for} \quad S_{PV} = \{n, n+1, \ldots\} \tag{6.33}$$

The expectation of (6.33) is computed as follows:

$$_{n|}\ddot{a}_x = \frac{v^n}{d}\sum_{k=n}^{\infty}(1 - v^{k+1-n})\,_kp_x\,q_{x+k} = \sum_{k=n}^{\infty} v^k\,_kp_x \tag{6.34}$$

In these common life annuities, if the level payments are valued at π, the present value function is multiplied by π, and multiplying the actuarial expectation by π yields the NSP. Two annuity computation examples that demonstrate both discrete and continuous calculations are now presented.

Example 6.10

An individual age 50 takes out a discrete 4-year temporary life annuity-due where $i = .05$ and the life table quantities listed in Table 5.1 hold. Following (6.32), we compute

$$\ddot{a}_{50:\overline{4}|} = \sum_{k=0}^{3} v^k\,_kp_x = 3.66903$$

Hence, if the yearly payments are $1,000, the expected present value, or NSP, of the annuity is

$$\text{NSP} = \$1,000(3.66903) = \$3,669.03$$

The NSP is the expected present value of the life annuity corresponding to an initial lump-sum payment base.

Example 6.11

For a continuous stochastic status model age x the future lifetime random variable is an exponential random variable as in Example 6.8. The survival function for T takes form ${}_t p_x = \exp(-t/\theta)$ for $t > 0$. For a continuous whole life annuity expectation, (6.27) computes as

$$\bar{a}_x = \int_0^\infty \exp(-t\delta)\exp(-t/\theta)\,dt = \frac{1}{\delta + \dfrac{1}{\theta}}\exp(-t(\delta + 1/\theta))\Big|_\infty^0 = \frac{\theta}{1+\delta\theta} \quad (6.35)$$

The expectation for a continuous n-year temporary annuity following (6.30) simplifies to

$$\bar{a}_{x:\overline{n}|} = \int_0^n \exp(-t(\delta + 1/\theta))\,dt = \left[\frac{\theta}{1+\delta\theta}\right](1 - \exp(-n(\delta + 1/\theta)))$$

Following in a similar manner, expectations for other standard types of continuous annuities can be computed.

An economic complication can arise in the case of a discrete life annuity in the decrement year. In this year, an overpayment may exist corresponding to the fraction of the time between status failure and the end of the year. In apportionable annuities, discussed in the next section, this overpayment is addressed.

6.5.2 Apportionable Annuities

In the case of a discrete life annuity with the payments made at the start of each year (x) survives, each full year is financed by the NSP expectation discussed in Section 6.5. For this reason, depending on the contract, there may be an annuity overpayment corresponding to the interval of time between status change signified by decrement and the end of the time period. In an apportionable annuity there is an adjustment or repayment to the discrete annuity present value associated with this fractional time period.

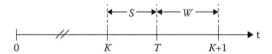

FIGURE 6.1
Apportionable annuity structure.

An analysis of apportionable annuities starts by considering a discrete life annuity for (x) with constant interest rate and level payments of π made at the start of each year. The underlying stochastic structure for the continuous life annuity is based on the future lifetime random variable T partitioned as $T = K + S$, where the UDD assumption holds. Utilizing the results of Section 5.4, K is the curtate future lifetime random variable, S is uniform on $[0, 1]$, and K and S are independent. In an apportionable annuity, a refund is paid to the insurer associated with the length of time between the status change or decrement and the end of the interval. This fractional time is denoted by W, $S \leq W \leq 1$, and is assumed to be a continuous uniform random variable on $[0, 1]$. This is depicted in Figure 6.1.

Based on present value computations of a continuous annuity, the amount of the refund is the ratio of

$$\int_{S}^{1} \exp(\delta w)\,dw \quad \text{to} \quad \int_{0}^{1} \exp(\delta w)\,dw$$

After integration this ratio can be written as the future value quantity

$$FV(T) = \frac{\exp(\delta) - \exp(\delta S)}{\exp(\delta) - 1} \tag{6.36}$$

Quantity (6.36) is the amount of the refund at the time the status fails, where $S = T - K$. Applying UDD through formula (6.5), the expectation of (6.36) can be written as

$$E\{PV(T)\} = \sum_{S_{PV}} v^{k}\,{}_{k}p_{x}\,q_{x+k} \int_{0}^{1} \frac{[\exp(\delta(1-s)) - 1]}{\exp(\delta) - 1}\,ds \tag{6.37}$$

and signifies the expected value of the adjustment payment. After simplification, expectation (6.37) becomes a function of the expectation associated with a continuous-status whole life insurance (6.13), written as

$$E\{PV(T)\} = \frac{i - \delta}{\delta d}\,A_{x} \tag{6.38}$$

The NSP or expectation of the present value of a unit payment apportionable annuity is found by subtracting (6.38) from the annuity expectation and after simplification becomes

$$\ddot{a}_x^{\{1\}} = \frac{i}{\delta}\ddot{a} - \frac{i-\delta}{\delta d} \tag{6.39}$$

The general form of (6.39) is $\ddot{a}_x^{\{m\}}$. A mathematical modeling example that demonstrates many of these calculations now follows.

Example 6.12

The curtate future lifetime associated with a stochastic status follows a geometric distribution as discussed in Example 1.19, where the pdf is $f(k) = pq^k$, where $S_K = \{0, 1, ...\}$. For an individual age x the whole life insurance expectation, (6.13), is

$$A_x = \sum_{k=0}^{\infty} e^{-\delta(k+1)} pq^k = pe^{-\delta}\sum_{k=0}^{\infty}(e^{-\delta}q)^k = \frac{p}{e^{\delta}-q}$$

For a discrete whole life annuity for an individual age x formula (6.28) gives the expectation

$$\ddot{a}_x = \sum_{k=0}^{\infty} e^{-\delta k} q^k = \sum_{k=0}^{\infty}(e^{-\delta}p)^k = \frac{e^{\delta}}{e^{\delta}-q}$$

In particular, if $\delta = .1$, $p = .95$, and $q = .05$, we compute $A_x = .900329$ and $a_x = 1.047366$. To compute the expectation of an apportionable annuity we first find that $i = e^1 - 1 = .1051709$, and $d = .1051709/1.1051709 = .0951625$. Using (6.39), the apportionable annuity expectation is

$$\ddot{a}_x^{\{1\}} = 1.051709\,(1.047366) - \frac{.0051709}{.00951625} = .5581484$$

Further, the expected amount of the refund, from (6.38), is

$$E\{PV(T)\} = .5433758(.900329) = .4892169$$

In this case, the adjustment payment made at status failure is substantial, as it is about ½ a unit amount.

6.6 Relating Risk Calculations

The expectation of the general present value function (6.2) associated with actuarial models, such as life insurance and life annuities, yields the NSV or NSP formula (6.5). Assuming UDD, this decomposition structure allows for relationship formulas among standard life insurance and life annuity expectations. For example, the present value for discrete and continuous whole life annuities (6.21) and (6.24) can be written as functions of the present values of discrete- and continuous-status whole life insurance, given by (6.13) and (6.12). Taking expectations, relationships among these actuarial NSP computations are realized. In our discussion, the force of interest is fixed at δ throughout the support of the stochastic status lifetime, and the relationships among the expectations of the standard actuarial statuses of Sections 6.4.1 and 6.5.1 are presented. Annuities pay a level benefit of $\pi = 1$, and life insurance has a unit benefit of $b = 1$.

6.6.1 Relations among Insurance Expectations

The first connections among life insurance formulas we investigate are associations between life insurance expectations. The expectation for life insurance, in both the continuous- and discrete-status cases, can be partitioned into segments designating the first n and subsequent future years. Expectations for term life insurance, (6.14) and (6.15), and those for both whole life, (6.12) and (6.13), and endowment insurance, (6.18), are related by

$$\bar{A}_x = \bar{A}^1_{x:\overline{n}|} + {}_nE_x\,\bar{A}_{x+n} \quad \text{and} \quad A_x = A^1_{x:\overline{n}|} + {}_nE_x\,A_{x+n} \qquad (6.40)$$

and

$$\bar{A}_{x:\overline{n}|} = \bar{A}^1_{x:\overline{n}|} + {}_nE_x \quad \text{and} \quad A_{x:\overline{n}|} = A^1_{x:\overline{n}|} + {}_nE_x \qquad (6.41)$$

Further, using the pure endowment mean ${}_n$Example as a time operator, deferred life insurance (6.19) and whole life insurance computations, (6.12) and (6.13), are related by

$${}_{n|}\bar{A}_x = {}_nE_x\,\bar{A}_{x+n} \quad \text{and} \quad {}_{n|}A_x = {}_nE_x\,A_{x+n} \qquad (6.42)$$

Understanding relationships among insurance formulas aids in overall comprehension of these types of insurance policies. An explanatory example follows.

Example 6.13

A continuous-status model associated with the exponential future lifetime random variable discussed in Example 6.8 is considered. Expectations for a continuous-status n-year term and continuous-status n-year endowment unit benefit insurance are related to whole insurance formulas using (6.40) and (6.41) as

$$\bar{A}^1_{x:\overline{n}|} = \bar{A}_x - {}_nE_x\,\bar{A}_{x+n} \quad \text{and} \quad \bar{A}_{x:\overline{n}|} = \bar{A}_x + {}_nE_x\,(1 - \bar{A}_{x+n}) \qquad (6.43)$$

Applying (6.43) and the exponential pdf (1.16) the expectation for continuous-status n-year term life insurance is

$$\bar{A}^1_{x:\overline{n}|} = \frac{1}{1+\theta\delta}(1 - v^n e^{-n/\theta})$$

The similar expectation computation for continuous-status n-year endowment follows directly and is left to the reader.

Relationships between the discrete- and continuous-status life insurance expectations exist based on yearly mortality assumptions. In Section 6.2.3, assuming UDD, the general net single premium formula for the mixed calculations is given in (6.6). For general unit benefit continuous-status life insurance, with support S_T and an UDD assumption, the net single premium is

$$\text{NSP} = b\,\frac{i}{\delta}\sum_{S_k} v^{k+1}\,{}_kp_x\,q_{x+k} \qquad (6.44)$$

where i and δ are related as in (3.5). Applying (6.44) to continuous-status whole life insurance (6.12) and continuous-status n-year term life insurance (6.14), expectations yield relationships

$$\bar{A}_x = \frac{i}{\delta}A_x \quad \text{and} \quad \bar{A}^1_{x:\overline{n}|} = \frac{i}{\delta}A^1_{x:\overline{n}|} \qquad (6.45)$$

Utilizing these concepts and formulas, expectations for a variety of different types of continuous-status life insurance can be computed using life table data, as discussed in Section 5.7. An example relating continuous- and discrete-status insurance follows.

Example 6.14

Consider a stochastic status model described by curtate future lifetime following a geometric distribution with pmf $f(k) = p\,q^k$ on $S_K = \{0, 1, \dots\}$

given in Example 1.19. The discrete-status whole insurance expectation (6.13) simplifies using summation (1.32), and the continuous-status life insurance expectation follows from (6.45) so that

$$A_x = \frac{p}{e^\delta - q} \quad \text{and} \quad \bar{A}_x = \frac{p}{\delta}\left(\frac{e^\delta - 1}{e^\delta - q}\right)$$

These concepts and formulas can be extended to life table distributions.

There are other relations among insurance moment computations. Relational formulas among life insurance risk calculations based on subsequent years can be derived. To observe this, consider a discrete-status whole life insurance policy for a person age x. Using (6.13), we find

$$A_x = v q_x + v p_x A_{x+1} \tag{6.46}$$

where v is the discount operator. This formula relates sequential-year whole life insurance expectations and is applied in a later section in the calculation of reserves. These formulas are not exhaustive and other relations within insurance and annuity modeling can be derived using the basic computational laws.

6.6.2 Relations among Insurance and Annuity Expectations

In this section we investigate the relationships between life insurance and life annuity expectations. Present value functions for standard life annuities are simplified and written in terms of life insurance present values. For whole life annuities, utilizing (6.28) in the discrete case and (6.27) in the continuous case, the expectations can be written as

$$\text{NSP} = \frac{1}{d} E\{1 - v^{K+1}\} \quad \text{or} \quad \text{NSP} = \frac{1}{\delta} E\{1 - v^T\} \tag{6.47}$$

Applying standard notations for annuities, (6.27) and (6.28), and for life insurance, (6.12) and (6.13), to (6.47) yields

$$\ddot{a}_x = \frac{1}{d}(1 - A_x) \quad \text{and} \quad \bar{a}_x = \frac{1}{\delta}(1 - \bar{A}_x) \tag{6.48}$$

Similarly, in both the discrete and continuous cases a direct relationship exists between the expectations for n-year temporary life annuities and n-year endowment insurance, and is given by

$$\ddot{a}_{x:\overline{n}|} = \frac{1}{d}(1 - A_{x:\overline{n}|}) \quad \text{and} \quad \bar{a}_{x:\overline{n}|} = \frac{1}{\delta}(1 - \bar{A}_{x:\overline{n}|}) \tag{6.49}$$

In the continuous-status setting, statistical distributions for the future lifetime random variable are required to apply (6.48) and (6.49). For the discrete annuity status, utilizations of formulas (6.48) and (6.49) require suitable life tables. In the next section, following a continuous lifetime example, relations between continuous and discrete life annuity expectations are presented.

Example 6.15

A continuous whole life annuity is considered for the stochastic status of Example 6.8. The future lifetime random variable is exponential with mean θ. Applying (6.48) with the result of Example 6.8, the NSP is

$$\bar{a}_x = \frac{1}{\delta}\left(1 - \frac{1}{1+\theta\delta}\right) = \frac{\theta}{1+\theta\delta}$$

This NSP can be directly computed using (6.27) and taking $_t p_x = \exp(-t\theta)$ for $t > 0$.

6.6.3 Relations among Annuity Expectations

In the spirit of Sections 6.6.1 and 6.6.2, relationships exist among standard life insurance expectations. In this section, both continuous and discrete life annuities are considered where whole life annuity expectations are partitioned into two parts. The first part corresponds to the first n future years with expectation consistent with an n-year temporary annuity computation. The expectation for the subsequent time takes the form of a whole life annuity. The formulas for whole life annuity expectations, utilizing temporary life annuity computations (6.30) and (6.32), are

$$\bar{a}_x = \bar{a}_{x:\overline{n}|} + {}_n E_x\, \bar{a}_{x+n} \quad \text{and} \quad \ddot{a}_x = \ddot{a}_{x:\overline{n}|} + {}_n E_x\, \ddot{a}_{x+n} \tag{6.50}$$

Following a similar development for discrete and continuous deferred life annuities, the expectation is a function of whole life annuity computations as

$$_{n|}\bar{a}_x = {}_n E_x\, \bar{a}_{x+n} \quad \text{and} \quad _{n|}\ddot{a}_x = {}_n E_x\, \ddot{a}_{x+n} \tag{6.51}$$

Combining these expectation formulas yields a multitude of computing possibilities. For example, to compute the expectation for a continuous whole life annuity, under an UDD assumption, the formula for a discrete whole life insurance expectation can be utilized as follows:

$$\bar{a}_x = \frac{1}{\delta}\left(1 - \frac{i}{\delta} A_x\right) \tag{6.52}$$

These relational formulas can be useful in computational problems where redundant work is eliminated. Examples of such applications exist in the context of life table applications discussed in Section 6.7.

For the case of an apportionable annuity, introduced in Section 6.5.2, expectation relational formulas exit. Formula (6.39) relates the expectations associated with an apportionable annuity with those of a discrete whole life annuity. Another formula for this computation utilizing continuous whole life annuity expectation (6.27) takes the form

$$\ddot{a}_x^{\{1\}} = \frac{\delta}{d} \bar{a}_x \tag{6.53}$$

Based on formula (6.53), we observe an apportionable annuity can be viewed, in terms of expectation computation, as a continuous annuity with a constant adjustment, namely, δ/d. An approximation to (6.53) follows from (6.52) and is

$$\ddot{a}_x^{\{1\}} = \frac{1}{d}\left(1 - \frac{i}{\delta} A_x\right) \tag{6.54}$$

The expectation for a whole life apportionable annuity (6.54) can be computed from the discrete whole life insurance computation in (6.13).

Aided by these relational formulas, annuity expectations can be computed using discrete group survivorship totals or mortality rates available in life tables. Another annuity relational formula, utilized in the calculations of reserves, discussed in Section 6.9, relates the expectation for discrete whole life annuities based on statuses associated with subsequent years. The expectation for a discrete life annuity for an individual age $x + 1$, using the expectation in (6.28), can be computed utilizing age x quantities as

$$\ddot{a}_x = 1 + v p_x \ddot{a}_{x+1} \tag{6.55}$$

Changes in annuity expectations with status age can be found using (6.55) and a suitable life table.

Our attention turns to expected present value computations related to various forms of life insurance and life annuities using group survivorship-based life tables as discussed in Section 5.7 tables. The formulas relating expectations can be useful in both computation and comprehension associated with these models.

6.7 Actuarial Life Tables

The mortality structures associated with stochastic status models are often presented in the form of life tables. As introduced in Section 5.7, group survivorship life tables list expected decrement totals or mortality and rates from ages $x = 0$ to limiting age ω, defined so that $p_{\omega-1} = 0$. For a discrete stochastic status model age x, probabilities associated with the relevant curtate future life random variable K are computed utilizing life table quantities and (5.53) and (5.54) as

$$P(K = k) = \frac{d_{x+k}}{l_x} \quad \text{and} \quad P(K \geq k) = \frac{l_{x+k}}{l_x} \tag{6.56}$$

for k in S_k. Utilizing probabilities (6.56), present value expectations associated with various types of discrete life insurance and discrete life annuities can be computed. For an individual age x, applying (6.56) in the expected present value formulas for discrete whole life insurance (6.13) and discrete whole life annuity (6.28) produces

$$A_x = \sum_{k=0}^{\omega-x-1} v^{k+1} \frac{d_{x+k}}{l_x} \quad \text{and} \quad \ddot{a}_x = \sum_{k=0}^{\omega-x-1} v^k \frac{l_{x+k}}{l_x} \tag{6.57}$$

Expectations (6.57) can be directly computed using life tables targeted to specific populations or risk strata.

For a target population, such as males, females, or smokers, an actuarial life table lists the expected present value for discrete whole life insurance and whole life annuity expectations by age defined in (6.57). The relational formulas of Section 6.6 are used to compute the net single premiums corresponding to a variety of life insurance and life annuities. An example follows demonstrating the application of actuarial tables to various actuarial computations.

Example 6.16

Associated with a stochastic status model, the group survivorship quantities listed in Table 5.1 for selected years hold and the yearly effective interest rate is fixed at $i = .06$. For the selected years an actuarial table listing the expectations in (6.57) is presented in Table 6.1.

Assuming UDD, the actuarial expectations in Table 6.1 can be used to compute NSP corresponding to a variety of life insurances and life annuities. Examples of these computations are now demonstrated.

TABLE 6.1

Actuarial Life Table for Selected Ages

x	A_x	$\ddot{a}_x$	x	A_x	$\ddot{a}_x$	x	A_x	$\ddot{a}_x$
0	.093957	16.0068	50	.211036	13.9384	70	.351232	11.46161
1	.095061	15.9873	51	.215813	13.8554	71	.361583	11.2787
2	.096187	15.9674	52	.220769	13.7664	72	.372357	11.0884
3	.097344	15.9469	53	.225915	13.6756	73	.383597	10.8898
4	.098526	15.9260	54	.231239	13.5814	74	.395300	10.6830

1. An individual age 50 takes out a life insurance policy that pays a benefit of $5,000 where:

   a. The benefit is paid at the moment of death. Using (6.45), the net single premium for this continuous-status whole life insurance is

   $$\text{NSP} = \$5000 \, \bar{A}_x = \$5000 \, \frac{i}{\delta} \, A_{50} = \$5000 \left(\frac{.06}{.058269} \right)(.211036)$$

   $$= \$1,086.53$$

   b. The benefit is paid at the end of the year of death if death occurs within 22 years. Using (6.41) the net single premium for this discrete term life insurance is

   $$\text{NSP} = \$5,000 \, A^1_{50:\overline{22|}} = \$5,000[A_{50} - {}_{22}E_{50} \, A_{72}]$$

   $$= \$5,000[.211036 - (.207668) \cdot (.372357)]$$

   $$= \$668.55$$

   where we note ${}_{22}E_{50} = (l_{72}/l_{50})(1.06^{-22}) = (52,915/70,711).2775051 = .207668.$

   c. The benefit is paid at the moment of death or if age 72 is reached. From (6.43) and (6.45) the net single premium for this continuous-status endowment insurance is computed as

   $$\text{NSP} = \$5,000 \, \bar{A}_{50:\overline{22|}} = \$5,000 \left[\frac{i}{\delta} \, A_{50} + {}_{22}E_{50} \left(1 - \frac{i}{\delta} \, A_{72} \right) \right]$$

   $$= \$5,000 \left[\frac{.06}{.058269} .211036 + .207668 \left(1 - \frac{.06}{.058269} .372357 \right) \right]$$

   $$= \$1,726.75$$

2. An annuity is comprised of payments valued at $1,000.

   a. A person age 50 receives a continuous annuity for life. From (6.45) and (6.48) the net single premium for this continuous whole life annuity is

   $$\text{NSP} = \$1,000 \, \bar{a}_{50} = \$1,000 \, \frac{1}{\delta} \left(1 - \frac{i}{\delta} \, A_{50} \right)$$

   $$= \$1,000 \left(\frac{1}{.058269} \right) \left(1 - \frac{.06}{.058269} (.211036) \right) = \$13,432.45$$

   b. The person is age 3 and the payments last for at most 50 years. From (6.30) the net single premium for this discrete temporary annuity is

   $$\text{NSP} = \$1,000 \, \ddot{a}_{3:\overline{50|}} = \$1,000 \, (\ddot{a}_3 - {}_{50}E_3 \, \ddot{a}_{53})$$

   $$= \$1,000 \, (15.9469 - (.052635)(13.6756)) = \$15,618.06$$

   where ${}_{50}E_3 = v^{50} \, l_{50}/l_3 = (1.06^{-50})68,557/70,711 = .052635.$

c.   A person age 50 receives payments starting at year 72. The net
     single premium for the deferred annuity is found using (6.51):

$$\text{NSP} = \$1,000 \; _{22|}\ddot{a}_{50} = \$1,000 \; _{22}E_{50} \, \ddot{a}_{72} = \$1,000(.207668)(11.08836)$$

$$= \$2,302.70$$

These net single premiums represent the estimated present value associated with the individual actuarial statuses utilizing a risk or expectation criteria. These computations are central to the determination of life insurance and annuity premiums, as demonstrated in the next section.

6.8 Loss Models and Insurance Premiums

The most basic challenge in the field of insurance is the setting of premiums. In a typical insurance policy, future possible benefits are financed by a series of annuity payments, called premiums. The premiums must be set so as to cover all associated costs, and therefore depend on many factors. To analyze these types of stochastic status models, a stochastic loss model using loss function (4.1) is constructed. In a typical insurance setting, the present values of the expenditure and the revenue correspond to the insurance benefit and the annuity premium payments, respectively. For a fixed rates set of δ, premiums are computed using either the *RC* and *EP* or *PC*(.25) as discussed in Sections 4.2.1 and 4.2.2, respectively. In practice, stochastic status models accounting for associated insurance costs are developed and analyzed, resulting in loaded premiums. The analysis of expense models is presented in Section 6.11. Two examples demonstrating stochastic loss model construction and premium evaluation are now presented.

Example 6.17

A person age x purchases a whole life insurance policy that pays a benefit of b at the moment of death, financed by continuous whole life premium payments. The present values corresponding to insurance (6.12), for the expenditures, and the premium annuity (6.27), for revenues, are

$$\text{PVE}(T) = b \, v^T \quad \text{and} \quad \text{PVR}(T) = \frac{\pi}{\delta}(1 - v^T) \tag{6.58}$$

for $T \geq 0$. The resulting stochastic loss function is a function of $T \geq 0$, taking the form

$$\text{LF}(T) = \left(b + \frac{\pi}{\delta}\right) v^T - \frac{\pi}{\delta} \tag{6.59}$$

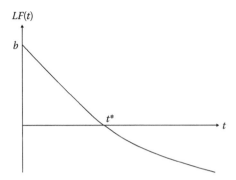

FIGURE 6.2
Loss function.

Stochastic loss function (6.59), viewed as a function of T, is represented in Figure 6.2, where it is zero at future time $t^* = \ln[(b\delta + \pi)/\pi]/\delta$. From Figure 6.2 we observe that as time increases the loss function decreases, reaching zero at t^*. Further, as interest rate δ increases, t^* decreases and more of the loss function becomes negative, implying a realized loss.

Both types of stochastic loss model criteria introduced in Section 4.2 can be applied to compute the premium π. Applying the $PC(.25)$ method we set $.25 = P(LF(T) > 0)$, resulting in

$$.25 = P(T < t^*) = -\frac{1}{\delta}\ln\left[\frac{\pi}{\delta}\left(b + \frac{\pi}{\delta}\right)^{-1}\right]$$

Utilizing the 25th percentile for the future lifetime random variable, $t_{.25}$, the resulting premium takes the form

$$\pi = b\delta\frac{v^{t_{.25}}}{1 - v^{t_{.25}}} \tag{6.60}$$

In the RC and EP method, imposed on (6.59), we set $E\{L(t)\} = 0$ and solve for the premium. Assuming $b = \$50{,}000$,

$$\pi = \$50{,}000\frac{\overline{A}_x}{\overline{a}_x} \tag{6.61}$$

These two techniques for evaluating premiums yield different results. This is due, in part, to the nonlinear nature of the stochastic loss function. The choice of criterion depends on the philosophy of the investigator or actuary.

Example 6.18

In a general setting, both the insurance policy benefits and premium payments may vary by time. A purely discrete setting is considered where whole life insurance with the benefit paid at the end of the decrement year is financed by premium payments made at the start of each year during the lifetime of the insured. The insurance benefit and premium payments associated with $K = k$ are denoted by b_k and π_k. The present values for expenditures and revenues are functions of the curtate future lifetime K and are

$$PVE(K) = b_K v^{K+1} \quad \text{and} \quad PVR(K) = \sum_{j=0}^{K} \pi_j v^j \qquad (6.62)$$

for $K = 0, 1, \ldots$. The discrete stochastic loss function in (4.1) is

$$LF(K) = PVE(K) - PVR(K) = b_K v^{K+1} - \sum_{j=0}^{K} \pi_j v^j \qquad (6.63)$$

In the case of level premiums where $\pi_k = \pi$ for $K \geq 0$, (6.63) simplifies by applying (6.21) as

$$LF(K) = b_K v^{K+1} - \frac{\pi}{d}(1 - v^{K+1}) \qquad (6.64)$$

for $K \geq 0$. Applying RC and EP on (6.64) $E\{LF(K)\} = 0$, and for fixed benefits, b_K for $K \geq 0$, a suitable premium π can be found.

In a stochastic loss model approach, both the present value functions for expenditures and revenues may correspond to standard types of life insurance in Section 6.4.1 and life annuities in Section 6.5.1. In this situation, when the insurance benefit and annuity premium are fixed through time, denoted by b and π, standard loss functions are formed and some are presented in Table 6.2.

For standard insurances, the net premium, which is based on desired benefits and ignoring expenses, is computed using RC and EP on the stochastic loss model. For special cases, such as those of Table 6.2, actuarial notations exist for the premium. This follows the precedent of notations for standard life insurance of Section 6.4.1 and standard life annuities of Section 6.5.1.

6.8.1 Unit Benefit Premium Notation

As demonstrated in the previous section, life insurance and premiums payable for life have been combined to form stochastic loss models used in the determination of premiums. For this purpose the RC method in conjunction with EP is commonly utilized. Assuming a constant interest rate for standard combinations of unit benefit life insurance and their level premiums, actuarial

TABLE 6.2

Standard Insurance Loss Function Constructions

Life Insurance	Premium Annuity	Loss Function
1. Continuous-status whole life insurance	Continuous whole life premiums	$LF(T) = bv^T - \dfrac{\pi}{\delta}(1-v^T)$ for $T > 0$
2. Discrete-status whole life insurance	Discrete whole life premiums	$LF(K) = bv^{K+1} - \dfrac{\pi}{d}(1-v^{K+1})$ for $K \geq 0$
3. Continuous-status n-year term insurance	Continuous premiums payable for min$\{T, h\}$ years, where $h \leq n$	$LF(T) = \begin{cases} bv^T - \dfrac{\pi}{\delta}(1-v^T) & \text{for} \quad 0 \leq T < h \\[2mm] bv^T - \dfrac{\pi}{\delta}(1-v^h) & \text{for} \quad h \leq T < n \\[2mm] -\dfrac{\pi}{\delta}(1-v^h) & \text{for} \quad T \geq n \end{cases}$
4. Discrete-status n-year endowment insurance	Discrete premiums payable for min$\{K, h\}$, where $h \leq n$	$LF(K) = \begin{cases} bv^{K+1} - \dfrac{\pi}{d}(1-v^{K+1}) & \text{for} \quad K = 0,1,...,h-1 \\[2mm] bv^{K+1} - \dfrac{\pi}{d}(1-v^h) & \text{for} \quad K = h,h+1,...,n-1 \\[2mm] bv^n - \dfrac{\pi}{d}(1-v^h) & \text{for} \quad K = n,n+1,... \end{cases}$

TABLE 6.3

Standard Unit Benefit Level Premium Notation

Expenditure	Revenue	EP Premiums
1. Continuous-status whole life insurance	Continuous whole life annuity	$\bar{P}(\bar{A}_x) = \dfrac{\bar{A}_x}{\bar{a}_x}$
2. Discrete-status whole life insurance	Discrete whole life annuity-due	$P_x = \dfrac{A_x}{\ddot{a}_x}$
3. Continuous-status n-year term insurance	Continuous h-year temporary life annuity	$_h P(\bar{A}^1_{x:\overline{n}|}) = \dfrac{\bar{A}^1_{x:\overline{n}|}}{\bar{a}_{x:\overline{h}|}}$
4. n-Year endowment	Discrete h-year temporary life annuity-due	$_h P(\bar{A}_{x:\overline{n}|}) = \dfrac{\bar{A}_{x:\overline{n}|}}{\ddot{a}_{x:\overline{h}|}}$
5. n-Year pure endowment	Discrete h-year temporary life annuity-due	$_h P(_n E_x) = \dfrac{_n E_x}{\ddot{a}_{x:\overline{h}|}}$
6. n-Year deferred whole life annuity-due	Discrete h-year temporary life annuity-due	$_h P(_{n|}\ddot{a}_x) = \dfrac{_{n|}\ddot{a}_x}{\ddot{a}_{x:\overline{h}|}}$

notations exist. Listed in Table 6.3 is the premium structure in terms of standard life insurance and life annuities expectations along with actuarial notations for a variety of standard life insurances. This collection is not all-inclusive but demonstrates a wide variety of possible combinations of insurance and annuity payments.

If the insurance benefit, b, differs from 1, then the corresponding premium is found by multiplying the standard premium by the value of b. Computational examples now follow that demonstrate the computation of premiums and the application of actuarial notations.

Example 6.19

Premiums for the discrete life insurance policy discussed in Example 6.7 are explored. For an individual age 50, the discrete 4-year term life insurance policy is financed by a discrete life annuity-due of level premiums π. The expectations corresponding to the unit benefit term life insurance and a 4-year temporary life annuity-due, computed in Examples 6.7 and 6.10, respectively, are

$$A^1_{50:\overline{4}|} = .036183 \quad \text{and} \quad \ddot{a}_{50:\overline{4}|} = 3.66903$$

From Table 6.3, the unit benefit premium, under *RC* and *EP*, is

$$_4 P(A^1_{50:\overline{4}|}) = \frac{A^1_{50:\overline{4}|}}{\ddot{a}_{50:\overline{4}|}} = \frac{.036183}{3.66903} = .009862$$

If the death benefit is \$100,000, the corresponding yearly premium is $\pi =$ \$100,000(.009862) = \$9,862.00. We remark that no loading to offset for expense costs is contained in this premium.

Example 6.20

Premiums for a continuous-status whole life insurance policy are determined based on an exponential future lifetime random variable with mean θ, as given in (1.16). Standard expectations are given in Examples 6.8 and 6.11, and the *RC* and *EP* unit benefit premium is found following Table 6.3 as

$$\bar{P}(\bar{A}_x) = \frac{\bar{A}_x}{\bar{a}_x} = \frac{\dfrac{1}{1+\theta\delta}}{\dfrac{\theta}{1+\theta\delta}} = \frac{1}{\theta}$$

In this case, the unit benefit yearly premium is independent of the interest rate and takes the form of the mean rate (inverse of the mean future lifetime) associated with the future lifetime random variable.

Applying the computed premium, the loss function is stochastic based on the future lifetime of the random variable. The analysis of observed loss functions at future times becomes important in assessing the future financial liability associated with the actuarial contracts. This concept is investigated in the discussion of reserves presented in Section 6.9 and follows a section exploring the variance of the stochastic loss function.

6.8.2 Variance of the Loss Function

In this section, variance computations of the stochastic loss function associated with life insurance models are explored. For standard life insurance loss functions, such as those in Table 6.2, where the insurance benefit is fixed at b and the *RC*- and *EP*-based level premium is π, variance formulas for the present value function are given. For two specific examples, one based on continuous statuses and one based on discrete statuses, the form of the loss function is simplified and variance formulas are produced.

Example 6.21

For an individual age x, a whole life insurance policy with payment b at the moment of death is financed by a continuous premium payable for life. The loss function is given in (1) in Table 6.2 and simplifies to

$$LF(T) = \left(b + \frac{\pi}{\delta}\right)v^T - \frac{\pi}{\delta} \tag{6.65}$$

for $t > 0$. The variance of (6.65) computes as

$$\text{Var}\{LF(T)\} = \left(b + \frac{\pi}{\delta}\right)^2 \text{Var}\{v^T\} = \left(b + \frac{\pi}{\delta}\right)^2 [^2\bar{A}_x - (\bar{A}_x)^2] \tag{6.66}$$

where, following basic second moment computation,

$$^2\bar{A}_x = \int_0^\infty e^{-2\delta t} {}_t p_x \, \mu_{x+t} \, dt \tag{6.67}$$

For example, if the lifetimes follow an exponential distribution with mean θ, then (6.67) computes as

$$^2\bar{A}_x = \int_0^\infty e^{-2\delta t} \frac{1}{\theta} e^{-t/\theta} \, dt = (1 + 2\delta\theta)^{-1} \tag{6.68}$$

Using the expectation of the continuous whole life insurance given in Example 6.8 and (6.68), the variance (6.66) becomes

$$\text{Var}\{LF(T)\} = \left(b + \frac{\pi}{\delta}\right)^2 [(1 + 2\delta\theta)^{-1} - (1 + \delta\theta)^{-2}] \tag{6.69}$$

Variances such as (6.69) measure the variability of loss functions associated with financial actions and can be applied in various statistical analyses.

Example 6.22

For an individual age x, a discrete-status whole life insurance policy with benefit b is financed by discrete premiums of π, payable for life. For this setting, the loss function is given in (2) of Table 6.2 and simplifies to

$$LF(K) = \left(b + \frac{\pi}{d}\right) v^{K+1} - \frac{\pi}{d} \tag{6.70}$$

for $K = 0, 1, \ldots$. The variance of (6.70) is

$$\text{Var}\{LF(K)\} = \left(b + \frac{\pi}{d}\right)^2 \text{Var}\{v^{K+1}\} = \left(b + \frac{\pi}{d}\right)^2 [^2A_x - (A_x)^2] \tag{6.71}$$

where

$$^2A_x = \sum_{k=0}^\infty v^{2(k+1)} {}_k p_x \, q_{x+k} \tag{6.72}$$

For exposition, consider the setting of Example 6.12 where curtate future lifetime is a geometric random variable, as in Example 1.19, with pmf $f(k) = p\,q^k$, where $p + q = 1$ and $S_K = \{0, 1, \ldots\}$. In this setting, (6.72) becomes

$$^2A_x = \sum_{k=0}^{\infty} e^{-2\delta(k+1)}\,pq^k = pe^{-2\delta}\sum_{k=0}^{\infty}(e^{-2\delta}q)^k = \frac{p}{(e^{2\delta}-q)}$$

and the variance (6.71) computes as

$$\mathrm{Var}\{LF(K)\} = \left(b+\frac{\pi}{d}\right)^2 \mathrm{Var}\{v^{K+1}\} = \left(b+\frac{\pi}{d}\right)^2 [p(e^{2\delta}-q)^{-1} - (p(e^{\delta}-q)^{-1})^2]$$

These computations can be utilized in asymptotic applications involving variances, such as prediction intervals for the aggregate of stochastic status loss functions. The use of Excel to compute net single premiums and variances of present value functions is outlined in Problem 6.14.

6.9 Reserves

The economic stability and monetary value of a company, such as an insurance company, is affected by the future financial obligations it has incurred. For this reason, life insurance and life annuity contracts must be analyzed at various times in their tenure. Similarly, the loss models associated with a continuous stochastic status age x are not only a function of the future lifetime random variable T, but also the time frame of the analysis. The benefit reserve or reserve is the expectation of the stochastic loss function relative to future age $x + t$. In this section, for both continuous and discrete standard life insurance, the reserve loss function is described and the expectation defining the reserve is discussed.

In the case of a life insurance status model where the computed premium is based on age x, analysis at age $x + t$ is based on a translated or reserve loss function. As is typical in stochastic status modeling, there are two primary settings, discrete and continuous. In the continuous case, the reserve loss function is a function of the translated future lifetime random variable $W = T - t$, while in the discrete case, it is a function of the curtate future lifetime random variable $U = K - [t]$, where $[t]$ is the greatest integer less than or equal to t. The reserve structure is demonstrated in Figure 6.3.

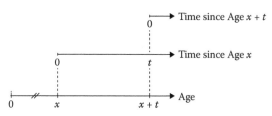

FIGURE 6.3
Reserve structure.

The expectation of the reserve loss function is referred to as the reserve and measures the expected future obligation associated with past life insurance policies. In general, the reserve is an increasing function in the future time variable. Reserve concepts and computations are demonstrated in the following examples.

Example 6.23

The reserve structure and computations for the continuous-status whole life insurance policy for an individual age x of Example 6.17 is explored. The analysis is done at age $x + t$, where $W = T - t$ represents the translated future lifetime random variable, and the relevant present value functions similar to (6.58) are

$$PVE(W) = b\,v^W \quad \text{and} \quad PVR(W) = \frac{\pi}{\delta}(1 - v^W) \qquad (6.73)$$

for $W \geq 0$. In (6.73) the premium π is computed based on the benefit b using the RC and EP approach at age x. Based on age $x + t$, the reserve loss function takes the form of (6.59), and for $W \geq 0$ is given by

$$_tLF(W) = \left(b + \frac{\pi}{\delta}\right)v^W - \frac{\pi}{\delta} \quad \text{for} \quad W \geq 0 \qquad (6.74)$$

The expectations of the present values in (6.73), using standard actuarial notations, are given by

$$E\{PVE(W)\} = b\,\bar{A}_{x+t} \quad \text{and} \quad E\{PVR(W)\} = \pi\,\bar{a}_{x+t} \qquad (6.75)$$

The reserve corresponding to age $x + t$ is the expectation of the loss function (6.74). Utilizing (6.75) and relation (6.48) the reserve can be simplified as

$$E\{_tLF(W)\} = b\,\bar{A}_{x+t} - \pi\,\bar{a}_{x+t} = \left(b + \frac{\pi}{\delta}\right)\bar{A}_{x+t} - \frac{\pi}{\delta} \qquad (6.76)$$

The EP approach implies that the reserve initially, corresponding to $t = 0$, takes the value zero. From (6.76) we note as t increases, since the whole life insurance expectation increases, the reserve increases.

Example 6.24

Reserve analysis for a 3-year endowment insurance policy for (x) is considered. The policy provides a benefit of \$1,000 paid at the end of the year of death, and premium payments are made at the start of each year during the lifetime of (x). The effective annual interest rate is $i = .15$ and the mortality probabilities are $q_x = .1$, $q_{x+1} = .1111$, and $q_{x+2} = .5$. Using the RC and EP the premium is $\pi = P_{x:\overline{3}|} = \$288.41$. Based on age $x + k$, for positive integer k,

TABLE 6.4

Year-by-Year Reserve Calculations

Time k	Outcome U	$_kLF(U)$	$P(U = u)$	$LF(U) \, P(U = u)$
0	0	581.16	.1	58.12
	1	216.94	$.1 = .9(.1111)$	21.69
	2	−99.76	$.8 = 1 − .1 − .1$	−79.81

The reserve is $E\{_0LF(U)\} = 0$ and $Var\{_0LF(U)\} = 46{,}445$.

1	0	581.16	.1111	64.57
	1	216.94	.8889	192.84

The reserve is $E\{_1LF(U)\} = 257.41$ and $Var\{_1LF(U)\} = 13{,}1001$.

2	0	581.16	1	581.16

The reserve is $E\{_2LF(U)\} = 581.16$ and $Var\{_2LF(U)\} = 0$.

the integer future lifetime random variable is $U = K − k$. For $U = u$ the present value of the premiums and reserve loss function are

$$\ddot{a}_u = \pi \sum_{r=0}^{u} v^{-r} \quad \text{and} \quad _kLF(U) = 1000 \, v^{U+1} − \pi \ddot{a}_u$$

For example, if $k = 0$, then

$$_0LF(0) = 1000(1.15)^{-1} − 288.41 = 581.1552$$

and

$$_0LF(1) = 1000(1.15)^{-2} − 288.41 − 288.41(1.15)^{-1} = 216.94$$

After computing yearly reserve loss function measurements, moment computations can be completed utilizing the pdf of U and a general discrete expectation (1.21). This is demonstrated through the information listed in Table 6.4.

As expected, $E\{_0LF(U)\} = 0$ and the conditions of the endowment policy, where a benefit is paid upon survival to future year $K = 2$, dictates $Var\{_2LF(K)\} = 0$. Approximate prediction interval estimates for the losses can be computed using the mean and variance calculations.

Example 6.25

In this example a person age x purchases a whole life insurance policy that pays a benefit of b at the end of the year of death. The policy is financed by premium payments at the start of year during the lifetime of (x), for at most h years. The RC and EP approach yields premium

$$\pi = b \, _hP_x \quad \text{where} \quad _hP_x = \frac{A_x}{\ddot{a}_{x:\overline{h}|}}$$

Analyzing this policy at future age $x + k$ requires the integer future lifetime random variable to be $U = K - k$, where the pdf takes the form ${}_u p_{x+k} \, q_{x+k+u}$ for positive integer k. The translated present value functions for the insurance benefit and the annuity payments are

$$\text{PVE}(U) = b \, v^{U+1} \quad \text{for} \quad U = 0, 1, \dots \tag{6.77}$$

and

$$\text{PVR}(U) = \frac{{}_h P_x}{d}(1 - v^{U+1}) \quad \text{for} \quad k < h,$$
$$= 0 \quad \text{for} \quad k \geq h \tag{6.78}$$

The reserve loss function based on additional time k takes the form

$$_k LF(U) = \left(b + \frac{{}_h P_x}{d} \right) v^{U+1} - \frac{{}_h P_x}{d} \quad \text{for} \quad k < h,$$
$$= b \, v^{U+1} \quad \text{for } k \geq h \tag{6.79}$$

The reserve is computed by taking the expectation of (6.79) over the disjoint regions, resulting in

$$E\{_k LF(U)\} = b \, A_{x+y} - {}_h P_x \, \ddot{a}_{x+k:\overline{h-k}|} \quad \text{for} \quad k < h,$$
$$= b \, A_{x+k} \quad \text{for} \quad k \geq h \tag{6.80}$$

For a target population, reserve computations (6.80) can be completed utilizing an actuarial life table as discussed in Section 6.7.

Analogous to the situations of standard life insurance, life annuities, and premiums discussed in Sections 6.4.1, 6.5.1, and 6.8.1, standard reserves for some standard insurance policies have defined actuarial notations. These notations are presented in the next section.

6.9.1 Unit Benefit Reserves Notations

As in the case of life insurance premiums, special notations have been developed for some standard reserves. In these life insurance loss models, the insurance benefit is taken to be 1 and the net level premium is evaluated utilizing the *RC* and *EP* approach, not additional costs. In Table 6.4, reserve computations and associated actuarial notations are listed for some standard life insurance models. These formulas give insight into the reserve structure.

TABLE 6.5

Reserves for Standard Life Insurance Policies

Expenditures and Revenue	Reserve	
	Symbol	Equation
1. Continuous-status whole life insurance and life annuity	$_t\bar{V}(\bar{A}_x)$	$\bar{A}_{x+t} - \bar{P}(\bar{A}_x)\bar{a}_{x+t}, \quad t \geq 0$
2. Discrete-status whole life insurance and life annuity	$_kV_x$	$A_{x+k} - P_x\ddot{a}_{x+k}, \quad k \geq 0$
3. Continuous-status n-year term insurance and h-year temporary life annuity	$_t^h\bar{V}(\bar{A}^1_{x:\overline{n}\mid})$	$\bar{A}^1_{x+t:\overline{n-t}\mid} - {_n}\bar{P}(\bar{A}^1_{x:\overline{n}\mid})\bar{a}_{x+t:\overline{h-t}\mid}, \quad 0 \leq t \leq h$
		$\bar{A}^1_{x+t:\overline{n-t}\mid}, \quad h < t$
4. n-Year continuous-status endowment and h-year temporary annuity	$_k^h\bar{V}(\bar{A}_{x:\overline{n}\mid})$	$\bar{A}_{x+k:\overline{n-k}\mid} - {_h}P(\bar{A}_{x:\overline{n}\mid})\ddot{a}_{x+k:\overline{h-k}\mid}, \quad 0 \leq k < h$
		$\bar{A}_{x+k:\overline{n-k}\mid}, \quad h \leq k < n$
5. n-Year pure endowment and h-year discrete-status temporary annuity	$_k^h V(_nE_x)$	$_{n-k}E_{x+k} - {_h}P(_nE_x)\ddot{a}_{x+k:\overline{h-k}\mid}, \quad 0 \leq k < h$
		$_{n-k}E_{x+k}, \quad h \leq k < n$
6. n-Year deferred whole life annuity and h-year temporary annuity	$_k^h V(_{n\mid}\ddot{a}_x)$	$_{n-k\mid}\ddot{a}_{x+k} - {_h}P(_{n\mid}\ddot{a}_x)\ddot{a}_{x+k:\overline{h-k}\mid}, \quad 0 \leq k < h$
		$_{n-k\mid}\ddot{a}_{x+k}, \quad h \leq k \leq n$

The development of the reserves listed in Table 6.5 is left to problems given at the end of the chapter, namely, Problem 6.12d–g. In these calculations, the mortality and survival structure of the stochastic lifetime random variable may be a statistical model based on or defined by a suitable life table. Further, the relationships between actuarial expectations presented in Section 6.6 are helpful in the calculation of many types of reserves. Moreover, there is a relationship between reserves based on subsequent years. These are explored in the next section and lead to the year-by-year approach to reserve computations presented in Section 6.9.3.

6.9.2 Relations among Reserve Calculations

Relations between consecutive year reserves exist and lead to an iterative approach to the computation of reserves. For example, for an individual age x, consider a unit benefit whole life insurance policy financed by discrete premiums for life. Applying *RC* and *EP*, the yearly premium from (2) in Table 6.3 is $\pi = P_x$, and the reserve formula for various years is listed in Table 6.6. For the first year,

$$0 = {_0}V_x = b\,A_x - \pi\,\ddot{a}_x \tag{6.81}$$

TABLE 6.6

Cash Flow Table for Life Insurance

k	l_{50+k}	$E\{Pr_k\}$	$E\{IF_k\}$	$E\{I_k\}$	$E\{B_k\}$	$E\{F_k\}$	l_{50+k+1}	$E\{_kLF(U)\}$
0	70,711	69,733	69,733	3,487	71,100	2,120	70,000	30.29
1	70,000	69,032	71,152	3,558	71,800	2,910	69,282	42.00
2	69,282	68,324	71,233	3,562	72,500	2,295	68,557	32.48
3	68,557	67,609	69,904	3,495	73,400	0000	67,823	0.00

Applying (6.46), formula (6.81) becomes

$$0 = {}_0V_x = b(v\,q_x + v\,p_x\,A_{x+1}) - \pi(1 + v\,p_x\,a_{x+1}) = v\,p_x\,{}_1V_x + b\,v\,q_x - \pi$$

Taking $b = 1$ and solving for the reserve at the end of year 1 gives

$$_1V_x = \frac{\pi}{vp_x} - \frac{q_x}{p_x} \tag{6.82}$$

Applying this approach to subsequent years, formulas relating consecutive reserves can be derived. From Bowers et al. (1997, p. 236) for time k and $k + 1$, $k = 1, 2, \ldots$,

$$_{k+1}V_x = \frac{(1+i)({}_kV_x + \pi)}{p_{x+k}} - \frac{q_{x+k}}{p_{x+k}} \tag{6.83}$$

Formula (6.83) defines the reserve as a function of year $k + 1$ for discrete whole life insurance. A graph of the reserve is given in Figure 6.4. In this

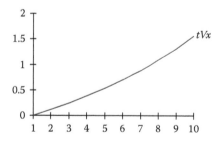

FIGURE 6.4
Reserve function.

situation the reserve increases as the year of analysis increases. Hence, a company with many older insurance contracts that has had adverse investment or mortality experience may have reserve obligations that can have a negative effect on their financial stability.

6.9.3 Survivorship Group Approach to Reserve Calculations

In this section discrete insurance status reserves are computed using a cash flow approach relating each year's reserve computations defined by (6.82) and (6.83). For a discrete life insurance status model with curate future lifetime K the life analysis notations hold, namely, l_{x+k} denotes the number of survivors to age $x + k$ and q_{x+k} is the 1-year mortality for individuals age $x + k$. Based on a group survivorship structure, the cash flow approach tracks the aggregate cash fund through time. For example, consider a whole life insurance for an individual age x and define the following aggregate expectations for future year k: $E\{Pr_k\}$ is the total expected premiums at the start of year k, $E\{IF_k\}$ denotes the expected initial fund at the start of year k, $E\{I_k\}$ is the expected interest earned during year k, $E\{B_k\}$ is the expected death benefits paid at the end of year k, and the expected fund at the end of year k is $E\{F_k\}$. Sequentially, in terms of the curtate future lifetime $K = k$ to mirror the collective cash flow, where the premium is π and benefit is b, we compute

$$E\{Pr_k\} = \pi \, l_{x+k}$$
$$E\{IF_k\} = E\{Pr_k\} + E\{F_{k-1}\}$$
$$E\{I_k\} = i \, E\{IF_k\}$$
$$E\{B_k\} = b \, l_{k\,1} q_{x+k}$$
$$E\{F_k\} = E\{IF_k\} + E\{I_k\} - E\{B_k\}$$

These quantities can be included in a sequential year-by-year table and comprise the information required for the reserve calculation. If $k = 0$, then using default value $E\{FE\}_{-1} = 0$ we have

$$E\{F_1\} = (1 + i) \, \pi \, l_x - b \, l_x \, q_x \tag{6.84}$$

Dividing (6.84) by $l_{x+1} = l_x \, p_x$ yields (6.82), the reserve for year 1. This expectation is

$$\frac{E\{F_1\}}{l_{x+1}} = (1 + i)\frac{\pi}{p_x} - b\frac{q_x}{p_x} = E\{{}_1 LF(U)\} \tag{6.85}$$

For subsequent years $k \geq 1$, $E\{FE\}_{k-1}$ may not be zero, and utilizing (6.83) we have the reserve formula

$$\frac{E\{F_k\}}{l_{x+k+1}} = \frac{(1+i)l_{x+k}(_kV_x + P_x) - bl_{x+k}\,q_{x+k}}{l_{x+k}\,p_{x+k}} = E\{_k LF(U)\} \tag{6.86}$$

A cash flow table consists of the yearly collective expectations and reserve (6.86). This is demonstrated in the following example.

Example 6.26

In this example consider the discrete 4-year term life insurance policy for a person age 50 as discussed in Examples 6.7 and 6.10, where premiums take the form of a discrete 4-year temporary life annuity and the benefit is $100,000. Table 5.1 holds, and for an individual age 50, the standard actuarial computations result in

$$A^1_{50:\overline{4|}} = .036183 \quad \text{and} \quad \ddot{a}_{50:\overline{4|}} = 3.66903$$

In terms of thousands of dollars, the RC and EP yearly premium is $\pi =$ 100,000(.0098617) = 986.17.

In Table 6.6, applying curtate future lifetime $K = k$, we give the cash flow approach where the reserve, $E\{_k LF(U)\}$, is in terms of dollars. For term life insurance, the reserve at the end of the term is fully exhausted and reaches the value zero. Another example of the cash flow approach using Excel is outlined in Problem 6.15.

6.10 General Time Period Models

For the discrete actuarial status models in this chapter it is assumed that the associated time period time is 1 year in length. In this section, the general time period structure introduced in Section 3.1.1 is utilized to construct investment present values where years are divided into m equal parts resulting in time intervals $[(h-1)/m, h/m)$ for $h = 1, 2, \ldots$. The effective annual rate of this model is referred to as a discrete monthly period model. Equating the future value of 1 unit under simple interest i with that of 1 unit with interest compounded following (3.3) at nominal rate $i^{(m)}$ gives $1 + i = (1 + i^{(m)}/m)^m$. Thus,

$$i^{(m)} = m((1+i)^{1/m} - 1) \tag{6.87}$$

The concepts and formulas associated with the financial and actuarial present value functions associated with the discrete 1-year time periods model are extended to the discrete monthly period model.

For a discrete stochastic status model based on the monthly period model, the time after H periods is $T = T_H = H/m$. Adapting (6.1), the present value associated with future value F_{T_H} discounted by $v_m = (1 + i^{(m)}/m)^{-1}$ for each time period is

$$\text{APV}(T_H) = F_{T_H} v_m^H \tag{6.88}$$

As in the case of yearly periods, single-status loss models are constructed using (6.88) and criteria such as RC or $PC(.25)$ are utilized in the analysis.

In life insurance settings the stochastic benefit is often financed by premium payments made at the start of each monthly period while the insured survives. If the period payments are level valued at $\pi^{(m)}$, applying (6.88) to each payment and summing yields the present value of the monthly period annuity:

$$PV(T_H) = \frac{\pi^m}{m} \sum_{r=0}^{H} \left(1 + \frac{i^{(m)}}{m} \right)^{-r} = \left(\frac{\pi}{m} \right) \frac{1 - v_m^{(H+1)}}{d^{(m)}} \tag{6.89}$$

where

$$d^{(m)} = m(1 - v_m) = \frac{i^{(m)}}{1 + \dfrac{i^{(m)}}{m}} \tag{6.90}$$

In the next section the expectation of monthly period present values for standard life insurance and standard life annuities is explored.

6.10.1 General Period Expectation

A discrete stochastic status model based on the monthly period model defines a monthly period curtate future lifetime random variable H where the endpoints of the periods are h/m for $h = 0, 1, \ldots$. Extending the general discrete distribution structure (5.12), the pmf of H can be written as $f(h) = {}_{h/m}p_x/{}_{1/m}q_{x+h/m}$ for $h = 0, 1, \ldots$. For both monthly period life insurance with fixed benefit $B_{T_h} = b$ and monthly period life annuities with level payments $\pi^{(m)}/m$, the present value expectations or net single premiums are modifications of previous computational formulas. Further, for standard life insurance and life annuities based on the discrete monthly period structure actuarial notation exists for the expectation. For example, based on structure (6.13) for a monthly period discrete whole life insurance with unit benefit the NSP is

$$A_x^{(m)} = \sum_{h=0}^{\infty} v^{h/m+1/m} \, {}_{h/m}p_x/{}_{1/m}q_{x+h/m} \tag{6.91}$$

In an analogous manner, adapting formula (6.28) for a standard monthly period discrete whole life annuity with yearly payments valued at 1, the NSP is

$$\ddot{a}_x^{(m)} = \frac{1}{m} \sum_{h=0}^{\infty} v_m^{h/m} \, {}_{h/m}p_x \tag{6.92}$$

The monthly period expectations can be directly computed only if the yearly mortality and survival expectations are available in terms of monthly periods. Since life tables are most often kept in terms of yearly, or longer, time periods, some adjustment or approximation is required. The adjustment of these monthly period formulas to account for yearly-based mortality and survival tables is the topic of the next section.

6.10.2 Relations among General Period Expectations

For some standard life insurance and life annuities, relations exist between monthly period and yearly period present value expectations. The basis in the construction of relational formulas is the decomposition of the future lifetime random variable $T = H/m + S/m$, for fixed positive integer m, and the UDD assumption applied to monthly length time periods. Hence, it is assumed that H and S are independent and S is a continuous uniform random variable defined on the unit interval. The support of monthly period curtate future lifetime H is denoted S_H. The required adjustments are presented through some standard actuarial models.

The relationship between the present value expectations corresponding to continuous whole life insurance and discrete monthly period whole life insurance, both with a unit benefit, is first derived. The derivation starts with the continuous whole life insurance expectation for a status age x given in (6.12). Following an analogous construction of (6.6), the net single premium can be written as

$$\bar{A}_x = \sum_{S_h} v^{(h+1)/m} \, {}_{h/m}p_x \, {}_{1/m}q_{x+h/m} \int_0^1 v^{(s-1)/m} \, ds \tag{6.93}$$

To simplify (6.93), we note $\exp(\delta/m) = (1 + i^{(m)}/m)$ and compute

$$\int_0^1 v^{(s-1)/m} \, ds = m(\exp(\delta/m) - 1)/\delta = \frac{i^{(m)}}{\delta}$$

Since $v = (1 + i^{(m)}/m)^{-m}$, (6.93) reduces to

$$\bar{A}_x = \left[\frac{i^{(m)}}{\delta} \right] A_x^{(m)} \tag{6.94}$$

Formula (6.94) associates a discrete monthly period whole life expectation with a continuous whole insurance expectation. Further, substituting (6.45) with (6.94) produces a computational relationship between discrete whole life expectations based on yearly and monthly period structures given by

$$A_x^{(m)} = \left(\frac{i}{i^{(m)}}\right) A_x \tag{6.95}$$

where $i^{(m)} = m((1+i)^{1/m} - 1)$. Formula (6.95) can be utilized in conjunction with actuarial life tables based on yearly intervals as discussed in Section 6.7 to compute NSP associated with monthly period whole life insurance.

Life annuity present value expectations involving monthly period payments can be related to life insurance and yearly payment life annuity present value expectations for some standard actuarial models. These relational formulas are derived in the context of specific examples. Utilizing (6.89),

$$\ddot{a}_x^{(m)} = \frac{1}{d^{(m)}} E\left\{1 - v_m^{H+1}\right\} = \frac{1}{d^{(m)}}(1 - A_x^{(m)}) \tag{6.96}$$

Applying (6.94), (6.45), and (6.48) to formula (6.96) produces a relationship between the expected present values of discrete whole life annuities based on the monthly period and yearly period structures. After simplification we have

$$\ddot{a}_x^{(m)} = \left(\frac{id}{d^{(m)}i^{(m)}}\right)\ddot{a}_x + \frac{i^{(m)} - i}{i^{(m)}d^{(m)}} \tag{6.97}$$

Hence, an approximation to the standard monthly period discrete whole life annuity expectation can be written as a linear function of the expectation for a standard discrete whole life annuity.

Relations for other discrete monthly period insurance and annuity present value expectations exist. For n-year term life insurance and n-year temporary annuity relations take the form

$$\bar{A}_{x:\overline{n}|}^1 = \frac{i^{(m)}}{\delta} A_{x:\overline{n}|}^{1(m)} \quad \text{and} \quad \ddot{a}_{x:\overline{n}|}^{(m)} = \frac{1}{d^{(m)}}\left(1 - A_{x:\overline{n}|}^{(m)}\right) \tag{6.98}$$

for $n > 0$. Further, combining the expectations in (6.98) and utilizing the discrete expectation in (6.40),

$$\ddot{a}_{x:\overline{n}|}^{(m)} = \frac{1}{d^{(m)}}\left[1 - \frac{i}{i^{(m)}}(A_x - {}_nE_x A_{x+n}) - {}_nE_x\right] \tag{6.99}$$

Present value expectations for monthly period actuarial models for both theoretical statistical models and life table data are demonstrated in the following two examples.

Example 6.27

For a discrete stochastic status model age x, benefit b is paid at the end of the month of death. Level premium payments are made at the start of every month during the lifetime of (x). The monthly premium payments for this discrete-status life insurance are determined using RC and EP and are

$$\pi = b \frac{A_x^{(12)}}{12\ddot{a}_x^{(12)}}$$

In particular, let the geometric distribution discussed in Example 6.12 be used to model the distribution of the curtate future lifetime where the UDD assumption holds within time periods. Utilizing (6.95), (6.96), and $i^{(m)} = m(e^{\delta/m} - 1)$, the expectations used in the premium computation simplify to

$$A_x^{(12)} = \frac{p(e^\delta - 1)}{12(e^{\delta/12} - 1)(e^\delta - q)} \quad \text{and} \quad \ddot{a}_x^{(12)} = \left(\frac{1}{d^{(12)}}\right)(1 - A_x^{(12)})$$

We remark that the premium payment is a function of the interest rate and the time period structure.

Example 6.28

For an individual age 50 the actuarial quantities given in Table 6.1 hold along with the UDD within years assumption. The effective interest rate is $i = .06$, and a \$50,000 whole life insurance policy is purchased where the benefit is payable at the moment of death. The level premium payments are made at the start of each month while the insured survives. Here $m = 12$, and from (6.87) and (6.90) we compute $i^{(12)} = 12(1.06^{1/12} - 1) = .058408$ and $d^{(12)} = .058408/1.00486 = .058128$. Using Table 6.1 for $x = 50$ and (6.45),

$$\bar{A}_{50} = \left(\frac{.06}{.0582689}\right).211036 = .217306$$

Applying relation (6.95),

$$\bar{A}_{50}^{(12)} = \left(\frac{.06}{.0584088}\right).211036 = .216785$$

and (6.96)

$$\ddot{a}_{50}^{(12)} = \left(\frac{1}{.0581251}\right)(1 - .211036) = 13.57355$$

Hence, using *RC* and *EP* the monthly premium payment is found to be

$$\pi^{(12)}/12 = \$50,000\left(\frac{1}{12}\right)\left(\frac{.216785}{13.57355}\right) = \frac{\$798.5383}{12} = \$66.54$$

An example utilizing these formulas is outlined in Problem 6.10. In practice the premium would be increased or loaded to account for cost factors associated with the insurance. This topic is discussed in Section 6.11.

6.11 Expense Models and Computations

In the administration and execution of financial and actuarial contracts many types of expenses, both fixed and variable, are encountered. The general loss model approach is easily adapted to this situation where the *RC* and *EP* can be utilized to solve for unknown constants. For an insurance company or an investment firm all expenses, including computational costs, maintenance fees, salaries, and profit, must be included in an estimation of expenditures. For insurance, this results in an increase of premiums over nonexpense *RC* and *EP* computations. This increase is referred to as a loading, and the resulting premiums are referred to as gross premiums. In this section we explore the different types of costs encountered where concentration is on life insurance models but the basic concepts and formulas can be extended to financial investment models. The list of cost types we give is not exhaustive and other costs may be encountered in practice. Further, accounting procedures associated with actuarial calculations will not be discussed, and for a more exhaustive review we refer to Bowers et al. (1976, Chapter 15).

For life insurance, common costs for the insurance company can be classified into the broad classifications of acquisition, maintenance, settlements, and general costs. Acquisition costs include selling expenses, such as advertising, risk classification, commissions, preparing of policies, and recording data. Maintenance costs include premium collection and accounting and other correspondence. Claim administration investigation and legal defense costs fall under settlement costs. General costs include costs associated with research, actuarial services, general accounting, taxes, and fees.

Costs can be divided into three classifications. There are fixed costs, costs related to the amount of the benefit, and costs related to the premium payments. Further, the costs associated with these different types may change from year to year. We follow with two examples of actuarial models that include different types of costs. These examples take the form of case study examples that combine different types of actuarial and financial computations.

Example 6.29

For an individual age x, a life insurance policy has a benefit of \$20,000 payable at the end of the year of death. Policy costs may be the same or vary from year to year. For example, the commission is 60% of the premium the first year and 5% of the premium in subsequent years, and there is a fixed maintenance cost valued at 6 units every year. The costs are divided into fixed costs, denoted by F, and proportions of the benefit (per \$1,000) and gross premium, denoted by B and G, respectively. In Table 6.7 the particular costs are defined.

A stochastic loss function is constructed where the present value of the expenditures includes the direct insurance payments and costs as defined in Table 6.7. Applying the RC and EP approach, the gross premium, G, can be found using

$$G\ddot{a}_x = \$20,000\,A_x + \left[40.5 + 5(20) + .87G\right] + \left[6 + .5(20) + .07G\right]_{1|}a_x$$

Solving for G we find

$$G = \frac{(\$20,000\,A_x + 140.5 + 10.6_{1|}\ddot{a}_x)}{(\ddot{a}_x - .87 - .07_{1|}\ddot{a}_x)} \qquad (6.100)$$

Utilizing the techniques of Section 6.9 a reserve can be computed based on future age $x + t$. The expected present value of the revenue and expenses is given by

$$E\{PVR(W)\} = G\ddot{a}_{x+t} \quad \text{and} \quad E\{PVE(W)\} = 20,000\,A_{x+t} + (16 + .07G)\ddot{a}_{x+t}$$

TABLE 6.7

Costs of Discrete Whole Life Insurance

Classification	First Year			Renewal Years		
	F	B	G	F	B	G
Acquisition	34.5	4.5	.85	0.0	0.0	.05
Maintenance	6.0	.5	.02	6.0	.5	.02
Total	40.5	5.0	.87	6.0	.5	.07

The reserve associated with age $x + k$ is

$$E\{_k LF(W)\} = 20,000 A_{x+k} + (16 - .93G)\ddot{a}_{x+k} \qquad (6.101)$$

For example, for an individual age $x = 50$ let the actuarial life table given in Table 6.1 hold and $i = .06$. From the table,

$$A_{50} = .211036, \qquad \ddot{a}_{50} = 13.9384 \qquad \text{and} \qquad \ddot{a}_{51} = 13.8554$$

Ignoring costs, the *RC* and the *EP* yearly premium is $\pi = \$20,000\ (.211036/13.9384) = \$302.81$. We compare this to the analysis including the expenses as expressed in Table 6.7. Using Table 5.1 the expectation of the 1-year deferred annuity is

$$_1|\ddot{a}_{50} = {}_1E_{50}\,\ddot{a}_{51} = 12.93907$$

Computing (6.100) the gross premium is found to be

$$G = \frac{20,000(.211036) + 140.5 + 10.6(12.93907)}{13.9384 - .87 - .07(12.93907)} = \$369.83$$

The gross premium in this hypothetical example is significantly larger than the nonexpense computed premium. The graph of the reserves against future years using (6.101) is presented in Figure 6.5. As expected, the reserves increase as the future time increases, and the expenses incurred cause the reserves to be negative for the first 2 years.

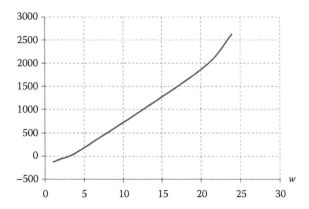

FIGURE 6.5
Reserve computations.

Example 6.30

An individual takes out a 30-year term life insurance policy that pays $50,000 at the end of the year of death. Premium payments are made at the start of each year while the insured survives up to age 50. The costs of the policy are both variable and fixed. For each year the sales commission is 7.5% and the taxes and fees are 3% of the premium. Other policy expenses are $25 the first year and $5 each renewal year, and there is a claims settlement fee of $35. Using the RC and EP approach, the gross premium, G, satisfies the equation

$$G\ddot{a}_{x:\overline{20|}} = 50,000\,A^1_{x:\overline{30|}} + 0.105\,G\ddot{a}_{x:\overline{30|}} + 25 + 5\,a_{x:\overline{29|}} + 35\,A^1_{x:\overline{30|}} \quad (6.102)$$

Solving for G we find

$$G = \frac{50,035\,A^1_{x:\overline{30|}} + 25 + 5\,a_{x:\overline{29|}}}{\ddot{a}^{(12)}_{x:\overline{20|}} - .105\,\ddot{a}_{x:\overline{30|}}}$$

Using appropriate actuarial life tables the loaded premium can be computed. Further, the reserves can be calculated corresponding to future year W. For $1 \le W < 20$, applying the computations in (6.102), the reserve is

$$E\{{}_k LF(W)\} = 50,035\ A^1_{x+k:\overline{30-k|}} - .895\,G\,a_{x+k:\overline{30-k|}}$$

For $20 \le w < 30$ we note there are no yearly premium payments and the reserve is computed as

$$E\{{}_k LF(W)\} = 50,035\,A^1_{x+k:\overline{30-k|}} + .105G\,a_{x+k:\overline{30-k|}}$$

In specific scenarios all these computations can be completed using appropriate actuarial life tables. A cost application is outlined in Problem 6.15.

Problems

6.1. An investment is to be sold for $5,000 at a future date where the rate of return is $\delta = .08$. The future time of sale is denoted by T. Find the expected present value where the distribution of T is (a) exponential with a mean of 3 years, (b) gamma with pdf $f(t) = 1/(\Gamma(\alpha)\beta^\alpha)t^{\alpha-1}e^{-t/\beta}$ where $\alpha = 1.5$ and $\beta = 2$, and (c) normal with mean 3 years and standard deviation given by 1 year.

6.2. In Problem 6.1 consider (c), where the future time of sale is a normal random variable. Construct a 95% prediction interval for the present value of the investment.

6.3. An investment is to be sold only at the end of a month for a value of $F_{K+1} = F$ where the financial return rate is estimated to be δ. The future time of sale has curtate lifetime K, which is a Poisson random variable with pmf $f(k) = e^{-\lambda}\lambda^k/k!$ where $S_k = \{0, 1, ...\}$. Find (a) the expected present value of the investment using (6.4) and (b) the variance of the present value function.

6.4. Consider the future sale described in Problem 6.3. Assume the UDD assumption and compute the NSV using (6.6).

6.5. In this problem the $PC(.25)$ method is applied to Problem 6.1. Find the NSP for (a) Problem 6.1a and (b) Problem 6.1c.

6.6. Let the lifetime associated with a status, X, have pdf $f(x) = 1/m$ where $S_x = (0, m)$ for constant $m > 0$. For a status age x give formulas for (a) the unit benefit insurances (i)$\bar{A}_x$, (ii)$\bar{A}^1_{x:\overline{n}|}$, and (iii)$\bar{A}_{x:\overline{n}|}$, and (b) the unit premium for annuities (i)$\bar{a}_x$, (ii)$\bar{a}_{x:\overline{n}|}$, and (iii)$_{n|}\bar{a}_x$.

6.7. An individual in a high-risk category has a curtate future lifetime given by

k	0	1	2	3	4	5	6	7	8	9
$f(k)$	.05	.05	.07	.1	.13	.1	.1	.15	.15	.1

The interest rate is given by $\delta = .05$, where UDD holds. Compute the expectation for the present value of (a) insurance paying a benefit of \$10,000 at the moment of death for (i) whole life insurance, (ii) 4-year term insurance, and (iii) 4-year endowment insurance; and b) an annuity paying \$1,000 at the start of each year for (i) whole life annuity, (ii) 4-year temporary annuity, and (iii) 4-year deferred annuity.

6.8. Let the curtate future lifetime associated with a status age x be a Poisson discrete random variable with pdf given in Problem 6.3. Give formulas for (a) the unit benefit insurance A_x, (b) the whole life annuity $\ddot{a}_x$, and (c) the yearly premium P_x using RC and EP.

6.9. An individual age 50 purchases life insurance with benefit of \$10,000 payable at the moment of death where Table 6.1 applies and $i = .06$. Further, the UDD assumption holds where the expectation criteria are used to compute premiums. Find the level premiums during the life of (50) for insurance where (a) the premiums are paid at the start of each surviving year, (b) the premiums are paid up to age 70, and (c) the premiums paid at the start of each year comprise a whole life approtionable annuity.

6.10. The conditions of Example 6.28 are considered where for a status age 50 the benefit b is associated with a continuous 22-year endowment insurance. The premium payments are made at the start of each month until age 72. Find the amount of monthly premium payment π.

6.11. Consider a status where the pdf of the continuous future lifetime random variable is exponential with pdf given by $f(t) = (1/\theta) \exp(-t/\theta)$ for $t > 0$. For a continuous whole life insurance policy where both the benefits and premiums are continuous find the needed premium using (a) the expectation criteria RC and (b) the $PC(.25)$ method using (6.60).

6.12. For a stochastic status let the UDD assumptions hold where $\delta > 0$. Give reduced formulas for the following:

(a) $P(\bar{A}_{50})$, (b) $\bar{P}(\bar{A}_{2:\overline{50}|})$, (c) $P(_{50|}\ddot{a}_2)$, (d) $_2V_{50}$, (e) $_2\bar{V}(\bar{A}_{50})$, (f) $_2^{50}V(A_{2:\overline{70}|})$

and (g) $_2^{50}V(_{70|}a_2)$.

6.13. Let Table 6.1 hold where $\delta = .06$ and the UDD assumptions hold. Compute (a) $P(\bar{A}_{50})$, (b) $\bar{P}(\bar{A}_{50:\overline{22}|})$, (c) $P(_{22}\ddot{a}_{50})$.

Excel Problems

6.14. The data in Problem 5.17 are utilized where $i = .05$. For a status age $x = 2$, discrete insurance with benefit b_k and discrete premium payments of π_k is analyzed using the general loss function approach and EP. Find the premium π using EP where the premium payments are level for (a) unit benefit whole life insurance and annuity premiums, (b) unit benefit 8-year term life insurance and 5-year temporary premium annuity, (c) life insurance with benefits $b_k = .5$ for $k = 0, 1$, $b_k = 1$ for $k = 2, 3, ..., 6$, and $b_k = 2$ for $k = 7$ and 8- and 5-year temporary premium annuity. Excel: Basic operations.

6.15. Apply the cash flow approach to Problem 6.14a and compute the reserves. Excel: Basic operations.

Solutions

6.1. $NSV = M_t(-\delta)$.

a. $NSV = \$5,000(1 + .08(3))^{-1} = \$4,032.26$.

b. $NSP = \$5,000(1 + .08(2))^{-1.5} = \$4,002.06$.

c. $NSP = \$5,000\exp(-3(.08) + .5(.08^2)) = \$3,945.75$.

6.2. 95% PI for T is $1.04 = 3 - 1.96 \le T \le 3 + 1.96 = 4.96$ yielding $\$3,362.34 \le PV(t) = \$5,000\exp(-.08t) \le \$4,600.84$.

6.3. $PV(K) = F\,v^{K+1}$.

    a. $NSV = F\,v\,\exp(\lambda(v-1))$.

    b. $\mathrm{Var}\{PV(K)\} = F^2v^2[\exp(\lambda(v^2-1)) - \exp(2\lambda(v-1))]$.

6.4. $NSV = F\,(d/\delta)\,\exp(\lambda(v-1))$.

6.5. $NSV = \$5{,}000\,\exp(-.08t_{.25})$.

    a. $t_{.25} = -3\ln(.75) = .86805$, so $NSV = \$4{,}666.43$.

    b. $t_{.25} = 3 - .6745 = 2.32551$, so $NSP = \$4{,}151.20$.

6.6. For $t \le m - x$ we have $_tp_x = (m-x-t)/(m-x)$.

    a. i. (6.12) gives $(1-\exp(-\delta(m-x)))/(\delta(m-x))$.

      ii. (6.14) gives $(1 - \exp(-\delta n))/(\delta(m-x))$.

      iii. (6.18) gives $(1 - \exp(-\delta n))/(\delta(m-x)) + \exp(-\delta n)\,(m-n-x)/(m-x)$.

    b. i. (6.27) gives $[(m-x)\delta + 1 - \exp(-\delta(m-x))]/[\delta^2(m-x)]$.

      ii. (6.30) gives $[(m-x)\delta + 1 - \exp(-\delta(m-x))(1 + \delta(m-x-n))]/[\delta^2(m-x)]$.

6.7. a. i. (6.13) and (6.45) give $NSP = \$10{,}000(.756316) = \$7563.16$.

      ii. (6.43) gives $NSP = \$10{,}000(.756316 - .597673(.862374)) = \$2{,}408.98$.

      iii. (6.43) gives $NSP = \$10{,}000(.756316 + .597673(1 - .826374)) = \$8{,}385.71$.

    b. i. (6.28) gives $NSP = \$1{,}000(5.38099) = \$5{,}380.99$.

      ii. (6.22) gives $NSP = \$1{,}000(3.432409) = \$3{,}432.41$.

      iii. (6.34) gives $NSP = \$1{,}000(1.948581) = \$1{,}948.58$.

6.8. a. Using (6.13) $A_x = v\,\exp(\lambda(v-1))$.

    b. (6.48) gives $[1 - v\,\exp(\lambda(v-1)]/(iv)$.

    c. $P_x = i\,v^2\,\exp(\lambda(v-1)/[1 - v\,\exp(\lambda(v-1))]$.

6.9. $NSP = \$10{,}000\,A_{50} = \$2173.06$.

    a. $\pi = \$2173.06/13.93836 = \$155.91$.

    b. Using (6.50) and Table 5.1 gives $\pi = \$2173.06/11.17015 = \$194.54$.

    c. (6.39) gives $\pi = \$2173.06/13.8273 = \$157.16$.

6.10. $\bar{A}_{50:\overline{22}|} = .345349$ and (6.99) gives $\ddot{a}^{(12)}_{50:\overline{22}|} = 11.26842$ and $\pi = \$50{,}000(.345349)/(11.26842(12)) = \$127.69$.

6.11. a. $P_x = (1 + \delta\theta)^{-1}/[\theta(1 + \delta\theta)^{-1}] = 1/\theta$.

    b. $t_{.25} = -\theta\ln(.75)$ so $V^{t.25} = .75^{-\delta\theta}$ and (6.60) gives $\pi = b\delta.75^{-\delta\theta}/(1 - .75^{-\delta\theta})$.

6.12. (a) $\dfrac{\bar{A}_{50}}{\ddot{a}_{50}}$, (b) $\dfrac{\bar{A}_{2:\overline{50}|}}{\ddot{a}_{2:\overline{50}|}}$, (c) $\dfrac{_{50|}\ddot{a}_2}{\ddot{a}_{2:\overline{50}|}}$, (d) $A_{52} - P_{50}\,\ddot{a}_{52}$, (e) $\bar{A}_{52} - \bar{P}(\bar{A}_{50})\,\bar{a}_{52}$,

    (f) $A_{4:\overline{68}|} - _{50}P(A_{2:\overline{70}|})\,\ddot{a}_{4:\overline{48}|}$, (g) $_{68|}\ddot{a}_4 - _{50}P(_{20|}\ddot{a}_2)\,\ddot{a}_{4:\overline{48}|}$

6.13. (a) $\dfrac{\overline{A}_{50}}{\ddot{a}_{50}} = .0155905$, (b) $\dfrac{\overline{A}_{50:\overline{22}|}}{\overline{a}_{50:\overline{22}|}} = .030738$, (c) $\dfrac{_{22|}\ddot{a}_{50}}{\ddot{a}_{50:\overline{22}|}} = .1978999$

6.14. a. $A_2 = .79843$, $a_2 = 4.232998$, $\pi = .18862$.

      b. $A^1_{2:\overline{8}|} = .763117$, $a_{2:\overline{5}|} = 3.637051$, so $\pi = .209818$.

      c. $\pi = 7.86734/3.637051 = .216311$.

6.15. 13184, .23917, .34943, .44190, .52204, .60519, .65944, .76374.

7

Advanced Stochastic Status Models

This chapter extends the analysis of stochastic status models of Chapter 6 to more complex stochastic structures. The theory and computations associated with two distinct actuarial stochastic status models and an important application are presented. In the first, discussed in Section 7.1, the survival probabilities associated with the stochastic status are functions of multiple future life random variables. Specifically, for two future lifetime random variables the cases of joint life, last survivor, and general contingent statuses are presented in Sections 7.1.1, 7.1.2, and 7.1.3, respectively. In the second, multiple-decrement stochastic models, where stochastic variables are present for both lifetime and mode of decrement, are explored in Section 7.2. Continuous and discrete multiple-decrement model cases are considered as well as single-decrement structures. The chapter ends with an introduction to pension models in Section 7.3. For pension models, topics concerning multiple types of benefits, contributions, and various types of yearly retirement structures are discussed.

7.1 Multiple Future Lifetimes

The lifetime associated with a stochastic status model may depend on more than one future lifetime random variable. For example, a husband and wife may take out an insurance policy that pays benefits based on the death of one or the other spouse. In the general setting, we have a collection of m individuals ages $x_1, x_2, \ldots,$ and x_m and corresponding future lifetime random variables $T(x_i) = T_i$ for $i = 1, \ldots, m$. The decomposition of the future lifetime random variable discussed in Section 4.4 yields $T_i = K_i + S_i$, where K_i is the curtate future lifetime and $0 \leq S_i \leq 1$ for $1 \leq i \leq m$. The future lifetime random variables T_i are assumed independent for $i \geq 1$, and their corresponding order statistics are given by $T_{(1)} < T_{(2)} < \ldots < T_{(m)}$ as discussed in Section 2.3.

In this discussion we concentrate on $m = 2$ future lifetime random variables where the stochastic status model is defined in terms of the multiple life functions introduced earlier in Example 1.5, namely, joint and last survivor status models. Other developments in the modeling of multiple lifetimes are given in Marshall and Olkin (1967, 1988) and Genest (1987).

7.1.1 Joint Life Status

In the general multiple life setting a stochastic status model is based on a function of a set of m independent future lifetime random variables denoted $T_1 = T(x_1)$, $T_2 = T(x_2)$, ..., $T_m = T(X_m)$. A joint life status (JLS) describes a stochastic status model where actions occur upon the first failure or decrement out of the m individuals. For JLS in the continuous random variable setting the future lifetime variable is the first order statistic

$$T_{(1)} = \min\{T_1, T_2, \ldots, T_m\} \tag{7.1}$$

In actuarial science notation $T_{(1)} = T(x_1, x_2, \ldots, x_m)$. The cumulative distribution function (cdf) of $T_{(1)}$ can be found applying (2.46) or utilizing the general theory of order statistics.

The case of two independent future lifetimes associated with ages x and y is examined in detail. Applying (1.6) and (2.46) for $t > 0$ and $m = 2$, the cdf of $T_{(1)}$ can be written as

$$F_{t(x,y)}(t) = P(T_{(1)} \le t) = 1 - {}_tp_x \, {}_tp_y \tag{7.2}$$

where the corresponding survival function follows, from (2.48), as

$$P(T_{(1)} > t) = {}_tp_{xy} = {}_tp_x \, {}_tp_y \tag{7.3}$$

To find the probability density function (pdf) of $T_{(1)}$ we take the derivative of (7.3) with respect to t:

$$\frac{d}{dt} {}_tp_{xy} = -{}_tp_x \, \mu_{x+t}({}_tp_y) - {}_tp_y \, \mu_{y+t}({}_tp_x)$$

Following (5.26) the pdf of $T_{(1)}$ can be written as the product of survival probability and a force of mortality:

$$f_{t(x,y)}(t) = {}_tp_{xy} \, \mu_{xy+t} \tag{7.4}$$

where the force of mortality for $T_{(1)}$ is the sum of the separate forces of mortality

$$\mu_{xy+t} = \mu_{x+t} + \mu_{y+t} \tag{7.5}$$

An example demonstrating the JLS for continuous future lifetimes is now given.

Example 7.1

For a population the mortality structure is that of Example 5.6 and holds where individual survival functions follow (5.28) and $_tp_x = [(100 - x - t)/(100 - x)]^{1/2}$ for $x + t \le 100$. Assuming independence, for a JLS for two people ages x and y, the survivor function, from (7.3), computes as

$$_tp_{xy} = \left[\frac{(100 - x - t)(100 - y - t)}{(100 - x)(100 - y)} \right]^{1/2}$$

Further, the force of mortality is found using (7.5) and is

$$\mu_{xy+t} = \frac{1}{2} \left[\frac{1}{100 - x - t} + \frac{1}{100 - y - t} \right]$$

In connection to JLS the force of mortality combines in a linear fashion, increasing the instantaneous failure rate over time.

Joint life statuses may be discrete in nature where the status conditions change is a function of independent curtate future lifetimes. For $m = 2$ independent future lifetimes associated with ages x and y, applying (7.3), the probability of failure of the first order statistic within the interval time interval $(k, k + 1]$, for nonnegative integer k, is

$$P(k < T_{(1)} \le k + 1) = {_kp_{xy}} - {_{k+1}p_{xy}} \tag{7.6}$$

The curtate future lifetime associated with $T_{(1)}$ is denoted by $K_{(1)}$, and for discrete $m = 2$ JLS the event $\{K_{(1)} = k\}$ means that either x or y, or both, will die within 1 year of k. Using independence on (7.6) we find the probability mass function (pmf) of $K_{(1)}$ takes the form

$$P(K_{(1)} = k) = {_kp_x} {_kp_y}(1 - p_{x+k} \, p_{y+k}) \tag{7.7}$$

A numerical example applying the discrete probability formulas follows.

Example 7.2

Consider two people ages $x = 55$ and $y = 50$ where the future lifetimes are independent and the pmf of $K_{(1)}$ follows (7.7). The probability the first death is after 5 years but before 10 years from (7.6) is

$$P(5 < T_{(1)} \le 10) = {_5p_{55}} {_5p_{50}} - {_{10}p_{55}} {_{10}p_{50}}$$

Other probabilities for JLS are computed following similar techniques.

Consider the JLS survival probabilities for fractional lifetimes as discussed in Section 5.4. In the case of $m = 2$ independent future lifetime random variables, the first order statistic is decomposed as $T_{(1)} = K_{(1)} + S_{(1)}$ and for fraction age s, for $0 < s < 1$, we have

$$P(K_{(1)} = k, S_{(1)} \le s) = {}_k p_{xy} - {}_{k+s} p_{xy} \tag{7.8}$$

Under the uniform distribution of death (UDD) assumption for both independent future lifetime random variables (5.35), after simplification, (7.8) reduces to

$$P(K_{(1)} = k, S_{(1)} \le s) = {}_k p_x \, {}_k p_y \, s(q_{x+k} + q_{y+k} - s \, q_{x+k} \, q_{y+k}) \tag{7.9}$$

Hence, (7.9) can be used to compute fractional lifetimes for JLS, when the future lifetimes are assumed independent and uniformly distributed within years.

7.1.2 Last Survivor Status

Following the stochastic status structure introduced in Section 7.1.1, we again consider m independent future lifetime random variables. In last survivor status (LSS), the conditions of the stochastic status model indicate an economic action upon failure or decrement of all of the m individuals. In the case of m continuous future lifetime random variables LSS is a function of the mth order statistic

$$T_{(m)} = \max\{T_1, T_2, \ldots, T_m\} \tag{7.10}$$

In actuarial science notation $T_{(m)} = T(x_1, x_2, \ldots, x_m)$. The cdf of (7.10) follows (2.47), and the general distributional theory for the maximum order statistics can be applied. An application of the general case of JLS is given in Problem 7.1.

The multiple lifetime setting consisting of two independent future lifetime random variables for individuals whose ages are given by x and y is explored in detail. The cdf of $T_{(2)}$ evaluated at $t > 0$ utilizing statistical independence is

$$F_{t(\overline{xy})}(t) = P(T_{(2)} \le t) = (1 - {}_t p_x)(1 - {}_t p_y) \tag{7.11}$$

and the survival function takes the form

$$P(T_{(2)} > t) = {}_t p_{\overline{xy}} = {}_t p_x + {}_t p_y - {}_t p_x \, {}_t p_y \tag{7.12}$$

To find the pdf of $T_{(2)}$ we take the derivative of (7.12) with respect to t. The resulting pdf for $T_{(2)}$ takes the form

$$f_{t(\overline{xy})}(t) = {}_t p_{\overline{xy}} \mu_{\overline{xy}+t} \tag{7.13}$$

where the force of mortality is given by

$$\mu_{\overline{xy}+t} = \frac{{}_tp_x\mu_{x+t} + {}_tp_y\mu_{y+t} - {}_tp_{xy}\mu_{xy+t}}{{}_tp_x + {}_tp_y - {}_tp_{xy}} \tag{7.14}$$

Probabilities for LSS can be computed using (7.11) and (7.12). For example, applying (7.12) the probability of observing $T_{(2)} = T(\overline{x,y})$ in future time interval $(k, k+1]$ for positive integer k is

$$P(k < T_{(2)} \le k+1) = {}_k p_{\overline{xy}} - {}_{k+1} p_{\overline{xy}} \tag{7.15}$$

From (7.12) we observe these LSS probabilities can be written as a function of JLS.

Similar to the $m = 2$ case for JLS of Section 7.1.1, distributions for integer future lifetimes associated with (7.10) can be developed. The curtate future lifetime of $T_{(2)}$ is denoted $K_{(2)} = K(\overline{x,y})$ and for nonnegative integer k, applying basic probability laws and assuming independence:

$$P(K_{(2)} = k) = P(K(x) = k, K(y) < k) + P(K(y) = k, K(x) < k) + P(K(x) = k, K(y) = k)$$

Thus, the pmf associated with $K_{(2)}$ takes the form

$$P(K_{(2)} = k) = (1 - {}_k p_y){}_k p_x\, q_{x+k} + (1 - {}_k p_x){}_k p_y\, q_{y+k} + {}_k p_x\, {}_k p_y\, q_{x+k}\, q_{y+k} \tag{7.16}$$

for $k \ge 0$. Probability construction for the LSS formula is demonstrated in the next example.

Example 7.3

Let the future lifetimes of people ages $x = 55$ and $y = 50$ be independent so that the pmf of $K_{(2)}$ follows (7.16). The probability the last death is after 5 years but before 10 years using (7.15) is

$$P(5 < T_{(2)} \le 10) = {}_5p_{55} - {}_{10}p_{55} + {}_5p_{50} - {}_{10}p_{50} - {}_5p_{55}\, {}_5p_{50} + {}_{10}p_{55}\, {}_{10}p_{50}$$

Alternative approaches to the development of such probability computations exist utilizing basic set operations and probability rules (see Section 1.1).

Fractional ages for LSS can be developed assuming UDD holds for the independent future lifetime random variables. Similar to the JLS case for $m = 2$, future lifetimes $T_{(2)} = K_{(2)} + S_{(2)}$, where we assume (5.35) holds for both x and y. For $0 < s < 1$,

$$P(K_{(2)} = k, S_{(2)} \le s) = {}_k p_{\overline{xy}} - {}_{k+s} p_{\overline{xy}} \tag{7.17}$$

Applying (7.12) and (5.35) we can rewrite (7.17) as

$$P(K_{(2)} = k, S_{(2)} \leq s) = {}_kp_x\,{}_kq_y\,s\,q_{x+k} + {}_kp_y\,{}_kq_x\,s\,q_{y+k} + {}_kp_x\,{}_kp_y\,s^2\,q_{x+k}\,q_{y+k} \qquad (7.18)$$

Fractional lifetime probabilities for LSS, assuming independence and UDD, are computed utilizing (7.18). An example demonstrating computations for both JLS and LSS is now presented.

Example 7.4

For a population let the survival function be $S_X(x) = (1/(1 + .015x))^2$, where the support is $S = \{x \geq 0\}$. The mean lifetime of a newborn, computed using the survivor function following (1.44), is

$$E\{X\} = \int_0^\infty S_X(x)\,dx = \int_0^\infty (1+.015x)^{-2}\,dx = \left. \frac{(1+.015x)^{-1}}{.015} \right|_0^\infty = 66.66$$

For a stochastic status age x the survival function for the future lifetime random variable T following (5.7) is

$${}_tp_x = S_X(x + t)/S_X(x) = ((1 + .015x)/(1 + .015x + .015t))^2$$

Further, the force of mortality associated with age x using (5.16) is

$$\mu_x = -\frac{d}{dx}\ln(S_X(x)) = \frac{2(.015)}{1+.015x}$$

This force of mortality is a decreasing function of x indicating an infant mortality structure. Multiple survival probabilities can be computed. For example, consider two statuses defined by ages $x = 20$ and $y = 25$ where the corresponding future lifetimes are independent. The probability both survive at least 10 years is an example of JLS application and using (7.3),

$$P(T_{(1)} > 10) = {}_{10}p_{20}\,{}_{10}p_{25} = (.8038)(.8129) = .65345$$

Also, the probability at least one survives past 50 years applies LSS formula (7.12):

$$P(T_{(2)} > 50) = {}_{50}p_{20} + {}_{50}p_{25} - {}_{50}p_{20}\,{}_{50}p_{25}$$

It is left to the reader to compute this survival probability.

In the last two sections specific forms of multiple life probabilities, namely, JLS or LSS, were presented. Life tables based on JLS and LSS can be constructed following the techniques in Section 5.7. In practice, the conditions of a stochastic status may depend on combinations of future lifetimes other than JLS or LSS. In the next section mortality rates for alternative conditioned models are introduced.

7.1.3 General Contingent Status

In the modeling of a stochastic status model based on multiple future lifetimes many possibilities exist. Actuarial and financial contracts may be written so that financial considerations, such as the amount of a benefit or annuity payments, may be contingent on the order of mortality of the people involved. In the simplest case, a stochastic status model depends on $m = 2$ individuals ages x and y, where the associated future lifetimes are continuous and assumed to be independent. Stochastic status model conditions may be defined in terms of contingent probabilities. For example, using the form of the pdf defined by (5.26) for both future lifetime random variables the probability that (x) dies before (y) within n years is

$$_n q^1_{xy} = \int_0^n \int_t^\infty {}_s p_y \mu_{y+s} \, {}_t p_x \mu_{x+t} \, ds \, dt = \int_0^n {}_t p_{xy} \, \mu_{x+t} \, dt \tag{7.19}$$

Similarly, the probability that (y) dies after (x) but before n years is

$$_n q^2_{xy} = \int_0^n \int_0^t {}_s p_x \mu_{x+s} \, {}_t p_y \, \mu_{y+t} \, ds \, dt = \int_0^n {}_t q_x \, {}_t p_y \, \mu_{y+t} \, dt \tag{7.20}$$

In specific cases, in order to compute the integrals in (7.19) and (7.20), numerical integration methods may be utilized. Many other contingent probabilities may be required in specific contracts, and this section ends with an example of a contingent mortality computation.

Example 7.5

Consider the setting of the two people ages x and y discussed in Example 5.5 where the force of mortality is the constant $\mu > 0$. In life insurance covering both people, where the individuals have different earning capabilities, the order of death may become important in that different benefits may be desired based on the order of individual decrement. The probability that (x) dies before (y) but within n years is

$$_n q^1_{xy} = \int_0^n \exp(-t\mu) \exp(-t\mu) \mu \, dt = \frac{1 - \exp(-2n\mu)}{2} \tag{7.21}$$

Problems concerning multiple lifetime models are outlined in Problems 7.1, 7.2, and 7.3.

7.2 Multiple-Decrement Models

In many stochastic status models both the time of an individual decrement and the particular cause of decrement are relevant. This is the case in pension and retirement plans where the benefits vary for termination, retirement, disability, and death. In these stochastic status models, survival and decrement probabilities are functions of two random variables. The first is the future active lifetime random variable and may be either discrete, denoted by curtate future lifetime K, or continuous, denoted by the future lifetime random variable T. The second is an indexing variable that designates the type of realized decrement. For m possible modes of decrement we utilize indicator random variable J with support $S_j = \{1, 2, \ldots, m\}$, where $J = j$ implies decrement by mode j.

In engineering and theoretical statistical modeling these models are commonly referred to as competing risk models. A history of multiple-decrement theory is given in Seal (1977). In the traditional setting the decrements are assumed to be independent (see Elandt-Johnson and Johnson, 1980; Cox and Oakes, 1990). In more recent work, the dependent decrement model such as the common shock model (see Marshall and Olkin, 1967) and techniques employing the copula function, as described by Genest and McKay (1986), have been proposed. The impact of dependent structures among the sources of decrement on actuarial calculations has been investigated by Gollier (1996).

Example 7.6

A person age x enters an employment program where decrement in the stochastic status model indicates exit from the program. There may be many causes, referred to as status decrements, for the person to leave this program. In a simplistic setting, there may be $m = 3$ different types of retirement. The indicator variable J is defined by $J = 1$ implies retirement at the normal retirement age, $J = 2$ implies retirement due to a disability, and $J = 3$ implies retirement before the normal retirement age. Different probabilities of survival and decrement due to the different modes need to be modeled since the future benefits paid may vary.

The theory and techniques associated with joint distribution of the future lifetime and the cause of decrement follow those introduced in Section 1.8. The distribution of the future lifetime may be either discrete or continuous, and a development of these models is given in the next section.

7.2.1 Continuous Multiple Decrements

For a stochastic status model age x with multiple causes of death, let the future lifetime random variable T be continuous and let decrement variable J

be discrete with support $S_j = \{1, 2, ..., m\}$. The joint mixed probability function (mpf) of T and J is of the mixed variety (see Section 1.2.3) and is denoted by $f(t, j)$. Statistical and mortality calculations fixed at $J = j$ are done by applying standard statistical concepts and formulas. For example, the probability of decrement due to cause $J = j$ on or before time $T = t$ is given by

$$_tq_x^{(j)} = P(T \leq t, J = j) = \int_0^t f(s, j)\, ds \qquad (7.22)$$

Other mortality and survival probabilities can be computed using the formulas presented earlier in Chapter 5.

The standard concepts and techniques associated with joint distributions as presented in Section 1.8 apply in the multiple-decrement setting. For a stochastic status age x the cdfs corresponding to T and J follow (6) in Section 1.8 and take the form

$$g(t) = \sum_{j=1}^{m} f(t, j) \quad \text{and} \quad h(j) = \int_0^{\infty} f(t, j)\, dt \qquad (7.23)$$

for $t > 0$ and $j \, \varepsilon \, S_j$. The probability that the future active lifetime takes values between a and b utilizes information from all failure modes and is computed as

$$P(a \leq T \leq b) = \sum_{j=1}^{m} \int_a^b f(t, j)\, dt \qquad (7.24)$$

Applying (7.23) and (7.24), it follows that the cdf of T is

$$G(t) = \int_0^t g(s)\, ds \qquad (7.25)$$

The cumulative probability (7.25) is general in that it accounts for all modes of decrement. The situations where some modes of decrement are eliminated is considered in Section 7.2.4.

In the multiple-decrement model the situation where all modes of decrement are active is important and requires special notation. For a stochastic status age x the probability of failure within future time $t > 0$ with all modes of decrement active, given by (7.25), is denoted

$$_tq_x^{(\tau)} = G(t) \qquad (7.26)$$

Similarly, the survival probability corresponding to t additional years is

$$_tp_x^{(\tau)} = P(T > t) = 1 - {}_tq_x^{(\tau)} \tag{7.27}$$

Also, utilizing the marginal distributions in (7.23), the overall decrement probability associated with t additional years can be written as

$$_tq_x^{(\tau)} = \sum_{j=1}^{m} {}_tq_x^{(j)} \tag{7.28}$$

We observe that the decrement probabilities due to separate modes of decrement sum to yield the total decrement probability. This concept is utilized in multiple mortality table constructions and applications as presented in Section 7.2.7. The concept of the instantaneous mortality rate, or force of mortality, in the multiple-decrement setting is now explored.

7.2.2 Forces of Decrement

In this section the concept of force of mortality is extended to the multiple-decrement setting. The causes of decrement in an employment system include, for example, termination, death, and retirement. Instantaneous failure rate structure (5.16) is modified by the condition of survival where all the forces of decrement are active. The force of decrement at age $x + t$ is defined by

$$\mu_{x+t}^{(\tau)} = \frac{f(t)}{1 - F(t)} = \left(\frac{1}{{}_tp_x^{(\tau)}} \right) \left(\frac{d}{dt} \right) {}_tq_x^{(\tau)} \tag{7.29}$$

where the pdf and cdf of T are $f(t)$ and $F(t)$, respectively. Following a similar rationale, if a single-decrement $J = j$ is considered, then the force of decrement becomes

$$\mu_{x+t}^{(j)} = \frac{f(t,j)}{1 - F(t)} = \left(\frac{1}{{}_tp_x^{(\tau)}} \right) \left(\frac{d}{dt} \right) {}_tq_x^{(j)} \tag{7.30}$$

Combining (7.29) and (7.30) we note the summation

$$\mu_{x+t}^{(\tau)} = \sum_{j=1}^{m} \mu_{x+t}^{(j)} \tag{7.31}$$

so that the separate forces of decrement sum to the overall force of decrement for the system. Condition (7.31) is useful in integrating combinations of decrement factors that act on a general status, such as an individual's lifetime.

As in the single-decrement setting, the joint distribution of T and J and the conditional distributions can be expressed using the force of decrement functions. The joint and marginal distributions are given by

$$f(t,j) = {}_tp_x^{(\tau)}\mu_{x+t}^{(j)}, \quad h(j) = {}_\infty q_x^{(j)} \quad \text{and} \quad f(t) = {}_tp_x^{(\tau)}\mu_{x+t}^{(\tau)} \tag{7.32}$$

The formulas in (7.32) define statistical distributions so that statistical concepts and techniques are applicable. For a stochastic status age x the decrement probability within the next t years due to the jth mode of decrement is

$$_tq_x^{(j)} = \int_0^t {}_sp_x^{(\tau)}\mu_{x+s}^{(j)}\, ds \tag{7.33}$$

Also of interest, the conditional pdf of J, given $T = t$, becomes the ratio

$$h(j|t) = \frac{f(t,j)}{f(t)} = \frac{\mu_{x+t}^{(j)}}{\mu_{x+t}^{(\tau)}} \tag{7.34}$$

The example that follows demonstrates these concepts and formulas and is similar to the two-decrement example given in Bowers et al. (1997, p. 312).

Example 7.7

For (x) there are three active modes of decrement indicated by forces of decrement at age $x + t$ defined as

$$\mu_{x+t}^{(1)} = \frac{t}{100}, \quad \mu_{x+t}^{(2)} = \frac{1}{100} \quad \text{and} \quad \mu_{x+t}^{(3)} = \frac{2}{100} \tag{7.35}$$

for $t \geq 0$. Applying (7.31), the total force of mortality computes as the sum of (7.35):

$$\mu_{x+t}^{(\tau)} = \frac{t+3}{100} \tag{7.36}$$

Modifying the formula (5.18), the overall survival probability for t additional years can be written in terms of the overall force of decrement (7.36):

$$_tp_x^{(\tau)} = \exp\left(-\int_0^t ((s+3)/100)\, ds\right) = \exp(-(t^2+6t)/200) \tag{7.37}$$

The joint mpf in (7.32) takes the multirule form

$$
f(t, j) = \begin{cases} \dfrac{t}{100} \exp(-(t^2 + 6t)/200) & \text{for} \quad j = 1 \\[2mm] \dfrac{1}{100} \exp(-(t^2 + 6t)/200) & \text{for} \quad j = 2 \\[2mm] \dfrac{2}{100} \exp(-(t^2 + 6t)/200) & \text{for} \quad j = 3 \end{cases} \tag{7.38}
$$

for $t > 0$. The marginal distributions are found using (7.32). First, for the future lifetime random variable T the pdf follows formula

$$
g(t) = \sum_{j=1}^{3} f(t, j) = \frac{3+t}{100} \exp(-(t^2 + 6t)/200) \tag{7.39}
$$

To find the marginal pmf corresponding to $J = j$, we consider each mode separately. Starting with $J = 2$, we have the marginal written as the integral

$$
h(2) = \int_0^\infty \frac{1}{100} \exp(-(t^2 + 6t) / 200) \, dt \tag{7.40}
$$

To simplify (7.40), we first complete the square in the exponent and then write the integral in terms of the normal random variable with mean -3 and standard deviation of 10. We then transform to a standard normal random variable. Letting $\Phi(z)$ denote the standard normal cdf, (7.40) reduces to

$$
h(2) = \frac{1}{10} \exp(9/200)(2\pi)^{1/2}(1 - \Phi(.3)) = .100 \tag{7.41}
$$

For the other modes of decrement the marginal pdfs are related by $h(3) = 2\,h(2) = .200$. Lastly, since probabilities sum to 1, $h(1) = 1 - .1 - .2 = .7$. The conditional pdf of J given future lifetime $T = t$ following (7.34) yields

$$
h(1|t) = \frac{t}{t+3} \quad \text{and} \quad h(2|t) = \frac{1}{t+3} \tag{7.42}
$$

A similar problem dealing with the construction of multiple-decrement distributions is given in Problem 7.4.

7.2.3 Discrete Multiple Decrements

In some cases a multiple-decrement stochastic status model may be a function of an integer-valued future lifetime. For a stochastic status age x let K be the curtate future lifetime associated with a decrement defined by type $J = j$. The resulting fully discrete joint pdf of K and J can be written as an integral in terms of the joint pdf in (7.32) as

$$P(K = k, J = j) = P(k \le T < k+1, J = j) = \int_k^{k+1} {}_t p_x^{(\tau)} \mu_{x+t}^{(j)} \, dt \qquad (7.43)$$

As before, the decomposition $T = K + S$ holds and the overall survival and decrement probabilities associated with t additional years, respectively, are

$$_t p_x^{(\tau)} = {}_k p_x^{(\tau)} {}_s p_{x+k}^{(\tau)} \quad and \quad q_{x+k}^{(j)} = \int_0^1 {}_s p_{x+k}^{(\tau)} \mu_{x+k+s}^{(j)} \, ds$$

Utilizing these basic forms for nonnegative integers k and j, (7.43) reduces to

$$P(K = k, J = j) = {}_k p_x^{(\tau)} q_{x+k}^{(j)} \qquad (7.44)$$

Joint pmf (7.44) can be used to compute other basic decrement probabilities. For example, the total probability of decrement, due to all causes or modes, for a person age $x + k$ within 1 year is

$$q_{x+k}^{(\tau)} = \int_0^1 {}_s p_{x+k}^{(\tau)} \mu_{x+k+s}^{(\tau)} \, ds = \int_0^1 {}_s p_{x+k}^{(\tau)} \sum_{j=1}^m \mu_{x+k+s}^{(j)} \, ds \qquad (7.45)$$

After simplification formula (7.45) reduces to

$$q_{x+k}^{(\tau)} = \sum_{j=1}^m q_{x+k}^{(j)} \qquad (7.46)$$

This formula is consistent with the additive of separate forces of decrement given in (7.28).

7.2.4 Single-Decrement Probabilities

For multiple-decrement stochastic status models, the different modes of decrement apply varying decrement stresses that can be modeled through the individual forces of decrement. In the study of the relative magnitudes of the

stresses associated with the prospective modes of decrement the hypothetical elimination of decrement modes, leading to marginal decrement structures, is sometimes useful. In the single- or absolute-decrement model, all
modes of decrement are eliminated except for the mode under consideration.
In general, the resulting general decrement mortality probabilities are not
identifiable through a unique probability distribution, as many probability
structures can produce an identical probability system. For general reference see Basu and Ghosh (1980) and Langberg et al. (1978). In this section the
upper and lower bounds on single-decrement rates as given by Borowiak
(1998) are presented. Further, under certain situations, as noted by many
sources (for example, Jordan, 1967), single- or absolute-decrement probabilities can be directly derived.

General survival probabilities are related to the force of interest by (5.18).
For a stochastic status model age x, the single or absolute survival probability
to age $x + t$, for decrement mode $J = j$, is defined by integrating the individual
force of decrement for $J = j$, denoted as

$$_t p_x^{s(j)} = \exp\left(-\int_0^t \mu_{x+s}^{(j)}\, ds\right) \tag{7.47}$$

The single-decrement rate before future time t is

$$_t q_x^{s(j)} = 1 - {}_t p_x^{s(j)} \tag{7.48}$$

These single survival and decrement probabilities are useful in the planning and modeling of future financial and actuarial systems where present
modes of decrement may be reduced or eliminated at a future date.

In general, we cannot directly observe single failure rates when all forces
of decrement are active. Single or absolute failure rates are not identifiable in
that they are not associated with a unique probability distribution. A lower
bound on these rates is easily attained by first noting for $t > 0$,

$$_t p_x^{(\tau)} = \exp\left(-\int_0^t \left(\sum_{j=1}^m \mu_{x+s}^{(j)}\right) ds\right) = \prod_{j=1}^m {}_t p_x^{s(j)} \tag{7.49}$$

It follows for all $j = 1, 2, \ldots, m$ that a natural bound is ${}_t p_x^{s(j)} \mu_{x+t}^{(j)} \geq {}_t p_x^{(\tau)} \mu_{x+t}^{(j)}$.
Realizing

$$_t p_x^{s(j)} \mu_{x+t}^{(j)} \geq {}_t p_x^{(\tau)} \mu_{x+t}^{(j)}$$

for $t > 0$ and utilizing the form of the joint pdf in (7.32), a lower bound on the single-decrement rate for $J = j$ associated with one additional year is

$$q_x^{s(j)} \geq q_x^{(j)} \tag{7.50}$$

for $j = 1, \ldots, m$. To explore further the relationship between single-decrement rates and the overall probabilities, distributional assumptions, such as the ones explored in the next section, are required.

7.2.5 Uniformly Distributed Single-Decrement Rates

Under certain assumptions, single failure rates defined by (7.47) and (7.48) can be directly computed. In this section, we assume each decrement has a uniform distribution of death (UDD) within each year. For $J = j$ and $0 \leq s \leq 1$, applying UDD (5.35) yields the fractional decrement rate

$$_s q_x^{(j)} = s q_x^{(j)} \tag{7.51}$$

Using definition (7.30), the force of decrement for decrement mode $J = j$ at age $x + s$ is

$$\mu_{x+s}^{(j)} = \frac{q_x^{(j)}}{1 - s \, q_x^{(\tau)}} \tag{7.52}$$

and the single-decrement rate for one additional year is computed as

$$q_x^{s(j)} = 1 - \exp\left(-\int_0^1 \frac{q_x^{(j)}}{1 - s \, q_x^{(\tau)}} \, ds \right) \tag{7.53}$$

The integral in (7.53) can be written as

$$\frac{q_x^{(j)}}{q_x^{(\tau)}} \int_0^1 \frac{q_x^{(\tau)}}{1 - s \, q_x^{(\tau)}} \, ds = -\frac{q_x^{(j)}}{q_x^{(\tau)}} \ln(1 - q_x^{(\tau)}).$$

Simplification yields single or absolute 1-year decrement and survival rates for $J = j$ and $x > 0$:

$$q_x^{s(j)} = 1 - (1 - q_x^{(\tau)})^{q_x^{(j)}/q_x^{(\tau)}} \quad \text{and} \quad p_x^{s(j)} = (1 - q_x^{(\tau)})^{q_x^{(j)}/q_x^{(\tau)}} \tag{7.54}$$

Formulas (7.54) define explicit relationships between single- and multiple-decrement quantities that are not possible without distributional assumptions. A computational example is given in the next section in Example 7.9, and an Excel computation of (7.54) is discussed in Problem 7.9.

Reversing the roles is possible where single-decrement rates are given and the multiple-decrement rates are then to be determined. It may be useful to combine single-decrement rates under assumption (7.51) to form overall decrement rates. If we invert (7.54) using the single-decrement rates as inputs we have

$$q_x^{(j)} = \left[\frac{\ln(p_x^{s(j)})}{\ln(p_x^{(\tau)})} \right] q_x^{(\tau)} \tag{7.55}$$

Formula (7.55) defines the multiple-decrement probabilities in terms of the single or absolute probabilities.

In practice, to apply formula (7.55), we switch the uniform distribution of deaths within interval assumptions from the multiple-decrement rates to the single mortality rates. For each $J = j$ and $0 \le t < 1$ we assume

$$_t q_x^{s(j)} = t \; q_x^{s(j)} \tag{7.56}$$

So that, for a fixed mode $J = j$,

$$q_x^{(j)} = \int_0^1 \prod_{i=1}^m {}_t p_x^{s(i)} \mu_{x+t}^{(j)} \, dt \tag{7.57}$$

From (7.56), $\frac{d}{dt} \left(-_t p_x^{s(j)} \right) = {}_t p_x^{s(j)} \mu_{x+t}^{(j)}$, and we can write (7.57) as

$$q_x^{(j)} = q_x^{s(j)} \int_0^1 \prod_{i \ne j} (1 - t \; q_x^{s(i)}) \, dt \tag{7.58}$$

This formula relates the single-decrement rates and multiple-decrement probabilities and is demonstrated in Example 7.8.

Example 7.8

Let $m = 2$ modes of decrement exist and assume a uniform distribution of deaths within years, so that (7.56) holds. For $J = j$, the rate of mortality for one-half additional year, given by (7.58), can be written as

$$q_x^{(j)} = q_x^{s(j)} \left(1 - .5 q_x^{s(i)} \right)$$

for i different from j. We note that the technique used to derive (7.58) can be extended to more than two modes of decrement.

7.2.6 Single-Decrement Probability Bounds

In general, without distributional assumptions, such as the UDD assumption, single-decrement rates cannot be computed exactly. In this section, bounds on the single-decrement rates, introduced by Borowiak (1998), are presented where there are m modes of decrement and no distributional assumptions on the survival or decrement rates. From (7.59) for mode $J = j$ we have the crude bound on 1-year rates of decrement:

$$q_x^{(j)} \leq q_x^{s(j)} \leq q_x^{(\tau)} \tag{7.59}$$

In (7.59) the lower bound is mathematically strict in that equality may hold in some theoretical settings. An alternative upper bound on the single-decrement rates is now derived.

Based on the definition of the force of decrement given in (7.30) for decrement $J = j$ evaluated at age $x + t$,

$$\mu_{x+t}^{(j)} = \left[-\frac{d}{dt} \ln({}_t p_x^{(\tau)}) \right] \frac{\mu_{x+t}^{(j)}}{\mu_{x+t}^{(\tau)}} \tag{7.60}$$

Using a Taylor series expansion and taking the derivative with respect to t we have

$$-\frac{d}{dt} \ln(1 - {}_t q_x^{(\tau)}) = \sum_{r=1}^{\infty} ({}_t q_x^{(\tau)})^{r-1} {}_t p_x^{(\tau)} \mu_{x+t}^{(\tau)} \tag{7.61}$$

Combining (7.60) and (7.61) produces the form of the single survival probability for one additional year;

$$p_x^{s(j)} = \exp\left(-\sum_{r=1}^{\infty} \int_0^1 ({}_t q_x^{(\tau)})^{r-1} {}_t p_x^{(\tau)} \mu_{x+t}^{(j)} \, dt \right) \tag{7.62}$$

If ${}_t q_x^{(\tau)} < 1$, then for some point in $(0, 1)$ the integral in (7.62) is bounded above by $(q_x^{(\tau)})^{r-1} q_x^{(j)}$. Hence, the single survival and decrement rate bounds corresponding to one additional year are given by

$$p_x^{s(j)} \geq \exp(-q_x^{(j)}/(1 - q_x^{(\tau)})) \quad \text{and} \quad q_x^{s(j)} \leq 1 - \exp(-q_x^{(j)}/(1 - q_x^{(\tau)})) \tag{7.63}$$

In practice, when the upper bound for the mortality rate in (7.63) is applied, it is assumed that it is less than the default upper bound $q_x^{(\tau)}$. As expected, the single-decrement rates under the uniform distribution of decrements within year assumption are within the lower and upper bounds. Further, in most cases it is observed that the range for the single-decrement rates defined by combining (7.59) and (7.63) is small. These concepts and formulas are demonstrated in the following example and investigated in terms of Excel work in Problem 7.10.

Example 7.9

A multiple-decrement stochastic status model containing $m = 3$ modes of decrement is considered where the 1-year failure rates are listed in Table 7.1. Both the single-decrement calculations assuming UDD given by (7.54) and the general upper bound, ub$^{(i)}$, given by (7.63) are computed with the results given in Table 7.2. First, we remark from the calculations in Table 7.2 that the bounds, namely, (7.59) and (7.63), hold for all years. Second, there is observed a close agreement between the UDD estimate and the upper bound given by (7.63). In fact, the narrowness of the upper and lower probability bounds indicates the efficiency of the UDD approximation for single-decrement rates.

TABLE 7.1

Multiple-Decrement Probabilities

Age	$q_x^{(1)}$	$q_x^{(2)}$	$q_x^{(3)}$	$q_x^{(\tau)}$
30	.001	.002	.002	.005
31	.002	.003	.004	.009
32	.002	.004	.004	.010
33	.003	.005	.006	.014

TABLE 7.2

UDD Multiple-Decrement Probabilities and Upper Bounds

Age	$q_x^{s(1)}$	ub$^{(1)}$	$q_x^{s(2)}$	ub$^{(2)}$	$q_x^{s(3)}$	ub$^{(3)}$
30	.001002	.001005	.002003	.002008	.002003	.002008
31	.002007	.002016	.003009	.003023	.004010	.004028
32	.002008	.002018	.004012	.004032	.004012	.004032
33	.003017	.003038	.005023	.005058	.006024	.006067

7.2.7 Multiple-Decrement Life Tables

Another area of application of group survivorship theory and notations involves the construction of life tables in the multiple-decrement setting of Section 7.2. In a multiple-decrement survivorship group, individuals leave the survivorship group, referred to as decrement, due to m causes or modes. In the continuous lifetime setting, the associated random vector is (T, J), where T is the future lifetime random variable and $J = j$ signifies the mode of decrement for $j = 1, 2, \ldots, m$. For a stochastic status age x, the force of decrement due to mode $J = j$ is $\mu_x^{(j)}$, the overall force of decrement is $\mu_x^{(\tau)}$, the joint and marginal pdfs are given in (7.32), and the overall survival probability associated with t additional years is denoted $_tp_x^{(\tau)}$. In the balance of this section, group survivorship notations are defined for the multiple-decrement setting leading to the construction of multiple-decrement tables.

As introduced in Section 5.6, the number of survivors to age x is denoted by the random variable $\mathcal{L}_x$. In multiple-decrement survivorship groups the expected number of survivors is a function of the overall force of mortality given by

$$l_x^{(\tau)} = E\left\{\mathcal{L}_x^{(\tau)}\right\} = l_0 \, _xp_0^{(\tau)} = l_0 \exp\left(-\int_0^x \mu_s^{(\tau)} ds\right) \tag{7.64}$$

The number of individuals leaving the group due to decrement $J = j$ between ages x and $x + n$ is denoted $_n\mathcal{D}_x^{(j)}$, where the associated expectation is

$$_nd_x^{(j)} = E\left\{_n\mathcal{D}_x^{(j)}\right\} = l_0 \int_x^{x+n} {}_tp_0^{(\tau)} \mu_t^{(j)} dt \tag{7.65}$$

In (7.65), taking $n = 1$ and assuming UDD yields the expected number of individuals leaving the group within 1 year due to mode $J = j$, and this can be written as

$$d_x^{(j)} = l_x^{(\tau)} q_x^{(j)} \tag{7.66}$$

From (7.66) we observe the expected number of decrements due to mode j in the year x is the product of the number of individuals alive at the start of the year times the jth yearly decrement rate. Further, the total number of decrements, due to all causes, between ages x and $x + n$ and corresponding expectations are

$$_n\mathcal{D}_x^{(\tau)} = \sum_{j-1}^{m} {}_n\mathcal{D}_x^{(j)} \quad \text{and} \quad _nd_x^{(\tau)} = E\left\{_n\mathcal{D}_x^{(\tau)}\right\} = \sum_{j-1}^{m} {}_nd_x^{(j)} \tag{7.67}$$

Simplifying (7.67), the expected total number of decrements between ages x and $x + 1$ is

$$d_x^{(\tau)} = l_x^{(\tau)} - l_{x+1}^{(\tau)} = l_x^{(\tau)} q_x^{(\tau)} \tag{7.68}$$

The construction and application of multiple-decrement life tables based on these concepts and formulas are explored in the next section.

Following the approach of Section 5.7, two types of multiple-decrement life tables can be constructed. In the first type, the multiple-decrement life table lists the number of survivors, l_x^{τ}, along with the number of decrements for each mode, $d_x^{(j)}$, for $j = 1, 2, \ldots, m$, for a set of ages. In the second type, specific and overall decrement probabilities defined as

$$q_x^{(j)} = \frac{d_x^{(j)}}{l_x^{(\tau)}} \quad \text{for} \quad j = 1, \ldots, m \quad \text{and} \quad q_x^{(\tau)} = \frac{d_x^{(\tau)}}{l_x^{(\tau)}} \tag{7.69}$$

are listed for various ages x. As before, the overall survival probability combines failure numbers of the various modes and is computed by

$$p_x^{(\tau)} = 1 - q_x^{(\tau)} \tag{7.70}$$

The construction and application of multiple-decrement life tables is demonstrated in Example 7.10. Problem 7.6 explores the use of Excel in the multiple-decrement life table analysis.

Example 7.10

In a survivorship group at age 50 there are of $l_{50}^{\tau} = 1,000$ survivors where two modes of death exist. A decrement life table listing the number of survivors along with decrement frequencies for 10 subsequent years is given in Table 7.3.

TABLE 7.3

Multiple-Decrement Life Table—Survivor and Decrement Expectations

x	l_x^{τ}	$d_x^{(1)}$	$d_x^{(2)}$	x	l_x^{τ}	$d_x^{(1)}$	$d_x^{(2)}$
50	1,000	10	15	55	859	15	20
51	975	11	16	56	824	16	21
52	948	12	16	57	787	16	23
53	920	13	17	58	748	18	25
54	890	13	18	59	705	20	27

Applying quantities in Table 7.3, basic probabilities can be computed. For example:

1. The probability an individual age 50 lives to age 55 is computed as

$$_5p_{50}^{(\tau)} = \frac{l_{55}^{(\tau)}}{l_{50}^{(\tau)}} = \frac{859}{1000} = .859$$

2. The probability an individual age 50 dies at age 54 by mode $J = 2$ is

$$_{4|1}q_{50}^{(2)} = {}_4p_{50}^{(\tau)} q_{54}^{(2)} = \frac{d_{54}^{(2)}}{l_{50}^{(\tau)}} = \frac{18}{1000} = .018$$

3. The probability an individual age 50 dies within 2 years by mode $J = 1$ is

$$_2q_{50}^{(1)} = \frac{d_{50}^{(1)} + d_{51}^{(1)}}{l_{50}^{(\tau)}} = .0210$$

The alternate form of the multiple-decrement life table listing the individual decrement-specific rates following (7.69) is presented in Table 7.4. Based on Table 7.4, basic decrement and survival probabilities can be computed. For example:

1. The probability an individual age 55 lives more than 2 years is computed by using the yearly decomposition

$$_2p_{55}^{(\tau)} = p_{55}^{(\tau)} p_{56}^{(\tau)} = .9592(.9551) = .9161$$

2. The probability an individual age 52 survives 2 years and dies in the third year by mode $J = 1$ is

$$_{2|}q_{52}^{(1)} = p_{52}^{(\tau)} p_{53}^{(\tau)} q_{54}^{(1)} = .0137$$

TABLE 7.4

Multiple-Decrement Life Table—Decrement Rates by Year

x	$q_x^{(1)}$	$q_x^{(2)}$	$p_x^{(\tau)}$	x	$q_x^{(1)}$	$q_x^{(2)}$	$p_x^{(\tau)}$
50	.0100	.0150	.9750	55	.0175	.0233	.9592
51	.0113	.0164	.9723	56	.0194	.0255	.9551
52	.0127	.0169	.9704	57	.0203	.0292	.9505
53	.0141	.0185	.9674	58	.0241	.0334	.9425
54	.0146	.0202	.9652	59	.0284	.0383	.9333

3. The probability an individual age 52 dies within 2 years by mode $J = 2$ is

$$_2 q_{52}^{(2)} = q_{52}^{(2)} + p_{52}^{(\tau)} \, q_{53}^{(2)}$$

In practice, Tables 7.3 and 7.4 might be merged into a larger multiple-decrement life table. Further, the overall decrement rate given in (7.69) could be listed in a multiple-decrement life table.

7.2.8 Single-Decrement Life Tables

In application, an actuarial modeler or insurance underwriter might want to consider the changes in mortality patterns under the hypothetical elimination of some modes of decrement. This section concludes by demonstrating the construction of life tables for single-decrement statuses as introduced in Section 7.2.4. In particular, utilizing the approaches and formulas of Sections 7.2.5 and 7.2.6, where the single rates are either estimated or bounded, life tables are constructed using the techniques explored in this section.

Assuming a uniform distribution of deaths within each year, or UDD, formula (7.54) yields exact mortality rates based on the decrement rates computed using (7.69). If no such assumptions are made concerning the mortality structure within years, the construction of lower and upper bounds on single rates can be utilized. In the following example these approaches and formulas are utilized to create a life table for single rates.

Example 7.11

For (x) the multiple-decrement life table given in Table 7.4 applies. A life table listing single mortality rates (7.54) along with the upper and lower bounds for the single rates using (7.59) and (7.63) is given in Table 7.5. The difference between the lower and upper bounds is small and, as

TABLE 7.5

Life Table for Single-Decrement Rates and Bounds

Year	Single-Decrement Rates		Upper Bounds	Lower Bounds
x	$q_x^{s(1)}$	$q_x^{s(2)}$	$J = 1$	$J = 2$
50	.0101	.0151	.0100–.0102	.0150–.0153
51	.0114	.0156	.0113–.0116	.0164–.0167
52	.0128	.0170	.0127–.0130	.0169–.0173
53	.0142	.0186	.0141–.0145	.0185–.0189
54	.0147	.0204	.0146–.0150	.0202–.0207
55	.0177	.0235	.0175–.0181	.0233–.0240

expected, captures the estimated rates. This implies the uniform assumption decrement rate bounds are a good estimate of the single-decrement rates as not much error is possible.

7.2.9 Multiple-Decrement Computations

The conditions in a life insurance or life annuity status model may include a multiple-decrement mortality structure as introduced in Section 7.2. In this setting, the stochastic structure includes the future lifetime along with the decrement cause or mode. For a continuous stochastic status model, the future lifetime random variable is T and the mode of failure is indicated by $J = j$ for $j = 1, 2, ..., m$. Examples of these models include insurance policies paying a variable benefit dependent on the cause of death and a pension system where future benefits depend on the form of decrement. Also applicable are retirement systems where the amount of monthly payments of a retirement annuity depends on the conditions of a retirement (explored in Section 7.3).

Multiple-decrement modeling concepts and statistical formulas presented in Sections 7.2 and 7.2.7 are applied to stochastic status counterparts. In the continuous model setting for (x) the random vector (T, J) takes joint mpf $f(t, j) = {}_t p_x^{(\tau)} \mu_{x+t}^{(j)}$ for $t \geq 0$ and $j = 1, 2, ..., m$. In complex structures, such as pension plans or insurance policies with multilevel benefits, the present value function takes the form of a general benefit function. The benefit function may depend on both the decrement time T and mode $J = j$ and is denoted by $b_t^{(j)}$. For (x) the present value expectation or net single premium is

$$\text{NSP} = \sum_{j=1}^{m} \int_{S_{PV}} b_t^{(j)} \, {}_t p_x^{(\tau)} \, \mu_{x+t}^{(j)} \, dt \tag{7.71}$$

where the support, based on the conditions of the stochastic status, is denoted S_{PV}. In the discrete-status setting, the curtate future lifetime is K and indicator variable J takes joint mpf $f(k, j) = {}_k p_x^{(\tau)} \, q_{x+k}^{(j)}$ for $k = 0, 1, ...$ and $j = 1, 2, ..., m$. The expectation associated with general benefit paid at the end of the year $b_{k+1}^{(j)}$ has a present value given by

$$\text{NSP} = \sum_{j=1}^{m} \sum_{S_{PV}} b_{k+1}^{(j)} \, {}_k p_x^{(\tau)} \, q_{x+k}^{(j)} \tag{7.72}$$

where, as in the continuous case, the support of the present value is S_{PV}. This technique is demonstrated in Example 7.12 dealing with the case of life insurance where the benefits are defined differently corresponding to particular decrement causes.

Example 7.12

A discrete whole life insurance policy contains a double-indemnity provision in the case of an accident. There are two modes of decrement where $J = 1$ implies death due to accident and $J = 2$ represents all other causes. The policy pays double if $J = 1$. From (7.71) the net single premium (NSP) for this insurance can be written as

$$\text{NSP} = 2\,A_x^{(1)} + A_x^{(2)}$$

where for $j = 1$ and 2

$$A_x^{(j)} = \sum_{k=0}^{\infty} v^{k+1}\,{}_k p_x^{(\tau)}\,q_{x+k}^{(j)}$$

The NSP computation is completed using either appropriate statistical or life table models.

One of the major applications of multiple-decrement models is in constructing pension models. This topic is broad in scope with many subtopics, and a brief introduction along with statistical model development is presented in the next section.

7.3 Pension Plans

An active actuarial science area for both research and applications is that of pension plan analysis and construction. Individuals in a retirement or pension system may be viewed as a survivorship group. For the individual, the stochastic status changes when the individual exits the group and a decrement is realized. A pension plan is a financial contract where the main pension benefit, in the form of a deferred life retirement annuity, is financed by pension contributions taken from current and future salary payments and employer contributions. The retirement benefit is commonly a function of different factors such as age, salary, and length of service. In the following, the individual enters the pension plan at age x at a source of employment where the earliest and latest retirement ages are denoted by α and β, respectively. There may be other benefits related to the pension plan, such as a death benefit paid to a beneficiary or a life annuity paid in the event of a disability. For this reason, a multiple-decrement model works well for pension benefit analysis where the decrements are associated with the causes for which an individual may leave the employment system. In this section, the multiple-decrement concepts and notations introduced in Section 7.2 are assumed to hold.

Pension benefits are financed through a series of payments referred to as contributions that may be a function of various variables, such as age, salary, inflation, and length of service. The loss model approach as presented in Section 4.1 applies to pension modeling where expenditures consist of the pension benefits and revenues correspond to the contributions. The *RC* and *EP* approach of Section 4.2.1 is applied to these stochastic loss models to equate the two balancing factors, contributions and benefits. In this introduction to pension plans the notations are taken from Bowers et al. (1997, Section 11.5).

7.3.1 Multiple-Decrement Benefits

In general, in a pension plan there are many different types of benefits possible, and an analysis requires an appropriate multiple-decrement model. Commonly, differing levels of benefits are associated with the different types of employment decrement. The typical structure of multiple-decrement benefits associated with a pension plan is introduced in the example form where the types of decrements are listed in Bowers et al. (1997, p. 350). In practice, other benefits may be added to the pension plan. In the pension status the lifetime variable example is the continuous future lifetime random variable of Section 5.1.

Example 7.13

Group survivorship theory is applied to a pension system consisting of four possible modes of decrement: withdrawal, death, disability, and age-based retirement. A demonstrational pension is discussed where in practice the individual structures may be more complex. In Table 7.6, the type of decrement and associated yearly probability of decrement and force of decrement notations are listed.

A possible actuarial pension table lists the multiple-decrement expectations of Section 7.2.7 along with associated whole life annuity expectations by age. For an individual age 55 such an actuarial pension table is given in Table 7.7 consisting of the expected number of decrements due to withdrawal, death, disability, and retirement, denoted by $d_x^{(w)}$, $d_x^{(d)}$,

TABLE 7.6

Pension System Modes of Decrement

Types of Decrement	Notation	Yearly Decrement Probability	Force of Decrement
General withdrawal	w	$q_x^{(w)}$	$\mu_{x+t}^{(w)}$
Death	d	$q_x^{(d)}$	$\mu_{x+t}^{(d)}$
Disability retirement	i	$q_x^{(i)}$	$\mu_{x+t}^{(i)}$
Age retirement	r	$q_x^{(r)}$	$\mu_{x+t}^{(r)}$

TABLE 7.7

Pension Actuarial Experience Table

x	$l_x^{(\tau)}$	$d_x^{(w)}$	$d_x^{(d)}$	$d_x^{(i)}$	$d_x^{(r)}$	$\ddot{a}_x^{(r)}$
55	100	2	0	0	0	14
56	98	3	0	1	0	13.333
57	94	3	0	1	0	12.727
58	90	4	1	0	0	12.174
59	85	2	2	0	0	11.667
60	81	1	0	1	0	11.200
61	79	1	1	0	0	10.769
62	77	1	1	2	0	10.370

x	$l_x^{(\tau)}$	$d_x^{(w)}$	$d_x^{(d)}$	$d_x^{(i)}$	$d_x^{(r)}$	$\ddot{a}_x^{(r)}$
63	73	0	1	1	0	10.000
64	71	0	1	1	0	9.655
65	69	0	2	0	10	9.333
66	57	0	2	0	15	9.032
67	40	0	3	0	10	8.750
68	27	0	4	0	10	4.485
69	13	0	5	0	8	8.235
70	0	0	0	0	0	

$d_x^{(i)}$, and $d_x^{(r)}$, along with the unit benefit whole life annuity expectations. In this demonstrational plan age-based retirements occur for ages 65 to 69. This is an experience table, and in practice the expected decrement totals would be graduated or smoothed.

The benefits for withdrawal and death consist of one lump-sum payment denoted by $b_{k+1}^{(w)}$ and $b_{k+1}^{(d)}$ paid at the end of the decrement year with associated net single values

$$\text{NSV}^{(j)} = \sum_{k=0}^{14} b_{k+1}^{(j)} v^{k+1} \, {}_k p_{55}^{(\tau)} \, q_{55+k}^{(j)} = \sum_{k=0}^{14} b_{k+1}^{(j)} v^{k+1} \frac{d_{55+k}^{(j)}}{l_{55}} \tag{7.73}$$

where j is either w or d. For disability and age-based retirement the benefits take the form of a standard discrete whole life annuity with yearly payments of $\pi_k^{(i)}$ and $\pi_k^{(r)}$. These net single values take the form

$$\text{NSV}^{(j)} = \sum_{k=0}^{14} v^{k+1} \, {}_k p_{55}^{(\tau)} \, q_{55+k}^{(j)} \, \pi_{k+1}^{(j)} \ddot{a}_{55+k+1} = \sum_{k=0}^{14} v^{k+1} \frac{d_{55+k}^{(j)}}{l_{55}} \pi_{k+1}^{(j)} \ddot{a}_{55+k+1} \tag{7.74}$$

for j being i or r. Let the yearly effective interest rate be $i = .05$, and using the decrements in Table 7.7, compute (7.73) for withdrawal and death with level yearly payments of $b^{(w)}$ and $b^{(d)}$:

$$\text{NSV}^{(w)} = .14209 \, b^{(w)} \quad \text{and} \quad \text{NSV}^{(d)} = .13335 \, b^{(d)} \tag{7.75}$$

Similarly, for disability and age-based retirement, (7.74) with level yearly payments of $\pi^{(i)}$ and $\pi^{(r)}$ produces expectations

$$\text{NSV}^{(i)} = .57857 \, \pi^{(i)} \quad \text{and} \quad \text{NSV}^{(r)} = 2.509605 \, \pi^{(r)} \tag{7.76}$$

The net single value of the total benefit expectations in (7.75) and (7.76) is

$$\text{NSV-B} = .14209 \, b^{(w)} + .13335 \, b^{(d)} + .57857 \, \pi^{(i)} + 2.509605 \, \pi^{(r)} \tag{7.77}$$

These benefits are financed by yearly payments based on salaries, referred to as contributions. In typical pension plans the size of the benefit often depends on tenure and/or salary.

Example 7.14

A person enters the pension system at age x where the types of decrements are listed in Table 7.6. The benefits are analyzed at age $x + h$ for $h \geq 0$, so that the future lifetime random variable T starts at age $x + h$. Typically, pension withdrawal taking a lump benefit of $b^{(w)}$ must be done prior to the lowest retirement age α, and therefore is modeled by term insurance. The unit benefit expectation takes the form associated with continuous term insurance based on age $x + h$ and is

$$\text{NSP}^{(w)} = \int_0^{\alpha - x - h} b_{h+t}^{(w)} v^t \, {}_t p_{x+h}^{(\tau)} \, \mu_{x+h+t}^{(w)} \, dt \tag{7.78}$$

The death benefit is payable at the moment of death, and hence corresponds to continuous-status whole life insurance. The unit benefit expectation for death is the continuous whole life insurance computation

$$\text{NSP}^{(d)} = \int_0^{\infty} b_{h+t}^{(d)} v^t \, {}_t p_{x+h}^{(\tau)} \, \mu_{x+h+t}^{(d)} \, dt \tag{7.79}$$

Normally there is no death benefit after age β, and this form of benefit is for expository purposes. Disability benefits take the form of a life annuity where the issue age $x + h + t$ must be before some upper age u. Per unit of payment the disability expectation follows:

$$\bar{a}_{x+h+t}^{(i)} = \frac{1}{\delta} \left(1 - \bar{A}_{x+h+t}^{(i)} \right) \tag{7.80}$$

for $t \le u - x - h$. In a more common structure disability benefits may only last until retirement benefits begin.

The expected age-based retirement benefit takes the form of a deferred annuity expectation, and per unit of payment is

$$_{t|}\bar{a}_{x+h}^{(r)} = \frac{1}{\delta} \left(\bar{A}_{x+h:\overline{t}|}^{(r)} - \bar{A}_{x+h}^{(r)} \right) \tag{7.81}$$

where $\alpha \le x + h + t \le \beta$. The computation of the NSP for the two retirement annuities at age $x + h$ requires the integration of (7.80) and (7.81) using the proper multiple-decrement distribution. The deferral portion of (7.81) may be based on different mortality rates than the retirement annuity. The net single value (NSV) for retirement due to disability is

$$\text{NSV}^{(i)} = \int_0^{u - x - h} v^t \, {}_t p_{x+h}^{(\tau)} \, \mu_{x+h+t}^{(i)} \, \pi_{h+t}^{(i)} \bar{a}_{x+h+t}^{(i)} \, dt \tag{7.82}$$

and for age-based retirement is

$$\text{NSV}^{(r)} = \int_{\alpha - x - h}^{\beta - x - h} v^t \, {}_t p_{x+h}^{(\tau)} \, \mu_{x+h+t}^{(r)} \, \pi_{h+t}^{(r)} \bar{a}_{x+h+t}^{(r)} \, dt \tag{7.83}$$

The aggregate benefit expectation is a linear combination of (7.78), (7.79), (7.82), and (7.83) and is written in terms of the net single value of the benefit:

$$NSV\text{--}B_{x+h} = NSP^{(w)} + NSP^{(d)} + NSP^{(s)} + NSP^{(r)} \tag{7.84}$$

Computations such as (7.78)–(7.83) can be simplified using the relational expectation formulas and techniques presented earlier in this chapter.

The preceding benefit calculations often are approximated based on appropriate multiple-decrement mortality tables. The contributions needed to finance these benefits are discussed in the next section.

7.3.2 Pension Contributions

Pension benefits are financed by contribution payments connected to an individual's salary levels continuing up to retirement or maximum retirement age, denoted by β. Applying the *RC* and *EP* approach, the magnitude of the pension contributions determines the amounts of pension benefits. There are two basic plans for contributions, where the first is a flat contribution, valued at c per year, which we denote by $FC(c)$. The second method is a proportion, defined by c, of the individual's salary for designated years, and it is denoted $PS(c)$. In this section $FC(c)$ is investigated, while the discussion of $PS(c)$ is left to the next section.

Example 7.15

The pension plan discussed in Example 7.13 is financed by $FC(c)$ contributions. For an individual age 55, the pension actuarial experience table given in Table 7.7 holds where the effective annual interest rate is $i = .05$. Flat contributions are valued at c and are made at the start of each year the individual is a plan participant and form a life annuity. The expectation, following (6.28), is given by

$$NSV\text{--}C = c \sum_{k=0}^{14} v^k {}_k p_{55}^{(\tau)} = c \sum_{k=0}^{14} v^k \frac{l_{55+k}^{(\tau)}}{l_{55}^{(\tau)}} = 8.1743\,c \tag{7.85}$$

Applying the *EP*, the net single premium for the benefits equals the expected contributions. Equating (7.77) and (7.85) yields the relationship

$$8.1743\,c = .14209\,b^{(w)} + .13335\,b^{(d)} + .57857\,\pi^{(s)} + 2.509605\,\pi^{(r)} \tag{7.86}$$

Fixing the expected benefits determines the contribution rate. For example, to achieve benefits, in units of $1,000, of $b^{(w)} = 2$, $b^{(d)} = 50$, $\pi^{(s)} = 15$, and $\pi^{(r)} = 30$, the required flat contribution from (7.86) is $c = 11.12$, corresponding to $11,120 per year.

Example 7.16

In this example, flat contributions are related to the pension benefits for the pension plan discussed in Example 7.14. For an individual age $x + h$, $FC(c)$ contributions take the form of a continuous life annuity with payment c. The standard decomposition $T = K + S$ is utilized to estimate the total contributions, where the curtate future lifetime K is independent of the uniform random variable S. At age $x + h$ the net single value of the contribution is

$$\text{NSV-}C_{x+h} = c \sum_{k=0}^{\beta-x-h-1} v^k \, {}_k p_{x+h}^{(\tau)} \int_0^1 v^s \, {}_s p_{x+h+k}^{(\tau)} \, ds \qquad (7.87)$$

One midpoint approximation for the integral part of (7.87) is

$$\int_0^1 v^s \, {}_s p_{x+h+k}^{(\tau)} \, ds = v^{1/2} \, {}_{1/2} p_{x+h+k}^{(\tau)} \qquad (7.88)$$

Substituting (7.88) into (7.87) the net single value of the contributions becomes

$$\text{NSV-}C_{x+h} = c \sum_{k=0}^{\beta-x-h-1} v^{k+1/2} \, {}_{k+1/2} p_{x+h}^{(\tau)} \qquad (7.89)$$

Expectation (7.89) can be viewed as the net single premium of a discrete annuity with payments made in the middle of each year. Applying the uniform distribution of deaths assumption, the approximation

$$_{k+1/2} p_{x+h}^{(\tau)} = {}_k p_{x+h}^{(\tau)} \left(1 - \frac{1}{2} q_{x+h+k}^{(\tau)} \right)$$

could be used in connection with formula (7.89). If we use the *EP* equating the net single premiums of benefits (7.84) and contributions (7.89), the relational formula between benefits and contributions is

$$c \sum_{k=0}^{\beta-x-h-1} v^{k+1/2} \, {}_{k+1/2} p_{x+h}^{(\tau)} = \text{NSV-}B_{x+h} = \text{NSP}^{(w)} + \text{NSP}^{(d)} + \\ \text{NSP}^{(s)} + \text{NSP}^{(r)} \qquad (7.90)$$

In (7.90) contribution c is related to the benefit quantities, with larger benefits requiring increased contributions. A pension system example and Excel worksheet are outlined in the problems. In more realistic models the individual benefits are more complex, but this model hints at the general relation that exists between contributions and benefits.

More complex components of pension models through other forms of benefits and contributions are introduced in the next section. In particular, benefits and contributions may be based on functions of yearly salaries.

7.3.3 Future Salary-Based Benefits and Contributions

In general, pension systems benefits and the contributions may be a function of an individual's past, present, and future year salaries. If the individual has attained age $x + h$, the actual salary is known and is denoted by AS_{x+h}. At future age $x + h + t$ the salary amounts must be estimated, denoted ES_{x+h+t}. To estimate future salaries, increases due to merit, seniority, and inflation must be modeled. The increase in salary through time can be approximated by applying exponential growth at rate γ to the actual salary defined by

$$ES_{x+h+t} = (1 + s)^t AS_{x+h} \tag{7.91}$$

where $(1 + s) = \exp(\gamma)$ for salary increase rate s. In Example 7.17 we consider $PS(c)$-type contributions that are based on past and future salaries.

Example 7.17

In $PS(c)$ methods contributions are a proportion, c, of the yearly salary and form a continuous life annuity. Utilizing (7.91), the expectation or net single value of the contributions is computed as

$$\text{NSV-}C_{x+h} = c\,AS_{x+h} \int_0^{\beta-x-h} (1+j)^{-t}\,{}_tp_{x+h}^{(\tau)}\,dt \quad \text{where} \quad j = \frac{i-s}{1+s} \tag{7.92}$$

If we apply the midpoint approximation (7.88), then (7.92) simplifies to

$$\text{NSV-}C_{x+h} = c\,AS_{x+h} \sum_{k=0}^{\beta-x-h-1} (1+j)^{-k-1/2}\,{}_{k+1/2}p_{x+h}^{(\tau)} \tag{7.93}$$

Computation (7.93) takes the form of a discrete annuity expectation where the payments are made in the middle of the year.

It is common in pension systems for benefit amounts to be functions of part or all of the future years' prospective salaries. We consider a general benefit structure for an individual who entered the pension plan at age x and at analysis is age $x + h$. A benefit retirement rate is a function of the status future lifetime at age $x + h + t$ and is denoted $R(x, h, t)$. This structure is general in nature, fitting many pension systems, and is demonstrated in Example 7.18.

Example 7.18

For an individual age $x + h$, the benefit rate associated with age-based retirement is $R(x, h, t)$, where retirement is at age $x + h + t$. Modifying (7.83), the expectation of the benefit corresponding to age-based retirement takes the general form

$$\text{NSV-}B_{x+h} \int_{\alpha-x-h}^{\beta-x-h} v^t \, _tp_{x+h}^{(\tau)} \mu_{x+t}^{(r)} R(x, h, t) \bar{a}_{x+h+t}^{(r)} \, dt \qquad (7.94)$$

To simplify (7.94), the standard decomposition $T = K + S$ is applied, yielding

$$\text{NSV-}B_{x+h} = \sum_{k=\alpha-x-h}^{\beta-x-h-l} v^k \, _kp_{x+h}^{(\tau)} q_{x+h+k}^{(r)} g(x, h, k) \qquad (7.95)$$

where

$$g(x, h, k) = \int_0^1 v^s R(x, h, k+s) \bar{a}_{x+h+k+s}^{(r)} \, ds$$

Applying a midpoint approximation to the integral in $g(x, h, k)$ simplifies (7.95) to

$$\text{NSV-}B_{x+h} = \sum_{k=\alpha-x-h}^{\beta-x-h-1} v^{k+1/2} \, _kp_{x+h}^{(\tau)} q_{x+h+k}^{(r)} R\left(x, h, k+\tfrac{1}{2}\right) \bar{a}_{x+h+k+1/2}^{(r)} \qquad (7.96)$$

These approximate expectation formulas allow for calculations based on actuarial life tables. In applications, the benefit rate function may be complex. In the next section the benefit rate function is related to yearly salaries.

7.3.4 Yearly Based Retirement Benefits

In typical pension systems, the yearly benefits paid by an age-based retirement annuity are a function of one or more of the individual's yearly salaries. In the simplest case, the retirement payments are a fraction, g, of the final salary. Based on salary projection (7.91) the benefit rate function is

$$R(x, h, t) = g \, AS_{x+h} \, (1 + s)^t \qquad (7.97)$$

TABLE 7.8

Benefit Rate Based on Final Year's Salary

	Retirement Age		
Benefit rate	60	65	70
$R(30, 5, t)$	$87,258.57	$112,042.23	$143,865.07

Rate formula (7.97) can be utilized in the net single premium formula (7.96). A demonstrative application is considered in the following example.

Example 7.19

In this example, an individual at age 35 possesses an income of $50,000 per year. The salary growth rate is taken to be $\gamma = .05$, and the retirement benefit is 50% of the final year's salary. In Table 7.8 the benefit rate function given by (7.97) is computed for three possible retirement ages: 60, 65, and 70. The benefit rate increases at a greater than linear rate with the number of additional years of service. This is primarily due to the exponential salary growth rate in the salary projection. Hence, the resulting pension benefits increase greatly as the number of years of service increases.

In more typical pension settings, the retirement payment rate is a function of the yearly salaries. If the initial age is x and the current age is $x + h$, then future salaries can be estimated using (7.91). In the final example of this section two basic rate functions are discussed.

Example 7.20

An individual started in the pension system at age x and is currently age $x + h$. Two retirement annuity benefit rates are explored. In the first, the retirement benefit rate is a fraction, g, of the final m-year salary average or mean given by

$$R(x, h, t) = g\, AS_{x+h} \int_{t-m}^{t} (1+s)^u \frac{du}{m} = g\, AS_{x+h} \frac{(1+s)^t(1-(1+s)^{-m})}{m\gamma} \tag{7.98}$$

where $m \leq t$. In some cases the previous inequality may not hold where actual salaries or projections can be utilized. Rate (7.98) is a function of the growth of salary through $1 + s$ and the number of years m. This rate increases in much the same way the final-year salary rate increases with years of service. In the second, the retirement benefit rate is based on the mean salary for the entire career. The total of the past salaries is a known quantity given by

$$TPS_{x+h} = \sum_{j=0}^{h-1} AS_{x+j} \tag{7.99}$$

For future service, corresponding to age $x + h$ to retirement, the salaries must be estimated. A formula for the total estimated future salary is obtained by adjusting (7.98) and is

$$TFS_{x+h} = AS_{x+h} (1+s)^t \frac{(1+s)^t - 1}{\gamma} \qquad (7.100)$$

where t is the time until retirement. If the retirement benefit rate is the fraction g of the mean of the salaries for all years, then using (7.99) and (7.100) the rate function is

$$R(x,h,t) = g \frac{(TPS_{x+h} + TFS_{x+h})}{h+t} \qquad (7.101)$$

The expectation for the retirement benefit due to age is computed or approximated by substituting (7.101) into (7.94) or (7.96).

Problems

7.1. The multiple future lifetime setting is considered where the individual lifetimes follow the distribution Problem 5.9.

    a. For JLS find the df and pdf.

    b. For LSS find the df and pdf.

7.2. Let two people ages $x = 30$ and $y = 25$ have independent future lifetimes where the individual pdfs are given in Problem 7.1.

    a. Find the probability that the first death is after 20 years.

    b. Find the probability the last death is after 40 years.

7.3. Consider the two individual multiple life examples of Example 7.5. Using (7.20) compute the probability that (y) dies after (x) but before n years.

7.4. For the multiple-decrement model setting where $m = 2$ find (a) $\mu_t^{(\tau)}$, (b) $_t p_x^{(\tau)}$, (c) joint mpf $f(t, j)$, and (d) the marginal pdfs $g(t)$ and $h(j)$ for forces of mortality given by (i) $\mu_t^{(1)} = 1/10$ and $\mu_t^{(2)} = t/10$ for $t \geq 0$ and (ii) $\mu_t^{(1)} = \lambda$ and $\mu_t^{(2)} = \alpha \exp(\beta t)$ for $t \geq 0$ and parameters $\lambda > 0$, $\alpha > 0$, and $\beta > 0$. The combined force of mortality corresponds to the exponential distribution for $J = 1$ and the Gompertz distribution of Section 1.6.4 for $J = 2$. The combined force of mortality $\mu_t^{(\tau)}$ corresponds to the Makeham distribution of Section 1.6.5.

Excel Problems

7.5. Consider a discrete stochastic status setting in Problem 5.17. Two independent statuses are considered where the respective ages are 2 and 3.

    a. Find the mean curtate future lifetime for JLS.

    b. Find the mean curtate future lifetime for LSS.

    Excel Problem 5.17 extension: Basic operations, SUMPRODUCT.

7.6. For a status model there are $m = 3$ modes of failure. A multiple-decrement survivorship table is listed below:

x	0	1	2	3	4	5	6	7
lx	100							
$d_x^{(1)}$	3	4	6	8	10	6	2	1
$d_x^{(2)}$	2	3	4	10	12	4	1	0
$d_3^{(3)}$	1	3	2	6	8	2	1	1

    a. Fill in the table and list the decrement mortality rates $q_x^{(j)}$ for $j = 1, 2, 3$, and $q_x^{(\tau)}$ for each year.

    b. Find the probability a status (i) age 2 dies by age 5 due to mode 2, and (ii) age 1 survives 2 years and then dies by mode 3.

    c. Find the expected future lifetime of an individual age $x = 1$. Excel: Utilize basic operations and SUM.

7.7. Apply the UDD assumption to the multiple-decrement setting given in Problem 7.6 to compute the single-decrement rates for each year and each of the three modes. Excel Problem 7.6 extension: Basic operations.

7.8. A multiple-decrement model with $m = 3$ independent modes of decrement follows the partial table given below:

Age = x	$d_x^{(1)}$	$d_x^{(2)}$	$d_x^{(3)}$
40	.011	.010	.015
41	.015	.018	.020
42	.020	.022	.031
43	.022	.025	.032

For ages 40 to 43 compute (a) the overall mortality rates $q_x^{(\tau)}$, and (b) the single mortality rates $q_x^{s(j)}$ for $j = 1, 2$, and 3 defined in (7.21). Excel: Basic operations.

7.9. In this problem the upper bounds for the single-decrement rates associated with Problem 7.8 are explored. Compute the single rates given by (7.63) and list general comments.

Excel Problem 4.14 extension: Basic operations.

7.10. A pension survivorship group is modeled starting at age 50. The types of pension decrement and notations are the ones of Example 7.13 where $i = .04$. The withdrawal and death benefits are end-of-the-year payments of $b^{(W)} = 3$ and $b^{(d)} = 55$. The disability and age-based retirement benefit is a discrete whole life annuity with unit expectation listed below where the yearly payments are $\pi^{(s)} = 20$ and $\pi^{(r)} = 35$. The contributions are c made at the start of each surviving year following Example 7.15.

x	l_x	$d_x^{(w)}$	$d_x^{(d)}$	$d_x^{(s)}$	$d_x^{(r)}$	$a_x^{(r)}$	x	l_x	$d_x^{(w)}$	$d_x^{(d)}$	$d_x^{(s)}$	$d_x^{(d)}$	$a_x^{(r)}$
50	100	2	0	0	0	14	58	73	0	1	1	0	10.000
51	98	3	0	1	0	13.333	59	71	0	1	1	0	9.655
52	94	3	0	1	0	12.727	60	69	0	2	0	10	9.333
53	90	4	1	0	0	12.174	61	57	0	2	0	15	9.032
54	85	2	2	0	0	11.667	62	40	0	3	0	10	8.750
55	81	1	0	1	0	11.200	63	27	0	4	0	10	8.485
56	79	1	1	0	0	10.769	64	13	0	5	0	8	8.235
57	77	1	1	2	0	10.370	65	0	0	0	0	0	0

Compute (a) NSV$^{(w)}$, (b) NSV$^{(d)}$, (c) NSV$^{(s)}$, (d) NSV$^{(r)}$, and (e) NSV-C using (7.85). (f) Solve for c as in (7.86).
Excel: Basic operations.

Solutions

7.1. a. (7.2) gives $1 - (x/(x + t))^2(y/(y + t))^2$, (7.5) gives $\mu_{(1)+t} = 2/(x + t) + 2/(y + t)$, and (7.4) gives $(x/(x + t))^2(y/(y + t))^2 [2/(x + t) + 2/(y + t)]$.

     b. (7.11) gives $(1 - (x/(x + t))^2)(1 - (y/(y + t))^2)$. Use (7.13).

7.2. a. $_{20}p_{xy} = (30/50)^2(25/45)^2 = .1111$.

     b. $_{40}p_{\overline{xy}} = (30/70)^2 + (25/60)^2 - (30/70)^2(25/60)^2 = .325397$.

7.3. Same as (7.21).

7.4. a. From (7.31) $\mu_t^{(\tau)} = (1 + t)/10$.

     b. As in (7.37) $_tp_x^{(\tau)} = \exp(-(t^2 + 2t)/20)$.

    c. Similar to (7.38) $f(t, j) = \exp(-(t^2 + 2t)/20)/10$ for $j = 1, = t \exp(-(t^2 + 2t)/20)/10$ for $j = 2$.

    d. As in (7.39) $g(t) = [(1 + t)/10] \exp(-(t^2 + 2t)/20)$. Following the steps in (7.40) $h(1) = (2\pi/10)^{1/2} \exp(1/20) \, \Phi(-1/10^{1/2}) = .313332$, and so $h(2) = .86668$.

  ii. a. $\mu_t^{(\tau)} = \lambda + \alpha \exp(\beta t)$.

7.5. a. Using (7.3) gives 2.145649.

    b. Using (7.12) gives 4.629832.

7.6. a. Following (7.69) we compute $q_x^{(j)}$ and $q_x^{(\tau)}$.

    b. i. $_5q_2^{(2)} = .309524$.

        ii. $_{2|\infty}q_1^{(3)} = .191489$.

    c. 2.446809.

7.7. Using (7.21) for $x = 0$, single-decrement rates are .03046, .02041, and .01026.

7.10. a. Following (7.73), .44123.

    b. Following (7.73), 8.142045.

    c. Following (7.74), 12.23917

    d. Following (7.74), 99.18147.

    e. 8.569023c.

    f. $c = 120.0039/8.569023 = 14.00439$.

8

Markov Chain Methods

Many texts, such as Scheerer (1969) and Cinlar (1975), present a basic introduction to the mathematical basics of Markov chains. Markov chain theory lends itself to a wide variety of applications. Allen (2003) applies the general Markov chain approach to biological and other settings. Recently, financial and actuarial applications have been analyzed using Markov chain approaches. One such as Daniel (1954) illustrates how a nonhomogeneous Markov chain can be applied to actuarial topics. In particular, mortality and survival probabilities, as well as cash flow problems, can be investigated through Markov chains.

In this chapter we discuss the theory and applications of discrete-time Markov chains as related to actuarial science. Both nonhomogeneous and homogeneous Markov chain approaches are applied to life table data to construct stochastic status chains that are introduced in Sections 8.2 and 8.3. Basic discrete-status model survival and mortality probabilities introduced in Chapter 5 are computed using Markov chain techniques under various actuarial settings, including single-decrement and multiple-decrement models. Analyses of multirisk strata structures are proposed in utilizing stochastic status chains in Section 8.2.4. Homogeneous stochastic status chain models, in Section 8.3, are developed to model entire life tables, leading to the direct computation of survival and mortality probabilities. Utilizing these models, techniques to compute the expected curtate future lifetimes are presented in Section 8.3.1. The Markov chain structure provides an efficient platform for stochastic status computations.

Further, applying the concept of a yearly discount function to model the effect of interest, actuarial chains are developed and utilized to compute expectations related to different types of discrete life insurance. In Sections 8.2.2 and 8.3.2 Markov chain techniques for the computation of discrete whole life and term life insurance net single premiums are presented and computational examples using actuarial chains are demonstrated. The topic of discrete actuarial expectations for individuals in high-risk strata is explored by the use of a multirisk strata actuarial chain approach.

The modeling of group survivorship systems by Markov chains is discussed in Section 8.4. For different mortality structures, including single-decrement, multiple-decrement, and multirisk strata models, these Markov chains can be utilized to track expectations for survival and mortality totals through a discrete-time structure. In this way the modeling of populations consisting of various strata through a discrete-time structure is accomplished.

8.1 Introduction to Markov Chains

In mortality modeling and actuarial applications we consider discrete-time Markov chains as described by Allen (2003, Chapter 2), where the set of the Markov state is finite and is denoted by $MS = \{1, 2, \ldots, s\}$. The Markov chain is a sequence of random variables $\{M_1, M_2, \ldots\}$, where each M_i takes values in MS. For fixed n the u-step transitions are the realization of $M_n = i$ and $M_{n+u} = j$ for $1 \leq i, j \leq s$. In a Markov chain the conditional single-step transitions probabilities, referred to as the transition probabilities, depend only on the previous state. Starting in M_n, the transition probabilities moving from state i to state j in one step are denoted

$$t_n(i, j) = P(M_{n+1} = j \,|\, M_n = i) \tag{8.1}$$

for $1 \leq i, j \leq s$. The transition matrix consists of the probabilities in (8.1) and is denoted by the matrix

$$T_n \text{ consisting of elements } t_n(i, j) \quad \text{for} \quad 1 \leq i, j \leq s \tag{8.2}$$

Here (8.1) implies

$$\sum_{j=1}^{s} t_n(i, j) = 1 \quad \text{for } 1 \leq i \leq s \text{ and } n \geq 1. \tag{8.3}$$

Matrix (8.2) under condition (8.3) is referred to as a stochastic matrix. States in MS are sometimes classified based on stochastic properties. For example, starting in M_n the state j from MS is an absorbing state if $t_n(j, j) = 1$. Absorption states are useful in modeling decrements.

For u-step, transition probabilities originating in M_n can be computed using the transition matrices of the form (8.2). Specifically, for positive integer u the u-step transition probabilities are defined by

$$t_n^{(u)}(i, j) = P(M_{n+u} = j \,|\, M_n = i) \tag{8.4}$$

for $1 \leq i, j \leq s$, and $n \geq 1$. Probabilities (8.4) are computed by multiplication of transition matrices as

$$\mathbf{T}_n^{(u)} = \prod_{i=1}^{u} \mathbf{T}_{n+i-1} \tag{8.5}$$

Here (8.5) is the u-step transition matrix where probability (8.4) corresponds to the ith row and jth column entry. Designating the initial state defines the nonhomogeneous Markov chain and lends itself to the modeling of various actuarial situations.

In a homogeneous Markov chain the transition probabilities (8.1) and associated transition matrix (8.2) are independent of the initial state M_n. In this case, the transition probabilities are denoted by $t(i, j)$ and the associated transition matrix is denoted

$$\mathbf{T} \text{ consisting of elements } t(i, j) \quad \text{for } 1 \leq i, j \leq s \tag{8.6}$$

State $J = j$ is an absorption state if $t(j, j) = 1$. For positive integer u the u-step transition probabilities are given by

$$t(i, j)^{(u)} = P(M_{1+u} = j \mid M_1 = i) \tag{8.7}$$

for $1 \leq i, j \leq s$. The u-step transition matrix, following procedure (8.5), is found by multiplication of (8.6) as

$$T^{(u)} = \prod_{i=1}^{u} T \tag{8.8}$$

for $u \geq 1$. Relationships between different u-step transition matrices known as the Chapman–Kolmogorov equations exist, and for discussion we refer the reader to Allen (2003, p. 44).

The theory and techniques of Markov chains have been applied to a wide variety of modeling problems. In the following sections, applications of Markov chains related to mortality tables and actuarial techniques are explored. In the next section, we discuss nonhomogeneous Markov chain modeling of actuarial problems. Many classifications of the Markov states have been discussed in the literature and include, as discussed by Allen (2003, p. 48), absorbing states.

8.2 Nonhomogeneous Stochastic Status Chains

In this section, applying the concepts associated with nonhomogeneous Markov chains, we develop stochastic status chains to model various discrete stochastic status structures presented in Chapter 5. Life tables discussed in Section 5.7 play a central role in the formulation of these models. For a status age x, the chain takes the form $\{M_x, M_{x+1}, ...\}$, where each M_x takes values in MS for $x \geq 0$. The form of MS depends on the type of stochastic status being modeled. The elements in the stochastic status chain transition matrices consist of survival and mortality rates for subsequent years. Single- and multiple-decrement models as well as multirisk strata are modeling. Further, modifying the transition matrices by an interest discount function, actuarial chains are constructed and utilized to compute the expectation associated with the present values of standard life insurances.

8.2.1 Single-Decrement Chains

Nonhomogeneous stochastic chain theory is now developed to model the discrete single-decrement setting discussed in Section 5.2. For a status age x, a life table lists mortality rates q_x and survivor rates defined in (5.5) for $x = 0, \ldots, \omega - 1$. Here ω is the limiting age where $q_{\omega-1} = 1$ and $p_{\omega-1} = 0$. In the stochastic chain modeling of a stochastic status age x the initial stage corresponds to age x and has two states, survival and decrement. Thus, $MS = \{1, 2\}$, where 1 corresponds to survival and 2 corresponds to decrement, and the decrement and survival rates are (5.2) and (5.3). For a stochastic status age x, the transition probabilities follow (8.1), so that the transition matrix, T_x, consisting of the nonzero entries, is

$$t_x(1, 1) = p_x, \quad t_x(1,2) = q_x \quad \text{and} \quad t_x(2,2) = 1 \tag{8.9}$$

Condition (8.3) holds and the transition matrix is a stochastic matrix. For nonnegative integer u, following (8.5), the u-step transition matrix is

$$T_x^{(u)} = \prod_{j=1}^{u} T_{x+j-1} \tag{8.10}$$

where the entries based on (8.4) are

$$t_x(1, 1)^{(u)} = {}_u p_x, \ t_x(1, 2)^{(u)} = {}_u q_x, \ t_x(2, 1)^{(u)} = 0 \quad \text{and} \quad t_x(2, 2)^{(u)} = 1 \tag{8.11}$$

for $u = 1, 2, \ldots, \omega - x$. The derivation of (8.11) is outlined in Problem 8.1, and Excel computations are discussed in Problem 8.7. A computational example now follows.

Example 8.1

For a discrete stochastic status age x a life table based on the curtate future lifetime is given in Table 8.1, where the limiting age is $\omega = x + 5$. The yearly survival and mortality rates are modeled using a nonhomogenous stochastic status chain.

Starting at age x, using matrix components (8.9), the single-step transition matrices are found to be

$$T_x = \begin{pmatrix} .9 & .1 \\ 0 & 1 \end{pmatrix} \ T_{x+1} = \begin{pmatrix} .8 & .2 \\ 0 & 1 \end{pmatrix} \ T_{x+2} = \begin{pmatrix} .7 & .3 \\ 0 & 1 \end{pmatrix} \ T_{x+3} = \begin{pmatrix} .3 & .7 \\ 0 & 1 \end{pmatrix}$$

$$\tag{8.12}$$

and

$$T_{x+4} = \begin{pmatrix} 0 & 1 \\ 0 & 1 \end{pmatrix} \tag{8.13}$$

TABLE 8.1

Life Table for Status Age x and Limiting Age $x + 5$

Duration	$k = 0$	$k = 1$	$k = 2$	$k = 3$	$k = 4$
p_{x+k}	.9	.8	.7	.3	0
q_{x+k}	.1	.2	.3	.7	1

The limiting age of $\omega = x + 5$ is indicated by the ones in the second column of matrix (8.13). The u-step probabilities for $u = 2$ and $u = 3$ are computed utilizing (8.12) in (8.10) as

$$T_x^{(2)} = \begin{pmatrix} .72 & .28 \\ 0 & 1 \end{pmatrix} \quad \text{and} \quad T_x^{(3)} = \begin{pmatrix} .504 & .496 \\ 0 & 1 \end{pmatrix} \qquad (8.14)$$

Continuing, we find

$$T_x^{(4)} = \begin{pmatrix} .1512 & .8488 \\ 0 & 1 \end{pmatrix} \quad \text{and} \quad T_x^{(5)} = \begin{pmatrix} 0 & 1 \\ 0 & 1 \end{pmatrix} \qquad (8.15)$$

Mortality and survival probabilities can be realized from formulas (8.11). For example, the probability the status age x fails within 3 years is found from (8.14) to be $_3q_x = .496$. Typical of Markov chains based on mortality tables, the limiting age causes the convergence, as in (8.15) for $u = 5$.

8.2.2 Actuarial Chains

The computation of actuarial expectations utilizing nonhomogeneous Markov chain techniques requires the definition of actuarial chains based on rate of interest considerations. This is more direct than the approach of Daniel (2004), where an additional matrix is interjected to model the force of interest. For yearly effective interest rate i following the economic modeling concept presented in (3.12), (3.13), and (6.1), the yearly discount function is defined as $v = (1 + i)^{-1}$. As in the previous section for a stochastic status age x the Markov states are $MS = \{1, 2\}$, where 2 is the decrement state. For a discrete life insurance for (x) an actuarial chain is defined in terms of actuarial transition matrix $\mathbf{R}_x$ consisting of nonzero entries:

$$r_x(1, 1) = v \, p_x, \; r_x(1, 2) = v \, q_x \quad \text{and} \quad r_x(2, 2) = 1 \qquad (8.16)$$

for $x \geq 0$. The actuarial transition matrix is not a stochastic matrix as property (8.3) does not hold. The components (8.16) can be interpreted as the yearly survival and mortality rates discounted for the effect of interest. For $u = 1, 2, \ldots, \omega - x$ the u-step actuarial transition matrix is defined as

$$\mathbf{R}_x^{(u)} = \prod_{j=1}^{u} \mathbf{R}_{x+j-1} \qquad (8.17)$$

where the nonzero entries correspond to standard unit benefit discrete annuity and insurance present value expectations:

$$r_x(1,1)^{(u)} = v^u {}_up_x = {}_uE_x, \quad r_x(1,2)^{(u)} = A^1_{x:\overline{u}|}$$

$$\text{and} \quad r_x(2,1)^u = 0 \quad \text{and} \quad r_x(2,2)^{(u)} = 1 \tag{8.18}$$

In (8.18) the expectations for a unit benefit pure endowment, defined in (6.16), and a u-term life insurance, given in (6.15), namely, ${}_uE_x = v^u {}_up_x$ and $A^1_{x:\overline{u}|}$, are revealed. An illustrative example now follows, while Excel instructions are discussed in Problem 8.8.

Example 8.2

In the single-decrement setting of Example 8.1 let $i = .06$. Following (8.16) the actuarial transition matrices are constructed as

$$\mathbf{R}_x = \begin{pmatrix} .8491 & .0943 \\ 0 & 1 \end{pmatrix} \quad \mathbf{R}_{x+1} = \begin{pmatrix} .7547 & .1887 \\ 0 & 1 \end{pmatrix}$$

$$\mathbf{R}_{x+2} = \begin{pmatrix} .6604 & .2830 \\ 0 & 1 \end{pmatrix}$$

For a status age x, applying (8.17), the u-step actuarial transitional matrices are computed as

$$\mathbf{R}_x^{(2)} = \begin{pmatrix} .6408 & .2544 \\ 0 & 1 \end{pmatrix} \quad \text{and} \quad \mathbf{R}_x^{(3)} = \begin{pmatrix} .4232 & .4359 \\ 0 & 1 \end{pmatrix} \tag{8.19}$$

for $u = 2$ and $u = 3$. From (8.19) we compute the present value expectations for a discrete 3-year pure endowment insurance as ${}_3E_x = .4232$, and for a 3-year term life insurance as $A^1_{x:\overline{3}|} = .4359$.

In the next two sections the concepts and techniques of nonhomogeneous Markov chains are extended to other actuarial settings, namely, multiple-decrement models in Section 8.2.3 and multirisk strata models in Section 8.2.4. Actuarial chain techniques for these models are developed and utilized.

8.2.3 Multiple-Decrement Chains

Nonhomogeneous Markov chain modeling is now applied to the multiple-decrement setting introduced in Section 7.2 with notations taken from Section 7.2.7. The discrete stochastic status model has m modes of decrement leading to a multiple-decrement chain consisting of $d + 1$ distinct states where the first is survival and decrement mode $J = j$ is designated by state $j + 1$. Here $MS = \{1, 2, \ldots, m + 1\}$. For a stochastic status age x the mode-specific decrement rates $q_x^{(j)}$, from (7.22), and the overall decrement rates $q_x^{(\tau)}$, from (7.28), as well as the survival probability $p_x^{(\tau)}$, from (7.27), are considered to be known table quantities. The transition matrix $\mathbf{T}_x$ is $(m + 1)$ by $(m + 1)$ consisting of nonzero entries

$$t_x(1,1) = p_x^{(\tau)}, \quad t_x(1,j) = q_x^{(j-1)} \quad \text{for } 2 \le j \le m+1, \quad t_x(i,i) = 1 \quad \text{for } i = 2, \ldots, m+1$$

$$(8.20)$$

Computing the u-step transition matrix (8.20), the nonzero components correspond to the multiple-decrement survival and decrement rates:

$$t_x(1,1)^{(u)} = {}_u p_x^{(\tau)}, \quad t_x(1,j)^{(u)} = {}_u q_x^{(j-1)} \text{ for } 2 \le j \le m+1 \quad \text{and} \quad t_x(i,i)^{(u)}$$

$$(8.21)$$

$$= 1 \text{ for } 2 \le i \le m+1$$

The nonnegative components follow stochastic property (8.3); that is, the rows sum to 1. We follow with an example taken from Chapter 4.

Example 8.3

For a discrete stochastic status there exits $m = 3$ modes of decrement with yearly decrement rates given in Table 7.1. For age $x = 30$ and $x = 31$ the transition matrices are

$$\mathbf{T}_{30} = \begin{pmatrix} .995 & .001 & .002 & .002 \\ 0 & 1 & 0 & 0 \\ 0 & 0 & 1 & 0 \\ 0 & 0 & 0 & 1 \end{pmatrix} \quad \text{and} \quad \mathbf{T}_{31} = \begin{pmatrix} .991 & .002 & .003 & .004 \\ 0 & 1 & 0 & 0 \\ 0 & 0 & 1 & 0 \\ 0 & 0 & 0 & 1 \end{pmatrix}$$

$$(8.22)$$

In (8.22), we realize survival rates $p_{30}^{(\tau)} = .995$, $p_{31}^{(\tau)} = .991$ and decrement rates $q_{30}^{(1)} = .001$, $q_{31}^{(3)} = .004$. For a status age $x = 30$ the u-step transition matrices for $u = 1$ and 2 are

$$\mathbf{T}_{30}^{(2)} = \begin{pmatrix} .9861 & .0030 & .0050 & .0060 \\ 0 & 1 & 0 & 0 \\ 0 & 0 & 1 & 0 \\ 0 & 0 & 0 & 1 \end{pmatrix}$$

$$(8.23)$$

$$\mathbf{T}_{30}^{(3)} = \begin{pmatrix} .9762 & .0050 & .0089 & .0099 \\ 0 & 1 & 0 & 0 \\ 0 & 0 & 1 & 0 \\ 0 & 0 & 0 & 1 \end{pmatrix}$$

Applying (8.21) to the components in (8.23) we find 2- and 3-year survival probabilities $_2p_{30}^{(\tau)} = .9861$ and $_3p_{30}^{(\tau)} = .9762$ for a status age 30. Also, for a status age 30 the 2- and 3-year decrement probabilities for the first failure mode $_2q_{30}^{(1)} = .003$ and $_3q_{30}^{(1)} = .005$.

The concept and techniques of actuarial chains introduced in Section 8.2.2 can be extended to the multiple-decrement setting as follows. For a stochastic status age x the actuarial transition matrix is $\mathbf{R}_x$, consisting of nonzero entries:

$$r_x(1,1) = v\,p_x^{(\tau)}, \quad r_x(1,j) = v\,q_x^{(j-1)} \quad \text{for } 2 \le j \le m+1 \quad \text{and}$$
$$r_x(i,i) = 1 \quad \text{for} \quad 2 \le i \le m \tag{8.24}$$

For $u \ge 1$ the u-step actuarial transition matrix $\mathbf{R}_x^{(u)}$ follows (8.17), where the nonzero entries correspond to unit benefit discrete:

$$r_x(1,1)^{(u)} = v^u\,{}_up_x^{(\tau)}, \, r_x(1,j)^{(u)} = A_{x:\overline{u}|}^{1(j-1)} \text{ for } 2 \le j \le m+1 \quad \text{and}$$
$$r_x(i,i)^{(u)} = 1 \text{ for } 2 \le i \le m+1 \tag{8.25}$$

Formula (8.25) can be used to realize the pure endowment with the expectation $_uE_x = v^u\,{}_up_x$, and $A_{x:\overline{u}|}^{1(j)}$ represents the expectation of the present value for a unit benefit u-year term life insurance where the benefit is paid only if failure is due to the jth mode of decrement. These concepts and formulas are now demonstrated.

Example 8.4

In this example actuarial chains are applied to Example 8.3 where the effective interest rate is $i = .06$ per annum. Following construction (8.24), the actuarial transition matrices are computed as

$$\mathbf{R}_{30} = \begin{pmatrix} .9387 & .0009 & .0019 & .0019 \\ 0 & 1 & 0 & 0 \\ 0 & 0 & 1 & 0 \\ 0 & 0 & 0 & 1 \end{pmatrix}$$

$$\mathbf{R}_{31} = \begin{pmatrix} .9349 & .0019 & .0028 & .0038 \\ 0 & 1 & 0 & 0 \\ 0 & 0 & 1 & 0 \\ 0 & 0 & 0 & 1 \end{pmatrix}$$

The $u = 2$ step actuarial transitional matrices are found to be

$$
\mathbf{R}_{30}^{(2)} = \begin{pmatrix}
.8776 & .0027 & .0045 & .0054 \\
0 & 1 & 1 & 1 \\
0 & 0 & 1 & 0 \\
0 & 0 & 0 & 1
\end{pmatrix}
$$

Utilizing the formulas of (8.18) for a status age 30 for a 2-year pure endowment, the present value expectation is $_2E_{30} = .8776$. Further, for 2-year term life insurance based on individual decrement modes the expectations are $A_{30:\overline{2}|}^{1(1)}.0027$, $A_{30:\overline{2}|}^{1(2)}.0045$, and $A_{30:\overline{2}|}^{1(3)} = .0054$.

8.2.4 Multirisk Strata Chains

In this section a new structure for discrete stochastic status modeling based on partitioning the group survivorship structure presented in Section 5.6 into subgroups is presented. Following the development given by Borowiak and Das (2007), at each age x individual statuses are grouped by distinct strata. The individual strata are associated with distinct survival and mortality patterns and lead to various levels of insurance risk and are referred to as risk strata. The computation of the multirisk strata survival and mortality rates as well as actuarial expectations is accomplished by applying nonhomogeneous Markov chain theory and techniques. Further, the relative sizes of the risk strata are modeled using standard matrix techniques. For similar techniques related to actuarial science we refer to Daniel (2004).

In general, we have s distinct risk strata index by $i = 1, \dots, s$. For a stochastic status model age x, the associated single-year survival probabilities associated with moving from risk strata i to risk strata j are $^{(i)}p_x^{(j)}$ for $1 \le i, j \le s$. The 1-year mortality rate associated with the ith risk stratum is $^{(i)}q_x$ for $1 \le i \le s$. Hence for a status model age x in the ith stratum the stochastic property (8.3) becomes

$$
1 = {}^{(i)}q_x + \sum_{j=1}^{s} {}^{(i)}p_x^{(j)}
\tag{8.26}
$$

The Markov chain corresponds to the risk strata and decrement so that $MS = \{1, 2, \dots, s, s + 1\}$, where $s + 1$ signifies decrement. The transition matrix based on a status age x is matrix $\mathbf{T}_x$ with nonzero probabilities:

$t_x(i, j) = {}^{(i)}p_x^{(j)}$ for $1 \le i, j \le s$, $t_x(i, s + 1) = {}^{(i)}q_x$ for $1 \le i \le s$ and
$t_x(s + 1, s + 1) = 1$

$$
\tag{8.27}
$$

The nonnegative components (8.27) satisfy property (8.3), yielding a stochastic matrix. For positive integer u the u-step transition matrix, $\mathbf{T}_x^{(u)}$, follows (8.10), where the components contain the m-step survival probabilities moving from the ith to the jth risk stratum,

$$t_x(i, j)^{(u)} = {}^{(i)}_u p_x^{(j)} \quad \text{for} \quad 1 \le i, j \le s \qquad (8.28)$$

the m-step decrements associated with the ith risk stratum,

$$t_x(i, s+1)^{(u)} = {}^{(i)}_u q_x \text{ for } 1 \le i \le s \qquad (8.29)$$

and the decrement associated u-step components,

$$t_x(s+1, j)^{(u)} = 0 \quad \text{for} \quad 1 \le j \le s \quad \text{and} \quad t_x(s+1, s+1)^{(u)} = 1 \qquad (8.30)$$

Modeling the multirisk strata survival transitions and decrements by Markov chains allows for the measuring and tracking of the aggregate population. An example of this follows.

The relative sizes of the risk strata can be tracked over sequential ages applying the Markov chain technique, and relative risk strata sizes can be found based on age. For a stochastic status model age x the strata proportions are ${}^{(i)}\pi_x$ for $1 \le i \le s$ and the strata proportion vector is

$$\pi_x = ({}^{(1)}\pi_x, {}^{(2)}\pi_x, \ldots, {}^{(s)}\pi_x, 0) \qquad (8.31)$$

The u-step transition matrix of (8.5) can be used to track the relative strata sizes. After u-steps the aggregate strata proportion vector is given by

$$\pi_{x+u} = \pi_x T_x^{(u)} = \pi_x \prod_{j=1}^{u} T_{x+j-1} = ({}^{(1)}\pi_{x+u}, {}^{(2)}\pi_{x+u}, \ldots, {}^{(s)}\pi_{x+u}, 0) \qquad (8.32)$$

for $u \ge 0$. This is demonstrated in Example 8.5, while Excel computations are considered in Problem 8.12.

Example 8.5

For a stochastic status model there exist two risk strata. Individuals in the first risk strata follow a wear-out mortality pattern where members die out quickly with a limiting age of $x = 3$. The second risk strata individuals follow an infant mortality structure. The profile for newborns consists of 10% from the first stratum and 90% from the second stratum. Individuals cannot move from status 2 to status 1, and the yearly survival and mortality rates are listed in Table 8.2.

TABLE 8.2

Multirisk Strata Mortality and Survival Rates

Age x	$^{(1)}q_x$	$^{(2)}q_x$	$^{(1)}p_x^{(1)}$	$^{(1)}p_x^{(2)}$	$^{(2)}p_x^{(1)}$	$^{(2)}p_x^{(2)}$
0	.4	.1	.5	.1	0	.9
1	.6	.08	.3	.1	0	.92
2	.8	.04	.1	.1	0	.96
3	1.0	.03	0	0	0	.97

For status model age $x = 0, 1, 2,$ and 3 the transition matrices are found following (7.41)–(7.43) as

$$T_0 = \begin{pmatrix} .5 & .1 & .4 \\ 0 & .9 & .1 \\ 0 & 0 & 1 \end{pmatrix} \quad T_1 = \begin{pmatrix} .3 & .1 & .6 \\ 0 & .92 & .08 \\ 0 & 0 & 1 \end{pmatrix} \quad T_2 = \begin{pmatrix} .1 & .1 & .8 \\ 0 & .96 & .04 \\ 0 & 0 & 1 \end{pmatrix}$$

$$T_3 = \begin{pmatrix} 0 & 0 & 1 \\ 0 & .97 & .03 \\ 0 & 0 & 1 \end{pmatrix}$$

(8.33)

The initial strata proportions are given in the vector $\pi_0 = (.1\ .9\ 0)$. For infants age $x = 0$ the transition matrix and strata proportion vector after 1 year are computed utilizing (8.32) as

$$\pi_1 = \pi_0 T_0 = (.05\ .82\ .13)$$

After 1 year only 5% of the newborns survive in the first stratum compared to 82% that survive in the second stratum. Further, the aggregate mortality rate over both strata after the first year is 13%. Following multiplication (8.8) the u-step transition matrices for infants are

$$T_0^{(2)} = \begin{pmatrix} .15 & .142 & .708 \\ 0 & .828 & .172 \\ 0 & 0 & 1 \end{pmatrix} \quad T_0^{(3)} = \begin{pmatrix} .015 & .1513 & .8337 \\ 0 & .7949 & .2051 \\ 0 & 0 & 1 \end{pmatrix}$$

(8.34)

$$T_0^{(4)} = \begin{pmatrix} 0 & .1468 & .8532 \\ 0 & .7710 & .2290 \\ 0 & 0 & 1 \end{pmatrix}$$

For newborns of the first stratum the proportion of decrements of newborns by year 2 is 70/8% and by year 4 is 85.32%. Further, following (8.32) the aggregate strata proportion vectors are

$$\pi_2 = (.015\ .7594\ .2256), \pi_3 = (.00150\ .7305\ .2680), \pi_4 = (0\ .7085\ .2914)$$

Thus, at age $x = 4$ we observe that 0% are in the first stratum, 70.86% in the second stratum, and 29.14% have failed.

The actuarial chain approach of Section 8.2.2 can be extended to the case of multirisk strata to compute present value expectations for standard types of life insurance. For a stochastic status model age x the actuarial transition matrix is denoted by $\mathbf{R}_x$ consisting of nonzero entries

$$r_x(i,j) = v^{(i)} p_x^{(j)} \quad \text{for} \quad 1 \le i, j \le s, \quad r_x(i, s+1) = v^{(i)} q_x \quad \text{for} \quad 1 \le i \le s$$

$$\text{and} \quad r_x(s+1, s+1) = 1$$

(8.35)

The components in (8.35) are the survival and mortality rates adjusted for the force of interest. For $u \ge 1$ the u-step actuarial transition matrix, $\mathbf{R}_x^{(u)}$, follows (8.17), where the nonzero entries correspond to unit benefit discrete insurance present value expectations:

$$r_x(i,j)^{(u)} = v^u \, {}_u^{(i)} p_x^{(j)} = {}_u^{(i)} E_x^{(j)} \quad \text{for} \quad 1 \le i, j \le s, \quad r_x(i, s+1)^{(u)} = {}^{(i)} A_{x:\overline{u}|}^1 \quad \text{for} \quad 1 \le i \le s$$

$$\text{and} \quad r_x(s+1, s+1)^{(u)} = 1$$

(8.36)

where for a pure endowment ${}_u^{(i)} E_x^{(j)} = v^u \, {}_u^{(i)} p_x^{(j)}$ the unit benefit u-year term life insurance present value expectation corresponding to status age x from the ith risk stratum is ${}^{(i)} A_{x:\overline{u}|}^1$. A computational example is now given. Actuarial chain computations using Excel are outlined in Problem 8.13.

Example 8.6

An actuarial chain is developed for the multirisk strata setting presented in Example 8.5 where the interest rate is $i = .06$. Applying (8.35), the nonnegative components of the actuarial transition matrices for ages 0, 1, and 2 are

$$\mathbf{R}_0 = \begin{pmatrix} .4717 & .0943 & .3774 \\ 0 & .8491 & .0943 \\ 0 & 0 & 1 \end{pmatrix} \quad \mathbf{R}_1 = \begin{pmatrix} .2830 & .0943 & .5660 \\ 0 & .8679 & .0755 \\ 0 & 0 & 1 \end{pmatrix}$$

$$\mathbf{R}_2 = \begin{pmatrix} .0943 & .0943 & .7547 \\ 0 & .9057 & .0377 \\ 0 & 0 & 1 \end{pmatrix}$$

The u-step actuarial transitional matrices with components (8.35) are found for $u = 2$ and $u = 3$, using (8.17), to be

$$\mathbf{R}_0^{(2)} = \begin{pmatrix} .1335 & .1264 & .6515 \\ 0 & .7369 & .1584 \\ 0 & 0 & 1 \end{pmatrix} \quad \text{and} \quad \mathbf{R}_0^{(3)} = \begin{pmatrix} .0126 & .1271 & .7570 \\ 0 & .6674 & .1862 \\ 0 & 0 & 1 \end{pmatrix}$$

(8.37)

From (8.37) the strata-dependent present value expectations for 3-year term life insurance are realized as $^{(1)}A^1_{0:\overline{3}|} = .7570$ and $^{(2)}A^1_{0:\overline{3}|} = .1862$. As expected, the estimated cost of insuring individuals of the first risk strata is greater than that of the second risk strata members.

8.3 Homogeneous Stochastic Status Chains

Homogeneous Markov chains can be used to model entire life tables. These stochastic status chains consist of a transition matrix comprised of yearly survival and mortality rates, and multiyear survival and decrement probabilities similar to those described in Section 8.2. In addition, for a discrete stochastic status with limiting age the computation of present value expectations for standard unit benefit whole life insurance, as well as the curtate future lifetime, is achieved using matrix techniques. This discussion is restricted to single-decrement models, and extensions to additional actuarial structures, such as multiple decrements or multirisk strata, are straightforward and left to the reader.

To construct a homogeneous Markov chain we start with a group survivorship model with mortality table listings for age x mortality rates q_x for $x = 0, \ldots,$ $\omega - 1$. For the mortality table the age ω is the limiting age and is signified by $q_{\omega-1} = 1$. For a discrete stochastic status age x let the associated curtate future lifetime be $K(x)$, where $0 \le K(x) \le \omega - x - 1$. The states in the stochastic status chain correspond to the range of curtate future lifetimes $0, 1, \ldots, \omega - 1$ and the decrement state denoted by d, so that $MS = \{0, 1, \ldots, \omega - 1\}$. Starting with an initial age x, the transition matrix is composed of survival rates and mortality rates for future ages denoted by **T**, where the nonzero components are

$$t(i, i + 1) = p_{x+i-1} \quad \text{and} \quad t(i, \omega + 1) = q_{x+i-1} \quad \text{for} \quad i = 1, 2, \ldots,$$
$$\omega - x - 1 \quad \text{and} \quad t(i, \omega + 1) = 1 \quad \text{for} \quad i = \omega - x, \omega - x + 1 \qquad (8.38)$$

The elements of **T** defined by (8.38) satisfy the stochastic requirement (8.3). For integer $u = 1, 2, \ldots, \omega$ the u-step transition probabilities are computed following (8.8) as

$$T^{(u)} = \prod_{j=1}^{u} T \quad \text{with elements denoted by} \quad t(i, j)^{(u)} \qquad (8.39)$$

The nonzero entries in (8.39) are

$$t(i, i + 1)^{(u)} = {}_u p_{x+i-1} \quad \text{and} \quad t(i, \omega + 2)^{(u)} = {}_u q_{x+i-1} \quad \text{for} \quad i = 1, 2, \ldots,$$
$$\omega - x - u \quad \text{and } t(i, \omega + 2)^{(u)} = 1 \quad \text{for} \quad i = \omega - x - u + 1, \ldots, \omega - x + 1 \quad (8.40)$$

The transition probability matrix **T** is denoted by $\mathbf{T}^{(u)}$ for $u = 1$ and can be partitioned into components defined by survival and decrement probabilities. For $1 \le u \le \omega - x - 1$ the u-step transition matrix (8.40) takes the general form

$$\mathbf{T}^{(u)} = \begin{pmatrix} \mathbf{0} & \mathbf{M}^{(u)} & \mathbf{N}^{(u)} \\ & & \\ \mathbf{0} & \mathbf{0} & \mathbf{1}^{(u)} \end{pmatrix} \tag{8.41}$$

where the components consist of the diagonal matrix

$$\mathbf{M}^{(u)} = \text{diag}({}_u p_x, {}_u p_{x+1}, \ldots, {}_u p_{\omega-u}) \tag{8.42}$$

the column vector

$$\mathbf{N}^{(u)} = ({}_u q_x, {}_u q_{x+1}, \ldots, {}_u q_{\omega-u})^T \tag{8.43}$$

a column vector consisting of all ones $\mathbf{1}^{(u)} = (1, 1, \ldots, 1)^T$, and matrices consisting of all zeros denoted by **0**. Since the limiting age is ω, we find that for $u = \omega - x$ the u-step transition matrix (8.41) converges to

$$\mathbf{T}^{(\omega-x)} = (0, 1) \tag{8.44}$$

The demonstration of homogeneous stochastic status chains is considered in Example 8.7, and Problem 8.14 outlines Excel computations.

Example 8.7

For a discrete stochastic status model age x, the single-decrement life table given in Table 8.2 applies where the limiting age is $\omega = x + 5$. The transition matrix (8.41) consists of components

$$\mathbf{M} = \begin{pmatrix} .9 & 0 & 0 & 0 \\ 0 & .8 & 0 & 0 \\ 0 & 0 & .7 & 0 \\ 0 & 0 & 0 & .3 \end{pmatrix} \quad \mathbf{N} = \begin{pmatrix} .1 \\ .2 \\ .3 \\ .7 \end{pmatrix} \tag{8.45}$$

The two- and three-step transition matrices follow by computing (8.39) using components (8.45) to give components in (8.41):

$$\mathbf{M}^{(2)} = \begin{pmatrix} .72 & 0 & 0 \\ 0 & .56 & 0 \\ 0 & 0 & .21 \end{pmatrix} \quad \mathbf{N}^{(2)} = \begin{pmatrix} .28 \\ .44 \\ .79 \end{pmatrix} \quad \text{and}$$

$$\mathbf{M}^{(3)} = \begin{pmatrix} .504 & 0 \\ 0 & .168 \end{pmatrix} \quad \mathbf{N}^{(3)} = \begin{pmatrix} .496 \\ .832 \end{pmatrix} \tag{8.46}$$

From (8.46), survival and decrement probabilities can be realized. Applying (8.42), the 2- and 3-year survival and decrement probabilities are $_2p_x = .72$ and $_3p_x = .504$ from (8.42) and $_2q_x = .28$ and $_3q_x = .496$ from (8.43).

8.3.1 Expected Curtate Future Lifetime

Stochastic status chains can be based on either homogeneous or nonhomogeneous Markov chains as outlined in Sections 8.2 and 8.3. One advantage of homogeneous chains is the stochastic status model age x; let the associated curtate future lifetime be denoted by $K(x)$, where the expected value is denoted $E\{K(x)\} = e_x$. For age less than the limiting age, $x \leq \omega - 1$, we have

$$e_x = 1 + p_x \, e_{x+1} \tag{8.47}$$

To write (8.47) in matrix form, first let

$$\mathbf{T}^* = \begin{pmatrix} 0 & M \\ 0 & 0 \end{pmatrix} \quad \text{and} \quad \mathbf{e}_x = (e_x, e_{x+1}, \ldots, e_\omega)^T \tag{8.48}$$

where $\mathbf{M} = \mathbf{M}^{(1)}$ as defined in (8.42). The expected curtate future lifetime satisfies relation

$$\mathbf{e}_x = 1 + \mathbf{T}^* \, \mathbf{e}_x \quad \text{with solution} \quad \mathbf{e}_x = (\mathbf{I} - \mathbf{T}^*)^{-1} \, 1 \tag{8.49}$$

Solution (8.49) represents a direct closed form for the computation of expected discrete future lifetimes in terms of component $\mathbf{M}$ and is demonstrated in Example 8.8. Excel computations of the expected curtate future lifetime are considered in Problem 8.11.

Example 8.8

The expected curtate future lifetime associated with the discrete stochastic status of Example 8.7 is computed. Component $\mathbf{M}$ is given in (8.45), and using matrix structure (8.48) gives

$$\mathbf{I} - \mathbf{T}^* = \begin{pmatrix} 1 & -.9 & 0 & 0 & 0 \\ 0 & 1 & -.8 & 0 & 0 \\ 0 & 0 & 1 & -.7 & 0 \\ 0 & 0 & 0 & 1 & -.3 \\ 0 & 0 & 0 & 0 & 1 \end{pmatrix} \tag{8.50}$$

Inverting (8.50) yields

$$(\mathbf{I} - \mathbf{T}^*)^{-1} = \begin{pmatrix} 1 & .9 & .72 & .504 & .1512 \\ 0 & 1 & .8 & .56 & .168 \\ 0 & 0 & 1 & .7 & .21 \\ 0 & 0 & 0 & 1 & .3 \\ 0 & 0 & 0 & 0 & 1 \end{pmatrix}$$

and the solution in (8.49) computes as the vector

$$\mathbf{e}_x = (3.2752, 2.528, 1.91, 1.3, 1.0)^T \qquad (8.51)$$

Thus, for a discrete stochastic status age x the expected curtate future lifetime is

$$E\{K(x)\} = 3.2752$$

Likewise, for subsequent ages the mean curtate future lifetimes are components in (8.51), namely, $E\{K(x+1)\} = 2.528$, $E\{K(x+2)\} = 1.91$, $E\{K(x+3)\} = 1.3$, and $E\{K(x+4)\} = 1$.

8.3.2 Actuarial Chains

Actuarial chains, introduced in Section 8.2.2, can be defined for stochastic status chains based on homogeneous structures and utilized to compute expected present values for standard life insurance. Analogous to Section 8.2.2, the discount factor is defined by $v = (1 + i)^{-1}$ for yearly effective interest rate i for $0 < i < 1$. For a stochastic status model age x, define the actuarial transition matrix $\mathbf{R}$ consisting of components denoted by $r(i, j)$ where the nonzero entries are

$$r(i, i+1) = E_{x+i-1}, \quad r(i, \omega) = A^1_{x:\overline{1}|} \quad \text{for } i = 1, 2, ..., \omega - x - 1$$

$$\text{and} \quad r(i, \omega + 1) = v \text{ for } i = \omega - x \quad r(i, \omega + 1) = 1 \quad \text{for } i = \omega - x + 1 \qquad (8.52)$$

where $E_{x+i-1} = v\, p_{x+i-1}$ and $A^1_{x:\overline{1}|} = v\, q_{x+i-1}$. For integer $u = 1, 2, ..., \omega - x$ the u-step actuarial transition matrix follows (8.17) and is

$$\mathbf{R}^{(u)} = \prod_{j=1}^{u} \mathbf{R} \qquad (8.53)$$

where the nonzero elements are

$$r_{i,r+1}^{(u)} = {}_uE_{x+i-1}, \quad r_{i,\omega+2}^{(u)} = A_{x+i-1:\overline{u}|}^1 \quad \text{for} \quad i = 1, 2, \ldots, \omega - x - u$$

$$r_{i,\omega+2}^{(u)} = A_{x+i-1} \quad \text{for} \quad i = \omega - x - u + 1, \ldots, \omega - x + 1 \quad \text{and} \quad r_{\omega+2,\omega+2}^{(u)} = 1 \qquad (8.54)$$

where we note that $A_\omega = v$. The elements defined in (8.54) correspond to present value expectations for term and endowment discrete life insurance.

Mirroring the notations and structures presented in (8.41), the actuarial matrix
R is denoted by $\mathbf{R}^{(u)}$, and for $1 \le u \le \omega - x$ these matrices take the general form

$$\mathbf{R}_x^{(u)} = \begin{pmatrix} \mathbf{0} & \mathbf{U}^{(u)} & \mathbf{V}^{(u)} \\ \mathbf{0} & \mathbf{0} & \mathbf{W}^{(u)} \end{pmatrix} \qquad (8.55)$$

where the nonzero components consist of the diagonal matrix

$$\mathbf{U}^{(u)} = \text{diag}({}_uE_{x}, {}_uE_{x+1}, \ldots, {}_uE_{\omega-u}) \qquad (8.56)$$

along with the two vectors

$$\mathbf{V}^{(u)} = (A_{x:\overline{u}|}^1, A_{x+1:\overline{u}|}^1, \ldots, A_{\omega-u-1:\overline{u}|}^1)^T \quad \text{and} \quad \mathbf{W}^{(u)} = (A_{\omega-u}, \ldots, A_{\omega-2}, v, 1)^T \quad (8.57)$$

If $u = \omega - x + 1$, then

$$\mathbf{R}^{(\omega-x)} = (\mathbf{0}, \mathbf{W}^{(\omega-x)}), \quad \text{where} \quad \mathbf{W}^{(\omega-x)} = (A_x, \ldots, A_{\omega-2}, v, 1)^T \qquad (8.58)$$

Thus, the multistep transition matrix approach can be utilized to compute
standard unit-valued discrete whole life insurance expectations, and an
example now follows.

Example 8.9

Actuarial computations for the single-decrement setting of Example 8.7
are explored where $i = .06$. The actuarial transition matrix with nonnegative components following (8.52) takes form (8.55) with components

$$\mathbf{U}^{(1)} = \begin{pmatrix} .8491 & 0 & 0 & 0 \\ 0 & .7547 & 0 & 0 \\ 0 & 0 & .6604 & 0 \\ 0 & 0 & 0 & .2830 \end{pmatrix} \quad \mathbf{V}^{(1)} = \begin{pmatrix} .0943 \\ .1887 \\ .2830 \\ .6604 \end{pmatrix}$$

$$(8.59)$$

$$\mathbf{W}^{(1)} = \begin{pmatrix} .9434 \\ 1 \end{pmatrix}$$

Multiplying (8.59) as in (8.53), the two- and three-step actuarial transition matrices take form (7.90) with components

$$
\mathbf{U}^{(2)} = \begin{pmatrix} .6408 & 0 & 0 \\ 0 & .4984 & 0 \\ 0 & 0 & .1869 \end{pmatrix} \quad \mathbf{V}^{(2)} = \begin{pmatrix} .2545 \\ .4023 \\ .7191 \end{pmatrix} \quad \mathbf{W}^{(2)} = \begin{pmatrix} .9274 \\ .9434 \\ 1.0 \end{pmatrix}
$$

$$(8.60)$$

and

$$
\mathbf{U}^{(3)} = \begin{pmatrix} .4232 & 0 \\ 0 & .1411 \end{pmatrix} \quad \mathbf{V}^{(3)} = \begin{pmatrix} .4359 \\ .7314 \end{pmatrix} \quad \mathbf{W}^{(3)} = \begin{pmatrix} .8954 \\ .9274 \\ .9434 \\ 1.0 \end{pmatrix}
$$

$$(8.61)$$

The computations in (8.60) give discrete life insurance present value expectations for a term life insurance $A^1_{x:\overline{2}|} = .25454$, for a whole life insurance $A^1_{x+3:\overline{2}|} = A_{x+3} = .9274$, and the discount quantity $v = .9434$. From (8.61), standard present value life insurance expectations are $A_{x:\overline{3}|} = .4359$ and $A_{x+2} = .8954$. Further, to demonstrate convergence (8.58) with $u = 5$ we compute

$$
\mathbf{W}^{(5)} = (.8283 \quad .8645 \quad .8954 \quad .9274 \quad .9434 \quad 1.0)^T \tag{8.62}
$$

Thus, whole life insurance present value expectations are realized as $A_x = .8283$, $A_{x+1} = .8645$, $A_{x+2} = .8954$, $A_{x+3} = .9274$, and $v = .9434$.

8.4 Survivorship Chains

In this section, nonhomogeneous Markov chain techniques are directly applied to a closed-group survivorship structure, as presented in Section 5.6, to produce survivorship chains. These chains are based on the stochastic chain structure of Section 8.2 and used to compute survivorship quantities associated with various actuarial structures, namely, single-decrement, multiple-decrement, and multirisk strata models. For these models, the application of survivorship chains allows for the modeling of the decrement pattern for target populations or individual risk strata. Combining multirisk strata, survivorship chains techniques to model the growth and decrement pattern for a general population are developed.

8.4.1 Single-Decrement Models

Consider the modeling of a discrete single-decrement stochastic status using group survivorship concepts and notations. The number of newborns is denoted l_0, while the mean number of survivors to age x, following (5.45), is denoted by l_x for $x > 0$. Associated with (x) the expected number of decrements within u years, following (5.47), is denoted by $_u d_x$ for nonnegative integers u and x. To model the single-decrement structure using a nonhomogeneous survivorship chain, we first define, for a stochastic status age x, the vector of population totals as

$$l_x = (l_x, 0)^T \tag{8.63}$$

For age x the u-step transition probabilities (8.11) and transition matrix (8.10) are found and utilized to compute the vector of expected survival and decrement totals for subsequent years. For a status age x the expected total after u additional years is computed as

$$l_{x+u}^T = l_x^T T_x^{(u)} = (l_{x+u}, \, _u d_x) \tag{8.64}$$

where $_u d_x = d_x + \ldots + d_{x+u-1}$ for $u \geq 1$. The individual years' mean decrement totals are computed as $d_{x+1} = {}_2 d_x - d_x = l_x - l_{x+1}$ for $x \geq 0$. An example demonstrating these computations is now given, and Excel computations are considered in Problem 8.16.

Example 8.10

For a population let the yearly decrement and survival rates be given in Table 5.2. To demonstrate the survivorship chain approach the surrogate population starts with $l_0 = 100{,}000$ newborns. The transition matrices with components (8.9) for the first 3 years are

$$\mathbf{T}_0 = \begin{pmatrix} .99499 & .00501 \\ 0 & 1 \end{pmatrix} \quad \mathbf{T}_1 = \begin{pmatrix} .99494 & .00506 \\ 0 & 1 \end{pmatrix}$$

$$\mathbf{T}_2 = \begin{pmatrix} .99489 & .00511 \\ 0 & 1 \end{pmatrix}$$

For newborns $x = 0$ and applying (8.5), the u-step transition matrices for two subsequent years are found to be

$$\mathbf{T}_0^{(2)} = \begin{pmatrix} .989955 & .010045 \\ 0 & 1 \end{pmatrix} \quad \mathbf{T}_0^{(3)} = \begin{pmatrix} .984897 & .015103 \\ 0 & 1 \end{pmatrix} \tag{8.65}$$

Utilizing initial vector (8.63), we have $l_0 = (100,000, 0)$, and applying (8.64) we compute vector mean totals for the first 3 years as

$$l_1 = (99499, 501)^T \quad l_2 = (98995.5, 1004.5)^T \quad l_3 = (98489.7, 1510.3)^T \quad (8.66)$$

The first components in the vectors of (8.66) are the mean survival totals

$$l_1 = 99499, l_2 = 98995.5, \text{ and } l_3 = 98489.7$$

while the second components are the expected yearly cumulative decrement totals

$$d_0 = 501, {}_2d_0 = 1004.5, \quad \text{and} \quad {}_3d_0 = 1510.3$$

The expected yearly decrement totals are computed as $d_1 = 1{,}004.5 - 501 = 503.5$ and $d_2 = 1{,}510.3 - 1{,}004.5 = 505.8$. Survivorship chain techniques can be utilized to compute various life insurance computations, such as present value expectations for discrete life insurance.

8.4.2 Multiple-Decrement Models

In this section, survivorship chains are developed for the multiple-decrement setting introduced in Section 7.2. Based on multiple-decrement chains of Section 8.2.3, survivorship chains are used to compute expected survival and multiple-decrement totals. For a stochastic status model age x, the expected number of survivors is denoted by l_x, and the mean number of decrements by mode j within n years, as before, is denoted ${}_nd_x^{(j)}$. For a stochastic status age x with m distinct modes of decrement, define the vector:

$$l_x = (l_x, 0, ..., 0)^T \tag{8.67}$$

The u-step transition matrix (8.10) with components (8.21) is used along with (8.67) to compute the vector of subsequent survival and decrement expectation totals. The vector associated with age $x + u$ is

$$l_{x+u}^T = l_x^T \, \mathbf{T}_x^{(u)} = \left(l_{x+u}, {}_ud_x^{(1)}, ..., {}_ud_x^{(m)} \right) \tag{8.68}$$

where ${}_ud_x^{(j)} = d_x^{(1)} + \cdots + d_{x+u-1}^{(j)}$ for $u \geq 1$ and $1 \leq j \leq m$. Individual decrement mean totals are computed as $d_{x+1}^{(j)} = {}_2d_x^{(j)} - d_x^{(j)}$ for $j = 1, ..., m$ and $x \geq 0$. A multiple-decrement computational example follows, and Excel computations are outlined in Problem 8.17.

Example 8.11

A survivorship chain is applied to the multiple-decrement setting with $m = 3$ distinct modes of decrement discussed in Example 8.3. For a stochastic status age $x = 30$, let $l_{30} = 50{,}000$ so that from (8.67) we have $l_{30} =$

(50,000, 0, 0, 0). Using the transition probability matrix for age $x = 30$ in (8.22), the vector of mean survivorship totals associated with age 31 is

$$l_{31}^T = l_{30}^T \, \mathbf{T}_{30} = (49,750, 50, 100, 100)$$

The expected number of survivors to age 31 is $l_{31} = 49,750$, and the individual decrement mean totals are $d_{30}^{(1)} = 50$, $d_{30}^{(2)} = 100$, and $d_{30}^{(3)} = 100$. Likewise, using the u-step transition matrices (8.23) in (8.68) we compute

$$l_{31} = (49,302, 149.5, 249.25, 299) \text{ and } l_{32} = (48,809.23, 248.10, 446.46, 496.21)$$

Thus, the expected number of survivors to age 32 is found to be $l_{32} = 48,809.23$. Individual decrement mean totals for age $x = 31$ are $d_{31}^{(1)} = 149.6 - 50 = 99.6$, $d_{31}^{(2)} = 249.25 - 100 = 149.25$, and $d_{31}^{(3)} = 299 - 100 = 199$. Also, for age $x = 32$ the individual yearly decrement expectations are $d_{32}^{(1)} = 248.10 - 149.5 = 98.6$, $d_{32}^{(2)} = 446.46 - 249.25 = 197.21$, and $d_{32}^{(3)} = 496.21 - 299 = 197.21$.

8.4.3 Multirisk Strata Models

In this section, the survivorship chain approach is applied to the multirisk strata structure introduced in Section 8.2.4, where there are s distinct risk strata at each age. The expected number of newborns from the ith strata is denoted by $^{(i)}l_0$, and the mean number of survivors to age x belonging to the ith strata is denoted by $^{(i)}l_x$ for $x > 0$ and for $1 \le i \le s$. For a stochastic status age x, the vector of strata mean totals is denoted

$$l_x = (^{(1)}l_x, \ldots, ^{(s)}l_x, 0)^T \tag{8.69}$$

and the transition probabilities are given by (8.27). For positive integer u the u-step transition probabilities (8.28), (8.29), and (8.30) and transition matrix (8.5) are used to compute the risk strata expected survival and cumulative decrement totals as

$$l_{x+u}^T = l_x^T \mathbf{T}^{(u)} = \left(^{(1)}l_{x+u}, \ldots, ^{(s)}l_{x+u}, \, _u d_x \right) \tag{8.70}$$

where $_u d_x = d_x + \ldots + d_{x+u-1}$ for $u \ge 1$ and $1 \le j \le m$. Applying (8.70) the relative expected strata sizes can be modeled through subsequent years, and a demonstrative example follows. The use of Excel in survivorship chain computation is discussed in Problem 8.18.

Example 8.12

In the multirisk strata setting of Example 8.5 let the newborn totals for the two strata be $^{(1)}l_0 = 100$ and $^{(2)}l_0 = 900$, so that $^{(\circ)}l_0 = 1,000$. From (8.69) we have the initial vector of strata totals $l_0 = (100, 900, 0)^T$, and using (8.70) compute subsequent yearly vectors:

$$l_1 = (50, 820, 130), \quad l_2 = (15, 759.4, 225.6) \quad \text{and} \quad l_3 = (0, 708.6, 291.4) \tag{8.71}$$

For the first stratum, we find expected sizes 50 and 15, and that vanishes in subsequent years. This is consistent with the wear-out mortality structure of the first stratum. For the second stratum, the survival mean totals are 820, 759.4, and 708.6. The expected cumulative mortality totals are $d_0 = 130$, $_2d_0 = 225.6$, and $_3d_0 = 291.4$, so that the yearly decrement expectations are $d_1 = 225.6 - 130 = 95.6$ and $d_2 = 291.4 - 225.6 = 65.8$. These yearly decrement expectation totals for the second stratum are consistent with an infant mortality force of mortality.

Problems

8.1. Consider a discrete nonhomogeneous stochastic status chain where $MS = \{1, 2, ..., s\}$. For a discrete-status model age x let the transition probabilities be given by (8.9) with transition matrix (8.10) and actuarial transition matrix $\mathbf{R}_x$ with components (8.10).

   a. By matrix multiplication demonstrate for $u = 1$ and $u = 2$ multistep transition probabilities (8.11).

   b. Use (8.17) to show (8.18).

8.2. Consider a discrete nonhomogeneous stochastic status chain for multiple decrements where $m = 2$ and the transition probabilities are given by (8.20) with transition matrix (8.10). The actuarial transition matrix $\mathbf{R}_x$ has components (8.24).

   a. By matrix multiplication demonstrate for $u = 1$ and $u = 2$ multistep transition probabilities (8.21).

   b. Show (8.25).

8.3. Consider a discrete nonhomogeneous stochastic status chain for multirisk strata with $m = 2$ and the transition probabilities given by (8.27). The actuarial transition matrix $\mathbf{R}_x$ has components (8.35). By matrix multiplication (a) demonstrate for $u = 1$ multistep transition probabilities (8.28)–(8.30), and (b) show (8.36).

8.4. Consider a discrete homogeneous stochastic status chain for ages 0, 1, 2 and $\omega = 4$. The transition probabilities and the actuarial transition probabilities are given by (8.38) and (8.54). The u-step transition matrix (8.39) has components defined by (8.40), while the u-step actuarial transition matrix is given by (8.55).

   a. In (8.41) find the components for $\mathbf{M}^{(2)}$ and $\mathbf{N}^{(2)}$.

   b. Show (8.44) holds.

   c. In (8.56) and (8.57) find $\mathbf{U}^{(2)}$, $\mathbf{V}^{(2)}$, and $\mathbf{W}^{(2)}$.

8.5. Consider a discrete homogeneous stochastic status chain for ages 0 and 1 where $\omega = 2$.

a. Construct (8.48).

b. Demonstrate $(\mathbf{I} - \mathbf{T}^*)^{-1}$ consists of nonnegative entries given by $a_{ii} = 1$ for $i = 1, 2, 3$, $a_{12} = p_0$, $a_{13} = p_0 p_1$, and $a_{23} = p_1$.

c. Compute the mean in (8.49).

8.6. Consider the survival group nonhomogeneous survivorship chain under the settings of Section 8.4. Show for (a) single-decrement setting (8.39), (b) multiple-decrement setting (8.43), and (c) multirisk strata setting (8.70).

Excel Problems

8.7. We consider a single-decrement setting and consider mortality and survival computations using nonhomogeneous stochastic status chains. The below mortality table holds:

Age x	q_x	Age x	q_x	Age x	q_x
0	.05	3	.34	6	.71
1	.15	4	.47	7	.88
2	.22	5	.52	8	1.0

a. Following (8.9) find $\mathbf{T}_0$ and compute $\mathbf{T}_0^{(2)}$, give $_2q_0$, $_2p_0$, and compute $_{1|1}q_0$.

b. Compute $\mathbf{T}_0^{(3)}$, give $_3q_0$, $_3p_0$, and compute $_{2|1}q_0$.

Excel: MMULT.

8.8. We consider an actuarial chain associated with Problem 8.7 where $i = .04$.

a. Compute $\mathbf{R}_0^{(2)}$ and find $_2E_0$ and $A^1_{0:\overline{2}|}$.

b. Compute $\mathbf{R}_0^{(3)}$ and find $_3E_0$ and $A^1_{0:\overline{3}|}$.

Excel: Problem 8.6 extension: Basic operations.

8.9. The setting of Problem 8.7 is analyzed using a homogeneous stochastic status chain. In this problem we compute mortality and survival rates.

a. Compute $\mathbf{T}_0^{(2)}$, give $\mathbf{M}^{(2)}$ defined by (8.42), and $\mathbf{N}^{(2)}$ defined by (8.43). In particular, what are $_2p_0$, $_2p_2$, $_2q_0$, and $_2q_2$?

b. Compute $\mathbf{T}_0^{(3)}$, give $\mathbf{M}^{(3)}$ defined by (8.42) and $\mathbf{N}^{(3)}$ defined by (8.43). In particular, what are $_3p_0$, $_3p_2$, $_3q_0$, and $_3q_2$?

Excel: MMULT.

8.10. The actuarial setting associated with Problem 8.7 is analyzed using a homogeneous actuarial chain. We let $i = .05$.

 a. Compute $\mathbf{T}_0^{(2)}$, find $\mathbf{U}^{(2)}$ given by (8.56) and $\mathbf{V}^{(2)}$ and $\mathbf{W}^{(2)}$ given by (8.57). In particular, what are $_2E_0$, $_2E_2$, $A_{0:\overline{1}|}^1$, $A_{5:\overline{2}|}^1$, and A_7?

 b. Compute $\mathbf{R}_0^{(3)}$, find $\mathbf{U}^{(3)}$, $\mathbf{V}^{(3)}$, and $\mathbf{W}^{(3)}$. In particular, what are $_3E_0$, $_3E_2$, $A_{1:\overline{3}|}^1$, and A_6?

 Excel: MMULT.

8.11. The computation of expected curtate future lifetimes associated with Problem 8.7 is explored using a homogeneous stochastic status chain.

 a. Use (8.49) to compute e_x.

 b. What are $E\{K(0)\}$, $E\{K(1)\}$, and $E\{K(3)\}$?

 Excel: Utilize (8.48), MINVERSE, and MMULT.

8.12. Investigating Problem 8.7 we find two modes of decrement exist. Using nonhomogeneous multiple-decrement chains, mortality and survival probabilities are found based on the below multiple-decrement table:

Age x	$q_x^{(1)}$	$q_x^{(2)}$	$q_x^{(\tau)}$	Age x	$q_x^{(1)}$	$q_x^{(2)}$	$q_x^{(\tau)}$
0	.03	.02	.05	3	.22	.12	.34
1	.10	.05	.15	4	.30	.17	.47
2	.14	.08	.26	5	.32	.20	.52

 a. Compute $\mathbf{T}_0^{(2)}$, and give $_2q_0^{(1)}$, $_2p_0^{(2)}$, and $_2p_0^{(\tau)}$.

 b. Compute $\mathbf{T}_0^{(3)}$ and give $_3q_0^{(1)}$, $_3p_0^{(2)}$, and $_3p_0^{(\tau)}$.

 Excel: MMULT.

8.13. Consider the actuarial computations associated with Problem 8.12 where $i = .04$.

 a. Compute $\mathbf{R}_0^{(2)}$ and find $_2E_0$, $A_{0:\overline{2}|}^{1(1)}$, and $A_{0:\overline{2}|}^{1(2)}$.

 b. Compute $\mathbf{R}_0^{(3)}$ and give $_3E_0$, $A_{0:\overline{3}|}^{1(1)}$, and $A_{0:\overline{3}|}^{1(2)}$.

 Excel: Problem 8.6 extension: MMULT.

8.14. Investigating Problem 8.7 we introduce two risk strata. Using nonhomogeneous multirisk strata chains, mortality and survival probabilities are found based on the below multiple decrement table:

Age x	$^{(1)}q_x$	$^{(2)}q_x$	$^{(1)}p_x^{(2)}$	$^{(1)}p_x^{(2)}$	$^{(2)}p_x^{(1)}$	$^{(2)}p_x^{(2)}$
0	.07	.03	.90	.03	.02	.95
1	.18	.12	.72	.10	.13	.75
2	.27	.25	.62	.11	.09	.66

a. Compute $\mathbf{T}_0^{(2)}$ and give $_2^{(1)}p_0^{(1)}$, $_2^{(2)}p_0^{(2)}$, $_2^{(1)}q_0$, and $_2^{(2)}q_0$.

b. Compute $\mathbf{T}_0^{(3)}$ and give $_3^{(1)}p_0^{(1)}$, $_3^{(1)}p_0^{(1)}$, $_3^{(1)}q_0$, and $_3^{(2)}q_0$.

Excel: MMULT.

8.15. In the multirisk strata setting of Problem 8.14 suppose there are 40% newborns in stratum 1 and 60% newborns in stratum 2.

a. Give π_0.

b. Using (8.32) find π_1.

c. Using (8.32) find π_3.

d. After 3 years what proportion of the population (i) is in stratum 1 and (ii) has failed?

Excel: MMULT.

8.16. We consider group survivorship quantities associated with Problem 8.7 in the spreadsheet of Problem 8.6. We have 100,000 newborns so that $l_0 = 100,000$.

a. Compute l_1. What are l_1 and d_0?

b. Compute l_2. What are l_2 and $_2d_0$?

c. Compute l_3. What are l_3 and $_3d_0$?

Excel: MMULT.

8.17. Consider group survivorship quantities associated with Problem 8.12 where there are 100,000 newborns so that $l_0 = 100,000$. Enter l_0^T in N1:P1 following (8.67).

a. Compute l_1. What are l_1, $d_0^{(1)}$, and $d_0^{(2)}$?

b. Compute l_2. What are l_2, $_2d_0^{(1)}$, and $_2d_0^{(2)}$?

c. Compute l_3. What are l_3, $_3d_0^{(1)}$, and $_3d_0^{(2)}$?

Excel: Problem 8.12 extension: MMULT.

8.18. Group survivorship quantities associated with Problem 8.14 are considered where there are 60,000 in stratum 1 and 40,000 in stratum 2.

a. Compute l_1. What are $^{(1)}l_1$, $^{(2)}l_1$, and d_0?

b. Compute l_2. What are $^{(1)}l_2$, $^{(2)}l_2$, and $_2d_0$?

Excel: Problem 8.14 extension: MMULT.

Solutions

8.1. a. $t_x(1, 1)^{(2)} = p_x\, p_{x+1}$, $t_x(1, 2)^{(2)} = p_x\, q_{x+1} + q_x, t_x(1, 1)^{(3)} = p_x\, p_{x+1}\, p_{x+2}$, $t_x(1, 2)^{(3)}$
$= p_x\, p_{x+1}\, q_{x+2} + p_x\, q_{x+1} + q_x$.

  b. $r_x(1, 1)^{(2)} = v^2\, p_x\, p_{x+1}$, $t_x(1, 2)^{(2)} = v^2\, p_x\, q_{x+1} + v\, q_x, t_x(1, 1)^{(3)} = v^3\, p_x\, p_{x+1}\, p_{x+2}$,
$t_x(1, 2)^{(3)} = v^3\, p_x\, p_{x+1}\, q_{x+2} + v^2\, p_x\, q_{x+1} + v\, q_x$.

8.2. a. $t_x(1, 1)^{(2)} = p_x^{(\tau)}\, p_{x+1}^{(\tau)}$, $t_x(1, 2)^{(2)} = q_x^{(1)} + p_x^{(\tau)}\, q_{x+1}^{(1)}$, $t_x(2, 2)^{(2)} = q_x^{(2)} + p_x^{(\tau)}$
$q_{x+1}^{(2)}$.

  b. $t_x(1, 1)^{(2)} = v^2\, p_x^{(\tau)}\, p_{x+1}^{(\tau)}$, $t_x(1, 2)^{(2)} = v\, q_x^{(1)} + v^2\, p_x^{(\tau)}\, q_{x+1}^{(1)}$, $t_x(2, 2)^{(2)} = v$
$q_x^{(2)} + v^2\, p_x^{(\tau)}\, q_{x+1}^{(2)}$.

8.3. a. $t_x(1, 1)^{(2)} = {}^{(1)}p_x^{(1)}\, {}^{(1)}p_{x+1}^{(1)} + {}^{(1)}p_x^{(2)}\, {}^{(2)}p_{x+1}^{(1)}$, $t_x(1, 2)^{(2)} = {}^{(1)}p_x^{(1)}\, {}^{(1)}p_{x+1}^{(2)} +$
${}^{(1)}p_x^{(2)}\, {}^{(2)}p_{x+1}^{(2)}$, $t_x(1, 3)^{(2)} = {}^{(1)}q_x + {}^{(1)}p_x^{(1)}\, {}^{(1)}q_{\ x+1} + {}^{(1)}p_x^{(2)}\, {}^{(2)}q_{x+1}$.

  b. $r_x(1, 1)^{(2)} = v^2\, {}^{(1)}p_x^{(1)}\, {}^{(1)}p_{x+1}^{(1)} + v^2\, {}^{(1)}p_x^{(2)}\, {}^{(2)}p_{x+1}^{(1)}$, $t_x(1, 2)^{(2)} = v^2\, {}^{(1)}p_x^{(1)}$
${}^{(1)}p_{x+1}^{(2)} + v^2\, {}^{(1)}p_x^{(2)}\, {}^{(2)}p_{x+1}^{(2)}$, $t_x(1, 3)^{(2)} = v\, {}^{(1)}q_x + v^2\, {}^{(1)}p_x^{(1)}\, {}^{(1)}q_{\ x+1} + v^2$
${}^{(1)}p_x^{(2)}\, {}^{(2)}q_{x+1}$.

8.4. a. $\mathbf{M}^{(2)} = \mathrm{diag}(p_0\, p_1, p_1\, p_2)$, $\mathbf{N}^{(2)} = (q_0 + p_0 q_1, q_1 + p_1\, q_2)^T$.

  b. $\mathbf{U}^{(2)} = \mathrm{diag}(v^2\, p_0\, p_1, v^2\, p_1\, p_2)$, $\mathbf{V}^{(2)} = (v\, q_0 + v^2\, p_0\, q_1, v\, q_1 + v^2\, p_1\, q_2)^T$,
and $\mathbf{W}^{(2)} = (v^2\, p_2 + v,\ v,\ 1)^T$.

8.5. c. $\mathbf{e}_0 = (1 + p_0 + p_0\, p_1, 1 + p_1, 1)^T$.

8.6. a. Using (8.63) and (8.11), $l_{x+u} = (l_{x\ u}p_x,\ l_{x\ u}q_x)^T$.

  b. Using (8.67) and (8.21), $l_{x+u} = (l_{x\ u}p_x,\ l_{x\ u}q_x^{(1)}, ...,\ l_{x\ u}q_x^{(m)})^T$.

  c. Using (8.69) and (8.28)–(8.30), $l_{x+u} = (\sum {}^{(i)}l_{x\ u}^{(i)}p_x^{(1)}, ..., \sum {}^{(1)}l_{x\ u}p_x^{(s)}$,
$\sum {}^{(1)}l_{x\ u}^{(1)}q_x)$.

8.7. a. $_2q_0 = .1925$, $_2p_0 = .8075$, and $_{1|1}q_0 = .1925 - .05 = .1425$.

  b. $_3q_0 = .37015$, $_3p_0 = .62985$, and $_{2|1}q_0 = .37015 - .1925 = .17765$.

8.8 a. $_2E_0 = .746579$ and $A_{0:\overline{2}|}^1 = .179826$.

  b. $_3E_0 = .559934$ and $A_{0:\overline{3}|}^1 = .337756$.

8.9 a. $\mathbf{M}^{(2)} = \mathrm{diag}(.8075, .663, .5148, .3498, .2544, .1392, .0348)$, $\mathbf{N}^{(2)} =$
$(.1925, .337, .4852, .7456, .8608, .9652)^T$, $_2p_0 = .8075$, $_2p_2 = .5148$, $_2q_0 =$
$.1925$, and $_2q_2 = .6502$.

  b. $\mathbf{M}^{(3)} = \mathrm{diag}(.6299, .4376, .2728, .1679, .0738, .0167)$, $\mathbf{N}^{(3)} = (.3702,$
$.5624, .7272, .8321, .9262, .9833)$, $_3p_0 = .6299$, $_3p_2 = .2728$, $_3q_0 =$
$.3702$, and $_3q_2 = .8321$.

8.10. a. $\mathbf{U}^{(2)} = \mathrm{diag}(.7466, .6130, .4760, .3234, .2352, .1287, .0322)$, $\mathbf{V}^{(2)} =$
$(.1798, .3171, .4567, .6137, .7067, .8151, .9186)^T$, $\mathbf{W}^{(2)} = (.9571, .9615)^T$,
$_2E_0 = .7466\ _2E_2 = .4760$, $A_{0:\overline{2}|}^1 = .1798$, $A_{5:\overline{2}|}^1 = .8151$, and $A_7 = .9571$?

  b. $\mathbf{U}^{(3)} = \mathrm{diag}(.5599, .3890, .2426, .1493, .0656, .0149)$, $\mathbf{V}^{(3)} = (.3378,$
$.5175, .6718, .7754, .8673, .9240)^T$, $\mathbf{W}^{(3)} = (.9496, .9571, .9615)^T$, $_3E_0 =$
$.5599$, $_3E_2 = .2425$, $A_{1:\overline{3}|}^1 = .5175$, and $A_6 = .9496$?

8.11. a. $\mathbf{e}_x = (4.1635, 3.3299, 2.7411, 2.2322, 1.8670, 1.6359, 1.3248, 1.12, 1)^T.$

 b. $E\{K(0)\} = 4.1635$, $E\{K(1)\} = 3.3299$, and $E\{K(3)\} = 2.2322.$

8.12. a. $_2q_0^{(1)} = .125$, $_2p_0^{(2)} = .0675$, and $_2p_0^{(\tau)} = .8075.$

 b. $_3q_0^{(1)} = .23805$, $_3p_0^{(2)} = .1321$, and $_3p_0^{(\tau)} = .62985.$

8.13. a. $_2E_0 = .74658$, $A_{0:2|}^{1(1)} = .11668$, and $A_{0:2|}^{1(2)} = .06315.$

 b. $_3E_0 = .55993$, $A_{0:3|}^{1(1)} = .21718$, and $A_{0:3|}^{1(2)} = .12058.$

8.14. a. $_2^{(1)}p_0^{(1)} = .6513$, $_2^{(1)}p_0^{(2)} = .48674$, $_2^{(1)}q_0 = .2356$, and $_2^{(2)}q_0 = .1476.$

 b. $_3^{(1)}p_0^{(1)} = .41430$, $_3^{(2)}p_0^{(2)} = .48674$, $(_3^{(1)}q_0 = .43974)$, and $_3^{(2)}q_0 = .36346.$

8.15. a. $\pi_0 = (.4, .6, 0).$

 b. $\pi_1 = (.372, .582, .046).$

 c. $\pi_3 = (.25560, .35043, .39397).$

 d. i. 2,556.

 ii. 39,397.

8.16. a. $l_1 = 95,000$, $d_0 = 5,000.$

 b. $l_2 = 80,750$, $_2d_0 = 19,250.$

 c. $l_3 = 62,985$, $_3d_0 = 37,015.$

8.17. a. $l_1 = 95,000$, $d_0^{(1)} = 3,000$, $d_0^{(2)} = 2,000.$

 b. $l_2 = 80,750$, $_2d_0^{(1)} = 12,500$, and $_2d_0^{(2)} = 6,750.$

 c. $l_3 = 62,985$, $_3d_0^{(1)} = 23,805$, and $_3d_0^{(2)} = 13,210.$

8.18. a. $^{(1)}l_1 = 54,800$, $^{(2)}l_1 = 39,800$, $d_0 = 5,400.$

 b. $^{(1)}l_2 = 44,630$, $^{(2)}l_2 = 35,330$, $_2d_0 = 20,040.$

9

Scenario and Simulation Testing

Many modern techniques in financial and actuarial modeling involve the construction and analysis of mathematical and statistical models under various hypothetical settings. In fact, the demonstration of the financial viability under various economic scenarios, such as changing interest rates or stochastic return rates, may be desired or even required for individual companies or investment plans. Scenario testing, the effect of changing parameters on financial and actuarial computations, is introduced in Section 9.1. The stability of these measurements is assessed through this type of approach. Scenario testing is demonstrated for both deterministic and stochastic status models in, respectively, Sections 9.1.1 and 9.1.2.

Many financial or actuarial models may be complex in nature and may not lend themselves to direct theoretical statistical evaluation. These problems are attacked using modern simulation techniques. Generally, simulation methods refer to a collection of techniques where proxy data sets for stochastic outcomes are constructed and used for statistical evaluation. Simulation methods include bootstrap sampling and simulation sampling discussed in, respectively, Sections 9.2.1 and 9.2.2. Using these techniques, simulation probabilities are introduced in Section 9.2.3 and simulation prediction intervals are developed in Section 9.2.4. These statistical simulation techniques can be applied to various financial and actuarial problems. In particular, the simulation analyses of investment pricing models and stochastic surplus model evaluations are discussed in Sections 9.3 and 9.4. In applications, scenario testing and simulation methods may be applied simultaneously. The chapter ends with a discussion of the future direction of simulation analysis in Section 9.5.

9.1 Scenario Testing

In scenario testing, economic and actuarial formulas are often examined for their sensitivity to changes in model parameters. For example, changes in the future value of an investment or premiums for life insurance are examined in terms of interest rate variations. In Example 3.5, the present value of a future quantity over a time interval in the presence of varying interest rates was computed and compared to present value computed using the average interest rate over the time period. In this scenario test it was found that the mean interest rate approximation underestimates the present value. This general characteristic is a result of Jensen's inequality presented in Bowers (1997, p. 9) as applied to the present value function.

Scenario testing for deterministic and stochastic status models is demonstrated in separate sections. The criteria are used in scenario comparisons, including comparisons of fixed financial computations, statistical expectations, and prediction intervals for actuarial statistics.

9.1.1 Deterministic Status Scenarios

In the simplest form, the formulas related to deterministic status models, such as the present or future value function, contain no stochastic components, and direct computations under varying conditions can be compared. An introductory example (Example 3.4) demonstrated the effect on the continuous percentage rate for changes in the effective annual interest rate. The situation of stochastic rates present in deterministic status models is considered in Section 9.1.3. A typical use of scenario testing is the demonstration of the soundness of an economic strategy under adverse conditions, such as high or low interest or return rates.

In the two examples that follow the effects of changing rates on present and future value calculations in deterministic models are explored. The collection of scenario rates are denoted by either δ_j or i_j with corresponding present and future values $PV(t, j)$ and $FV(t, j)$ for $1 \leq j \leq m$. In the first example the present value of an investment is considered, and in the second the growth of an investment is discussed in terms of future values.

Example 9.1

The goal of an investment is to realize $1,000 in 5 years. The present value corresponding to this future amount follows formula (3.9) and is dependent on the continuous rate, δ, of a hypothetical investment that is held for 5 years. In Table 9.1 the present value of the investment, denoted $PV(5, j)$, is computed for return rates $\delta_j = .06 + .02j$ for $1 \leq j \leq 5$.

The changes in the present values related with hypothesized continuous rates are not linear, and overestimating the financial rate will cause the present value to be badly underestimated.

TABLE 9.1

Present Value Investment Scenario Testing Example

Return rate δ_j	.08	.10	.12	.14
$PV(5, j)$	$740.81	$670.32	$606.53	$496.58

TABLE 9.2

Future Value Annuity Scenario Testing Example

Interest rate i	.024	.036	.048	.060	.072
$FV(4, j)$	$7,549	$7,731	$7,920	$8,114	$8,315

Example 9.2

Scenario testing is applied to the discrete annuity model in Example 3.8 where the effect of changing annual interest rates on future values is investigated. A sum of $150 is deposited at the end of each month for 4 years. The future values of these deposits, $FV(4, j)$, are computed using (3.18) for interest rates $i_j = .024 + .012j$ for $0 \leq j \leq 4$ and are given in Table 9.2. From the future values listed, even minor changes in the interest rate affect the future value of the discrete annuity. As expected, the effect of changing rates is more dramatic for annuities as opposed to single-installment investments.

The previous two examples involve deterministic status models where the return and interest rates are fixed over time. Scenario testing is useful when comparing situations where the rate is not fixed over time but follows some variable pattern, and this concept is demonstrated in Problem 9.1. Our attention now turns to stochastic scenario models.

9.1.2 Stochastic Status Scenarios

Scenario testing can be applied to stochastic status models where the rates are varied over a range of values. The lifetime of the status is either the future lifetime random variable or the curtate future lifetime random variable and used to form relevant present value functions and related loss functions. In the examples that follow scenario testing is used to measure the effect of changing rates on net single values or insurance premium computations using the risk criteria (*RC*) and equivalence principal (*EP*) approach.

Example 9.3

For a unit benefit discrete whole life insurance policy, the curtate future lifetime follows a Poisson distribution with mean λ and probability mass function (pmf) $f(k) = e^{-\lambda}\lambda^k/k!$ for $k = 0, 1, \ldots$. The interest rates utilized in scenario testing are $i_j = .02j - .01$ for $j = 1, 2, 3, 4$, and 5. The present value expectations utilize $v_j = (1 + i_j)^{-1}$, and the net single values are

$$PV(K, j) = v_j^{K+1} \quad \text{and} \quad NSV(j) = {}^jA_x = \sum_{k=0}^{\infty} v_j^{k+1} e^{-\lambda}\lambda^k/k! = v_j \exp(\lambda(v_j - 1))$$

$$(9.1)$$

TABLE 9.3

Poisson Curtate Future Lifetime Whole Life Insurance

Interest rate i_j	.02	.04	.05	.06	.08
NSV(j)		.6624	.4455	.3614	.3041	.2105

for $1 \leq j \leq 5$. For mean lifetime $\lambda = 20$, expectation (9.1) is computed for the prospective interest rates and listed in Table 9.3.

The computed net single value is sensitive to interest rate changes, with greater interest rates leading to lower associated net single values.

Example 9.4

A whole life insurance policy pays a unit benefit at the moment of death where the future lifetime random variable follows an exponential distribution with mean $\theta > 0$ and probability density function (pdf) $f(t) = (1/\theta)$ $\exp(-\theta t)$ for $t > 0$. The expectation of the present value, or net single value, based on interest rate δ_j is denoted by $_j$NSV and computed as

$$NSV(j) = {}^j\bar{A}_x = \int_0^\infty (1/\theta)\exp(-\delta_j t)\exp(-t/\theta)\,dt = (1 + \delta_j\theta)^{-1} \qquad (9.2)$$

for $j \geq 1$. For mean lifetime $\theta = 20$ and five interest rates between .02 and .08, expectation (9.2) is computed and listed in Table 9.4.

As expected, interest rate increases are associated with decreases in the net single value for unit benefit whole life insurance.

Example 9.5

In this example scenario testing is applied to Example 6.7. For a status age 50, a discrete 4-year term life insurance with unit benefit is examined where possible interest rates are defined by $i_j = .02j - .01$ for $j = 1, 2, 3, 4,$ and 5. The unit benefit present value expectation associated with interest rate i_j and discount function $v_j = (1 + i_j)^{-1}$ is computed as

$$NSV(j) = {}^jA^1_{50:\overline{40}|} = \sum_{k=0}^{40-1} v_j^{k+1} d_{50+k}/l_{50} \qquad (9.3)$$

TABLE 9.4

Exponential Future Lifetime Whole Life Insurance

Interest rate δ_j	.02	.04	.05	.06	.08
NSV(j)		.71429	.55556	.50000	.45454	.38462

TABLE 9.5

Life Table-Based Term Life Insurance Scenario Testing

Interest rate i_j	.01	.03	.05	.07	.09
NSV(j)	.03984	.03794	.03618	.03456	.03304

for j = 1, 2, ..., 5. The expectations computed using (9.3) are listed in Table 9.5.

For term life insurance over a short time period the effect of varying interest rates on the computed expectation is minimal. If the term period is extended, the effect of changes in interest rates on expectations may be more profound.

Example 9.6

Associated with a stochastic status model age x the curtate future lifetime follows the pmf given by $P(K = k) = 1/60$ for $k = 0, 1, ..., 59$. For interest rates $i_j = .02j - .01$ and $v_j = (1 + i_j)^{-1}$ for $1 \le j \le 5$, the computations associated with fully discrete unit benefit whole life insurance are investigated. Namely, the present value expectations for discrete whole life insurance, whole life annuity, and EP premium are

$$^jA_x = \sum_{k=0}^{59} v_j^{k+1}/60 = (v_j - v_j^{61})/[60(1 - v_j)]^j\ddot{a}_x, = (1 - {}^jA_x)/d_j \text{ and } {}^jP_x = {}^jA_x/{}^j\ddot{a}_x$$

(9.4)

where $d_j = i_j/(1 + i_j)$ for $j \ge 1$. Quantities (9.4) are computed for the four effective annual interest rates and are given in Table 9.6.

We observe that the premium computation is less sensitive to changes in interest rate, in contrast to either life insurance or life annuity expectations separately. Further, as the interest rate increases the RC- and EP-based premium decreases.

From the previous examples we note that both life insurance and life annuity expectations, and to a lesser extent premiums, are affected by assumed interest rates. Our attention now turns to scenario testing in the case of stochastic rates where both deterministic and stochastic status models are considered.

TABLE 9.6

Fully Discrete Life Insurance Premium Scenario

Interest rate i_j	.01	.03	.05	.07
$^jA_x/{}^j\ddot{a}_x$	.7492/25.326	.4613/18.497	.3155/14.375	.2340/11.709
^jP_x	.02959	.02494	.02195	.01998

9.1.3 Stochastic Rate Scenarios

In this section scenario testing is demonstrated for both deterministic and stochastic status models where the rate is considered to be a random variable. Under various scenarios the distributional parameters are changed. In the first example, scenario testing is applied to an investment deterministic status where the return rate is a normal random variable. The sensitivity of the future value expectation in terms of changes in the mean parameter is examined. In the second example, life insurance net single computations are scenario tested where the interest is modeled by a uniform random variable. Prediction intervals for the future value are computed for the range of parameter values and utilized in sensitivity comparison. Excel usage in scenario testing of life insurance expectations is given in Problems 9.13 and 9.14.

Example 9.7

Scenario testing is applied to the investment setting of Example 3.12. The amount \$10,000 is invested for a period of 5 years where the monthly rate is a normal random variable defined by (3.28). The expected future value after 5 years, or $n = 60$ periods, ${}^{j}E\{FV(5)\}$, for various parameter vectors $\theta_j = (\gamma_j, \beta_j^2)$, follows (3.35), taking form

$$ {}^{j}E\{FV(5)\} = \$10,000 \ \exp(60(\gamma_j + (1/2)\ \beta_j^2)) \qquad (9.5) $$

for $j \geq 1$. For specific parameter combinations (9.5) is computed and the results are listed in Table 9.7.

Changes in the hypothetical parameters affect the expected future value calculations, where the mean has a greater impact than the standard deviation. Further, scenario prediction intervals for the future value are constructed for sets of hypothetical parameters. The variable monthly mean γ_j is varied, but standard deviation is fixed at $\beta = .01$. Following formula (3.30), a 90% prediction interval for the future value is ${}^{j}LB(FV(5)) \leq {}^{j}FV(5) \leq {}^{j}UB(FV(5))$, where the bounds are

$$ {}^{j}LB(FV(5)) = \$10,000 \ \exp(60\gamma_j - 1.645\ (60^{1/2})\ \beta_j) \qquad (9.6) $$

TABLE 9.7

Expected Future Value for Normal Rate Random Variable

Mean parameter	$\gamma_1 = .005$	$\gamma_2 = .008$	$\gamma_3 = .010$
Standard deviation	$\beta = .008/.012$	$\beta = .008/.012$	$\beta = .008/.012$
${}^{j}E\{FV(5)\}$	\$13,520/\$13,550	\$16,190/\$16,230	\$18,250/\$18,330

TABLE 9.8

Prediction Interval for Normal Rate Random Variable

Mean Rate γ_j	$^jLB(FV(5))$	$^jE\{FV(5)\}$	$^jUB(FV(5))$
.005	$11,884	$13,539	$15,333
.010	$16,041	$18,276	$20,697
.020	$29,230	$33,301	$37,712

and

$$^jUB(FV(5)) = \$10,000 \exp(60\gamma_j + 1.645\,(60^{1/2})\,\beta_j) \tag{9.7}$$

for $j \geq 1$. For mean rates .005, .008, and .010 quantities (9.6) and (9.7) are computed and listed in Table 9.8.

Scenario testing results can be used in application assessments. For example, if the monthly mean rate is .005, then the future value is at least $11,884, while this minimum increases to $16,041 for a mean return rate of $\delta = .01$.

Example 9.8

The unit benefit continuous whole life insurance policy introduced in Example 9.4 is examined where the future lifetime random variable is exponentially distributed with mean $\theta = 20$. Uncertainty about future interest rates is manifested by assuming a uniform distribution for the interest rate variable between positive constants a and b. For scenario testing, we compare unit benefit expectations using parameter (a_j, b_j), where $a_j = .01(j - 1)$ and $b_j = .11 - .01j$ for $1 \leq j \leq 5$. Using RC and EP gives (9.2), and integrating with respect to the interest rate random variable produces the expected net single value

$$\text{ENSV}(j) = E\{^j\bar{A}_x\} = \int_{a_j}^{b_j} (1 + \delta\theta)^{-1}\,d\delta/(b_j - a_j) = \ln[(1 + \theta b_j)/(1 + \theta a_j)]/[\theta(b_j - a_j)]$$

$$\tag{9.8}$$

for $1 \leq j \leq 5$. Further, prediction intervals for the net single value (NSV) are formed utilizing the distribution of the interest rate. For a 90% prediction interval (PI) based on equal tail probabilities the uniform distribution implies

$$.05 = P(\delta < .95a_j + .05b_j) = P(\delta > .05a_j + .95b_j) \tag{9.9}$$

A 90% PI for the NSV(j) defined by (9.2) is

$$(1 + .05a_j + .95b_j)^{-1} = {}^jLB(\text{NSV}) \leq \text{NSV}(j) \leq {}^jUB(\text{NSV}) = (1 + .95a_j + .05b_j)^{-1} \tag{9.10}$$

TABLE 9.9

Uniform Interest Rate Whole Life Insurance

Parameter (a, b)	(.00, .10)	(.01, .09)	(.02, .08)	(.03, .07)	(.04, .06)
ENSV$_j$	.5493	.5296	.5159	.5068	.5017
^jLB(NSV)	.3448	.3676	.3937	.4237	.4587
^jUB(NSV)	.9091	.7813	.6849	.6098	.5495

The expectations given by (9.8) as well as the *PI* given in (9.10) are listed in Table 9.9 for the sets of parametric values.

Applying the mean of the uniform interest rate $\delta = .05$ from Table 9.4, we find NSV = .5. As uncertainty grows, reflected by increasing the support of the interest random variable, the NSV expectation increases.

9.2 Simulation Techniques

The effect of modern computing power on modern mathematics and statistics can be seen in the number of new calculation-intensive modeling estimation and evaluation techniques. Some of these methods have become quite accepted, with examples being the Quenouille–Tukey jackknife, discussed in Quenouille (1949), Tukey (1958), and Miller (1974), along with the bootstrap procedures of Efron (1979). Modern procedures, such as data mining techniques (see Hand et al., 2000), hold great promise in the modeling of random variables with less structure, such as interest and financial return rates. Breiman (2001b) has suggested that two separate statistical approaches exist. The first is the standard statistical model and distribution-based inference approach, while the second assumes much less about the structure of the data distribution and relies on algorithmic techniques. The basis of modern algorithmic models, such as random forests (see Breiman, 2001a) and support vector analysis (see Christianini and Shawe-Taylor, 2000) is modern computational techniques. For a discussion on the use of computing power in statistics we refer to Diaconnis and Efron (1983).

Financial and actuarial models based on stochastic statuses, by their very nature, often contain one or more stochastic components. The analysis of such compounding of stochastic actions may be mathematically complex. For this reason traditional statistical techniques and evaluation methods are not always available. Common theoretical statistical methods may not exist or may be too restrictive, in terms of their accompanying assumptions, to be employed. In this case, modern simulation methods provide a useful path for the analysis of these often complex models.

In this section, an introduction to simulation is presented in the form of two types of procedures that produce typical or proxy data sets. The first approach discussed in Section 9.2.1 is based on a bootstrap procedure where the observed sample serves as the basis set for the construction of future simulation samples. In the second technique, referred to as simulation sampling and considered in Section 9.2.2, typical data sets are generated from a hypothetical distribution. In these simulation approaches, the proxy data sets mirror the theoretical distributions, leading to the basis for robust statistical techniques. Statistical analysis, including probability estimation in Section 9.2.3 and prediction intervals in Section 9.2.4, which are based on these techniques, are considered.

9.2.1 Bootstrap Sampling

The first data generation method considered is a general bootstrap sampling method, denoted BS. The starting point is a sample of observed random variables denoted Y_j for $j = 1, 2, \ldots, n$, producing observed values $\{y_1, y_2, \ldots, y_n\}$. A sample from the observed is drawn at random without replacement to construct m simulated samples each of size n $\{y_{1i}, y_{2i}, \ldots, y_{ni}\}$ for $i = 1, 2, \ldots, m$. Statistics, such as future and present value functions, loss functions, or reserves, can be computed for each of the generated samples. For the ith sample, let the associated statistics be denoted by W_i for $i = 1, 2, \ldots, m$. In this way a proxy sampling distribution associated with these statistics is constructed, leading to statistical BS inference techniques. Two examples of BS applications follow, where the first concerns insurance claim amounts, while the second investigates investment future values.

Example 9.9

The claim values in Example 2.24 are analyzed using a BS procedure. From Table 2.5, the claim values are given by 1.5, 1.7, 1.8, 1.8, 1.9, 2.0, 2.1, 2.3, 2.5, and 2.8. The statistic to be computed based on each set of 10 simulated samples is the aggregate sum of the claims, denoted by W_i for $i = 1, 2, \ldots, m$. Using $m = 10$, simulated samples are drawn and the observed aggregate sum is computed and listed in increasing order in Table 9.10.

The observed mean and standard deviation of the BS sample are $\overline{W}_{BS} = 20.34$ and $S^2_{W,BS} = 1.099272$. An approximate normal distribution based on $(1 - \alpha)100\%$ PI for W is found, modifying (2.16), to be

$$\overline{W}_{BS} - z_{(1-\alpha/2)}[S^2_{W,BS}]^{1/2} \leq W \leq \overline{W}_{BS} + z_{(1-\alpha/2)}[S^2_{W,BS}]^{1/2} \tag{9.11}$$

TABLE 9.10

BS Claim Sum

18.1	18.8	19.8	20.4	20.5	20.7	20.8	20.9	21.6	21.8

TABLE 9.11

BS Future Values of Stock Prices

5.91	7.91	9.95	10.45	10.48	10.50	10.58	10.84	10.92	11.50	12.20	12.33
12.55	12.57	13.11	13.45	13.49	13.88	13.91	14.71	15.92	17.11	18.16	18.20 20.09

For example, a 50% *PI* for the aggregate sum *W* is found using (9.11) to be $19.599 \leq W \leq 21.082$. Normality concerns raise questions about the validity of (9.11) and alternate approaches are presented in Section 9.2.4.

Example 9.10

The *BS* procedure is applied to the stock prices observed in Example 3.16. Starting with initial value of 10.5, measured after the first period, the future value after the next 10 periods is to be estimated. Using Table 3.2, the observed values of the rates are given by .091, −.072, .156, .000, −.024, .024, .162, −.035, −.112, and .062. The statistic to be computed, based on each set of 10 simulated samples, is the future value utilizing (3.10) and (3.11), so that $W_i = 10.5 \exp(\psi_i)$ for $i = 1, 2, ..., m$. Using $m = 25$ simulated samples, the observed future values are given in increasing order in Table 9.11.

The mean and standard deviation of the *BS* future values are found to be 12.29 and 3.217, respectively. Hence, an approximate, normality-based, 90% prediction interval for the future value after an additional 10 periods is calculated following (9.11) as

$$6.998 = 12.29 - (1.645)(3.217) \leq FV(11) \leq 12.29 + (1.645)(3.217) = 17.582$$

Uncertainty of the appropriateness of the normal approximation to the aggregate variables leads to a validity problem for the *PI* given in (9.11). In Sections 9.2.3 and 9.2.4, non-normal-based prediction intervals for simulation data are presented.

9.2.2 Simulation Sampling

In the second simulation method a theoretical model is assumed for random variables present in the model, and distributional associated samples are drawn. Modern random data generation methods are employed to obtain the typical data sets. In this chapter, simulated sampling (*SS*) utilizes typical financial and actuarial-based distributions such as the normal, exponential, or uniform distribution to reproduce *m* replicated samples of size *n*. Using the notation of *BS*, the simulated samples are given by $\{y_{1i}, y_{2i}, ..., y_{ni}\}$ with associated statistics W_i, for $i = 1, 2,..., m$. Candidates for the sampling statistics are statistics of interest used in financial or actuarial modeling, such as present or future value functions.

The *SS* method produces a simulation sample from a specified distribution. To do this we utilize an algorithm to generate a uniform random variable, U, over $(0, 1)$, where the observed values are u_i for $i = 1, ..., m$. Let the random variable of interest be Y with known pdf $f(y)$ over support S and cumulative distribution function (cdf) $F(y)$. If Y is a continuous random variable, then $F(y)$ is monotone increasing with associated inverse of F denoted by F^{-1} over the support S. To generate observed data y_i from desired pdf and cdf, we utilize the uniform sample values u_i and apply the transformation

$$y_i = F^{-1}(u_i) \qquad (9.12)$$

for $i = 1, ..., m$. If the random variable to be simulated is discrete, the transformed method using (9.12) can be adjusted to yield samples taking values in the discrete-type support with the correct proportions. Normal random variables play a central role in the statistical modeling of financial and actuarial data, and their use in simulation is demonstrated in the following example.

Example 9.11

A sample is generated from a normal distribution with mean μ and standard deviation σ where the cdf $F(x) = \Phi((x - \mu)/\sigma)/\sigma)$ given by (1.19). Applying inversion (9.12), normal variable observations are generated using

$$y_i = \mu + \sigma \, \Phi^{-1}(u_i) \qquad (9.13)$$

where u_i are realized values of the uniform random variable on the unit interval for $i = 1, 2, ...$. The SS technique is demonstrated on the observed return rates discussed in Example 3.16. The $AR(1)$ model analysis based on the normal distribution is applied with point estimates

$$\hat{\phi} = -.2124887 \quad \text{and} \quad \hat{\delta} = .006996$$

Here $\delta_1 \sim N(\delta, \sigma^2)$, and for $j > 1$,

$$\delta_j \sim N(\hat{\delta}, \hat{\phi}(\hat{\delta}_{j-1} - \delta))$$

Simulating $m = 25$ samples of 10 periods each, the computed statistics are the future values $W_i = FV(11)$ for $i = 1, ..., 25$. The ordered SS values are given in Table 9.12.

TABLE 9.12

Future Values Based on SS AR(1) Modeling

5.52	6.22	7.17	8.12	8.21	8.25	8.40	8.43	8.69	8.81	9.15	9.98	10.19	10.7
10.77	10.80	10.85	11.25	11.29	11.32	11.85	14.23	16.97	17.51	17.85			

The simulated mean and variance are computed to be 10.504 and 3.233, respectively, and the approximate 95% *PI* for the future value (9.11) is computed as

$$4.167 = 10.504 - (1.96)(3.233) \leq FV(11) \leq 10.504 + (1.96)(3.233) = 16.841$$

As previously mentioned, the nonnormal nature of the future value statistics makes this type of *PI* suspect.

The Box–Muller transformation (see Hogg et al., 2005, p. 177) presents an alternate method used to generate outcomes from normal random variables. In the next example, SSS samples are drawn from the collection of nonnegative random variables discussed in Section 1.6, namely, the Pareto, lognormal, and Weibull.

Example 9.12

For an investment stochastic status, the future lifetime random variable is modeled using a nonnegative random variable. The *SS* techniques applying the Pareto, lognormal, and Weibull distributions with cdfs given by (1.46), (1.51), and (1.55), respectively, are used to simulate future lifetimes. For $u_i \sim$ iid uniform (0, 1) random variables, for $i = 1, \ldots$, the transformations (9.12) are listed in Table 9.13. Also listed in Table 9.13 are the assumed parameters, chosen so that the median is equal to 20, used in the *SS* for each particular distribution.

The uniform variables along with the *SS* for the corresponding future lifetimes following the Pareto, lognormal, and Weibull distributions are given in Table 9.14.

The statistic of interest is the future lifetime $W_i = FV(t_i) = \exp(\delta t_i)$ for $1 \leq i \leq 10$. For fixed rate $\delta = .05$, these values are listed in Table 9.15.

TABLE 9.13

SS for Nonnegative Random Variables

	Pareto	Lognormal	Weibull
$F^{-1}(u_i)$	$\beta (1 - u_i)^{-1/\alpha}$	$\exp(\mu + \sigma \, \Phi^{-1}(u_i))$	$\beta (-\ln(1 - u_i))^{1/\alpha}$
Parameters	$\alpha = 2, \beta = 14.142$	$\mu = 2.996, \sigma = .5$	$\alpha = 2, \beta = 24.022$

TABLE 9.14

An SS for Nonnegative Random Variable Future Lifetimes

u_i	.0145	.1007	.1386	.3820	.4074	.5965	.8632	.8846	.8991	.9585
Pareto	14.25	14.91	15.24	17.99	18.37	22.26	38.24	41.63	44.52	69.39
Lognormal	6.71	10.56	11.62	17.22	17.80	22.60	34.59	36.42	37.87	47.59
Weibull	2.90	7.83	9.28	16.66	17.38	22.89	33.88	35.30	36.38	42.85

TABLE 9.15

Future Value SS for Nonnegative Random Variables

Pareto	2.03	2.11	2.14	2.46	2.51	3.04	6.77	8.02	9.26	32.12
Lognormal	1.40	1.70	1.79	2.37	2.44	3.10	5.64	6.18	6.64	10.80
Weibull	1.15	1.48	1.59	2.30	2.38	3.14	5.44	5.84	6.17	8.52

The nonlinearity of the future lifetime statistic causes the normality-based *PI* of the form (9.11) to be inefficient. This is demonstrated in Problem 9.9, where two of the normal-based prediction intervals contain negative values. An alternate approach to the construction of efficient *PI* is presented in the next two sections.

9.2.3 Simulation Probabilities

The simulation techniques presented in Sections 9.2.1 and 9.2.2 form a platform for estimating the sampling distribution for financial and actuarial statistics. Order statistics discussed in Section 2.3 are utilized for the empirical estimation of the sampling distribution for relevant statistics. Either simulation method *BS* or *SS* is used to generate computed statistics corresponding to the order statistics $W_{(1)} < W_{(2)} < \ldots < W_{(m)}$. For constant w the empirical or sample-based cumulative distribution function is defined by

$$F(w)_S = P(W \le w)_S = \sum_{i=1}^{m} I(W_{(i)} \le w)/m \qquad (9.14)$$

where $I(A) = 1$ if A holds and 0 otherwise. Simulated percentiles for W can be solved for a fixed number of probabilities. For example, the simulated median is defined as the value of w in (9.14) producing a simulated distribution function value of .5. For general percentiles interpolation methods are required.

Example 9.13

The *SS* procedure is used to analyze the portfolio of 25 short-term insurance policies of Example 2.13. The probability of a claim during the time period is .1, where the claim amounts are distributed as normal random variables with mean \$1,000 and standard deviation \$200. Applying (9.13), normal claim amounts are generated conditioning on a probability of .1 to form an *SS* of the aggregate of the 25 policies' claims. The probability that the sum of claims exceeds \$5,000 is to be estimated. The Haldane type *A* approximation (HAA) and saddlepoint approximation (SPA) methods, discussed in Sections 2.4.2 and 2.4.3, produce a reliability approximation close to .05. Based on $m = 50,000$ samples, where the statistic of interest is the sum of claims W using (9.14),

$$P(W > \$5,000)_S = 1 - F(5,000)_S = .06616$$

Thus, there is a 6.6% probability that the claim sums exceed $5,000. Based on the HAA and SPA computations, a 1% discrepancy in this reliability demonstrates the efficiency of the HAA and SPA approximations.

Example 9.14

A collection of 25 insurance policies is analyzed over a series of discrete-time periods. In each time period the probability of a claim is .1 and the claim amounts are assumed to be exponential random variables with a mean of 100. The insurance policies are tracked for 1 year where the interest rate is .01 per month and claims are paid at the end of the month. The individual present values follow (3.13) and (3.12) for $\delta_j = .01(j)$ for $j = 1, \ldots,$ 12. The statistic of interest is the sum of the 25 present values, denoted by W, and the exponential claim amounts are generated using the Weibull formulas in Table 9.15 with $\alpha = 1$ and $\beta = 100$. Using SS-based 1,000 repetitions, the mean and standard deviations for the present value function are computed and found to be

$$\bar{W}_{SS} = \$28,114.18 \quad \text{and} \quad S^2_{SS} = 7,156.129$$

Applying (9.11), the approximate 95% *PI* for the present value is computed to be

$$\$14,088.167 \le PV(1) \le \$42,140.193 \tag{9.15}$$

One of the important usages of simulation techniques is to check the accuracy of a large sample or asymptotic statistics. Employing the empirical cumulative distribution function given in (9.14), the coverage probability for *PI* (9.15) can be approximated. Taking $m = 1,000$ in (9.14), we find $F(14,088)_S = .009$ and $F(42140)_S = .965$, so that the simulation coverage probability is computed to be $.965 - .009 = .956$, slightly larger than desired. For this reason the normal-based *PI* may not be the best choice. For large SS, in a straightforward manner the empirical or sample cdf (9.14) can be used to construct prediction intervals. Applying (9.14) based on 1,000 simulated samples, we find $F(43,500)_S = .975$ and $F(16,000)_S = .025$ and a 95% *PI* is

$$\$16,000 \le PV(1) \le \$43,000 \tag{9.16}$$

Comparing (9.15) and (9.16), we note that the width of the simulation *PI* is less than that of the normality-based *CI*.

Empirical-based statistic measurements, such as distribution function and reliability computations as well as interval estimations, work well for large samples. Based on the empirical inference techniques introduced in Section 2.3, the construction of prediction intervals for moderate sample sizes is developed in the next section.

9.2.4 Simulation Prediction Intervals

Statistical techniques on order statistics can be used to construct prediction intervals for financial and actuarial statistics utilizing either *BS* or *SS*. A financial or actuarial statistic, *W*, is computed for the generated samples, producing order statistics $W_{(1)} < W_{(2)} < \ldots < W_{(r)}$. To construct prediction intervals for *W*, the symmetric prediction interval (SPI) and the bias correcting prediction interval (BCPI) approaches, presented in Section 2.3, are applied to these order statistics. Both methods rely on the empirical distribution function (9.14) and an associated interpolation method utilizing percentile ranks. The first method is symmetric with respect to percentile ranks and uses equal tail probabilities to select the order statistics, while the second is not and is influenced by the shape of the underlying distribution.

The symmetric simulation prediction interval (SSPI) is based on SPI approach and utilizes percentile ranks of the form (2.50) defined by

$$a = (\alpha/2)(r+1) \quad \text{and} \quad b = (1 - \alpha/2)(r+1) \tag{9.17}$$

For statistic of interest *W*, the lower bound from (2.51) takes the form

$$LB(W) = W_{[a]} + (a - [a])(W_{[a]+1} - W_{[a]}) \tag{9.18}$$

while the upper bound based on (2.52) is

$$UB(W) = W_{[b]} + (b - [b])(W_{[b]+1} - W_{[b]}) \tag{9.19}$$

Thus, the $(1 - \alpha)100\%$ simulated sample prediction interval or SSPI is given by

$$LB(W) \leq W \leq UB(W) \tag{9.20}$$

This procedure based on order statistics is demonstrated in Example 9.15.

Example 9.15

Based on the *BS* sample of Example 9.10 consisting of 25 future values a 90% SSPI for the future value after 11 time periods is computed. Following (9.17), $a = .05(26) = 1.3$ and $b = .950(26) = 24.7$, and utilizing (9.18) and (9.19), the lower and upper bounds for $FV(11)$ are

$$LB(FV(11)) = 5.91 + .3(7.91 - 5.91) = 6.51$$

and

$$UB(FV(11)) = 18.2 + .7(20.09 - 18.2) = 20.083$$

Thus, the SSPI is computed as

$$6.51 \le FV(11) \le 19.523 \qquad (9.21)$$

SSPI (9.21) is 22.9% wider than the normal-based *PI* computed in Example 9.10. This is a result, in part, of the nonnormal nature of the future values.

Alternate interval estimates to normal-based symmetric intervals have been proposed by many authors. One such approach is the bias-correcting percentile method given by Efron and Tibshirani (1986). This method is demonstrated to be useful when the statistic is not a linear function of an approximate normal random variable. The BCPI technique of Section 2.3 is applied to the simulation sample. The resulting prediction interval method, denoted by SBCPI, starts out by defining the simulation mean and corresponding (9.14) empirical cumulative probability

$$\bar{W}_S = \sum_{i=1}^{r} W_{(i)}/r \quad \text{and} \quad F(\bar{W}_S)_S = \sum_{i=1}^{r} I(W_{(i)} \le \bar{W}_S)/r \qquad (9.22)$$

where $F(s)_s$ is the sample cdf. The centering and tail areas similar to (2.55) are given by

$$z_0 = \Phi^{-1}(F(\bar{W}_S)), \quad \alpha_L = \Phi(2z_o + z_{(\alpha/2)}), \quad \text{and} \quad \alpha_U = \Phi(2z_o + z_{(1-\alpha/2)}) \qquad (9.23)$$

Based on the tail areas in (9.23), the percentile ranks are defined by

$$a = \alpha_L (r + 1) \quad \text{and} \quad b = \alpha_U (r + 1) \qquad (9.24)$$

The SBCPI for W takes the form (9.20), with lower and upper bounds defined by (9.18) and (9.19). If the simulation mean coincides with the empirical median, both defined in (9.22), then in (9.23) $z_o = 0$ and the BCPI reduces to the SSPI. This method is demonstrated in the examples that follow, and Excel computations are considered in Problems 9.17, 9.18, and 9.19.

Example 9.16

In this example the SBCPI is contrasted to the 90% SSPI for the future value after the 11 periods of Example 9.15. The *BS* in Table 9.13 is utilized where the mean of simulation future values is 12.29, and in (9.22) $F(12.29)_s$ = 11/25. The quantities in (9.23) are computed as

$z_o = \Phi^{-1}(11/25) = -.15097$ and $\alpha_l = \Phi(2(-.15097) - 1.645) = .0258$ and $\alpha_u = \Phi(2(-.15097) + 1.645) = .9104$

The percentile ranks (9.24) are $a = .0258(26) = .67$ and $b = .9104(26) = 23.669$. Since $[a] = 0$ we modify the lower bound (9.18) to coincide with the minimum order statistic and the upper bound follows (9.19). Thus,

$LB(FV(11)) = 5.91$ and $UB(FV(11)) = 18.16 + .669(18.2 - 18.16) = 18.187$

TABLE 9.16

Simulation 50% Future Value Prediction Intervals

Distribution	SSPI	SBCPI
Pareto	$2.133 \leq FV(11) \leq 8.330$	$2.498 \leq FV(11) \leq 25.125$
Lognormal	$1.768 \leq FV(11) \leq 6.295$	$2.424 \leq FV(11) \leq 9.527$
Weibull	$1.858 \leq FV(11) \leq 4.768$	$2.614 \leq FV(11) \leq 5.560$

and the 90% SBCPI for the future value is

$$5.91 \leq FV(11) \leq 18.187 \qquad (9.25)$$

In comparing the SBCPI given by (9.25) to the SSPI of (9.21), a shift to the left is noticed. This is due to the skewness of the simulation sample where the simulated mean is less than the empirical median.

Example 9.17

The future value of the investment after 11 time periods, discussed in Example 9.12, is estimated using the *PI* methods of this section. The *SS* values for $FV(11)$ under Pareto, lognormal, and Weibull future lifetimes are listed in Table 9.16 and used to construct prediction intervals. For each distribution, 50% prediction intervals SSPI and SBCPI are computed.

For each distribution the simulated mean exceeds the simulated median and the SBCPI are shifted to the right of the SSPI. For the Pareto distribution the shift picks up the long right tail, producing a substantially wider interval.

The nature of financial and actuarial stochastic status models often leads to the construction of complex scenarios and challenging statistical model analysis. In the analysis of such models all stochastic complexities must be considered. The future lifetime random variable associated with a stochastic status, as well as other variables such as the financial rate, can be modeled using simulation techniques. This section is concluded with a simulation analysis of a fully continuous whole life insurance contract.

Example 9.18

A continuous whole life insurance policy for an individual age x is analyzed where the future lifetime random variable takes survival function (5.28) or $S(t) = {}_tp_x = [(100 - x - t)/(100 - x)]^{1/2}$. Based on (9.12) an SS is constructed where the lifetimes are

$$t_i = (100 - x)(1 - u_i^2) \qquad (9.26)$$

for $1 \leq i \leq m$. The benefit paid is variable with time and is given by $t^{1/2}$ for $t \geq 0$. From the basic model (6.1) the present value takes the form

$$PV(T) = T^{1/2} \exp(-\delta T) \quad \text{for} \quad 0 \leq T \leq 100 - x \qquad (9.27)$$

Present value computation (9.27) applied to the sample lifetimes (9.26) produces an SS where the statistic of interest $W_i = PV(T_i)$ for $1 \leq i \leq r$. This whole life insurance policy is financed by a continuous premium and its associated whole life annuity with present value given by

$$PV(T) = \pi \, \bar{a}_{\overline{T}} = (\pi/\delta(1 - e^{-\delta T})) \qquad (9.28)$$

Similarly, the SS of future lifetimes (9.26) is used to generate a simulation sample of annuity present values using (9.28). For example, for a status age $x = 25$ the interest rate is fixed at $\delta = .08$ and the replication number is $m = 1,000$. The simulation mean expectations corresponding to the present values of whole life insurance and whole life annuity along with the EP-based premium are

$$^s\bar{A}_x = .289133, \; ^s\bar{a}_x = 11.42158, \quad \text{and} \quad ^s\bar{\pi}_x = {^s\bar{A}_x}/{^s\bar{a}_x} = .0253146 \qquad (9.29)$$

For a benefit of $10,000, the premium payments would be $253.15 annually. In this example, added complexity arises from two sources. One is the stochastic nature of the claim amounts, and the other is the force of interest over time.

The application of simulation techniques is broad and may be applied to the analysis techniques and stochastic status financial and actuarial models presented in Chapter 6. Our discussion of simulation analysis of stochastic status models continues in two previously introduced settings. Simulation techniques and applications are applied to investment pricing models in Section 9.3 and surplus models in Section 9.4.

9.3 Investment Pricing Applications

Simulation methods are useful in the analysis of investment stochastic status models. The basic concepts and formulas for the investment pricing model are taken from Section 4.3.2, and simulation techniques are applied to analyze these often complex models. In the simplest investment pricing setting, the stochastic status defines the future lifetime associated with the sale of an asset not related to the sale price. This is demonstrated in Example 9.19.

Example 9.19

An investment of value F_0 is made where the guaranteed force of interest is $r = .02$. The present value of this investment's value at future time $T > 0$ follows (4.30)

$$PV(T) = F_0 \exp(-rT + \delta_T) \qquad (9.30)$$

The investment has a stochastic yearly return rate taken to be uniform between .06 and .18. Hence, $\psi_T = T\delta$, where δ is distributed as a uniform random variable on $(.06, .18)$ and the guaranteed interest rate is assumed to be $r = .04$. The length of time the investment is held depends on outside factors and is taken to be a normal random variable with mean 2 years and standard deviation of .5 of a year. Based on an *SS* procedure with repetitions of size 10,000, the statistic of interest, W, is given by present value (9.30) where the simulated mean and variance are

$$W_s = 122.4159 \quad \text{and} \quad S_{W,SS}^2 = 117.1523$$

The normality-based 95% *PI* for the investment pricing formula, following (9.11), is

$$101.20 = 122.415 - 1.96(10.8237) \le PV(T) \le 122.415 + 1.96(10.8237) = 143.63$$

The present value of this investment is estimated to be between 101.20 and 143.63. Depending on the setting or whether the investor is buying or selling, either the lower or upper bound in the prediction interval may be important to the investor.

In a more realistic and complex setting, the future time of sale may be dependent on factors related to either the future price or present value of the investment. To demonstrate a discrete pricing model using the monthly period model setting and notations, actions, such selling or buying, may be done after j periods or at future time $t_j = j/m$ for $j \ge 1$. The present value of the investment where the period force of interest is $r_{1/m}$, adapted from (4.30), sold at future time T_j is

$$PV(T_j) = F_0 \exp(-r_{1/m}\, j + \psi_j), \quad \text{where} \quad \psi_j = \sum_{i=1}^{j} \delta_i \tag{9.31}$$

A financial investment strategy depends on the future sales price. For a fixed sales price, denoted K, at an unspecified future time, the present value of the pricing model measures the difference in the future value and the selling price. Similar to option pricing present value (4.35), the stochastic status investment pricing model takes the present value function

$$PV(T_j, K) = e^{-rj}(F_0 \exp(\psi_j) - K) \tag{9.32}$$

for $j = 1, 2, \ldots$. In the normal rate setting, it is assumed that the period return rate δ_i is distributed iid $N(\gamma, \beta^2)$ for $j \ge 1$. An application of simulation analysis applied to a discrete-time period stochastic investment pricing model follows in the next example.

Example 9.20

Simulation techniques are applied on the investment scenario analyzed by the Black-Scholes method in Example 4.12. An investment of $F_0 = 100$ is considered where the expiry or date of sale is at 1 year and the guaranteed force of interest is $r = .04$. Normally distributed rates are assumed where over 1 year $\sigma = .02$. The period is 1 month long, implying in each period $r_{1/12} = .04/12 = .003333$ and the variance is $\beta^2 = .0004/12$. Using the conservative estimate of (4.33), with monthly mean $\gamma = .0033166$, an SS based on 10,000 samples is run. The simulation statistic of interest, W, is the present value (9.32) truncated so that the negative values are replaced by zeros. The simulated mean is found to be .4250254, which compares closely with the Black–Scholes computation of .4364. Simulation prediction intervals for the present value, namely, SSPI and BCPI, could be found leading to a more complete picture and a reliability measure of the present value point estimate.

The generality of the SS method is a great advantage in financial investment analysis since there are so many prospective variables an investor may want to consider. Applying the present value asset pricing formula in (9.32) in conjunction with SS to a variety of settings produces many possibilities for analysis. One of these is the analysis of the more complex American call option model as outlined in Section 4.3.3. In the following, two examples of investment particular strategies dealing with the European call option model are analyzed using simulation methods.

Example 9.21

In this example, an American call option, where the asset may be purchased for fixed strike price K on or before the expiry date, is analyzed using an SS approach. This model is more complex than the European call option model in that it has an added stochastic component. The sale will be made the first time period or month the future value of the investment exceeds the price K, up to 1 year. An SS consisting of 10,000 samples is conducted for various choices of K, and the simulated mean of statistic (9.32) is found, where the negative values are truncated at zero. The results are given in Table 9.17.

The American call option, as defined in this example, yields a smaller expectation than the corresponding European call option. This is due to the strategy of buying the stock the first time the price exceeds K, thereby stop-

TABLE 9.17

An SS American Call Option Analysis by Strike Price

Strike price K	100	101	102	103	104	105
Simulated mean	.9808	1.0266	.68049	.3214	.1184	.0422

TABLE 9.18

An *SS* Asset Pricing Analysis by Present Value Price

Present value price K_0	100	101	102	103	104
Simulated mean	1.1279	1.5787	1.6278	1.2050	.7087

ping its growth. Hence, a comparison of different option strategies depends on the exact financial actions, along with time triggers, to be defined.

Example 9.22

In this example a stock purchasing strategy, similar to the previous ones, with a different twist is analyzed using simulation. The investment strategy is to purchase the stock the first time, within 1 year, the investment present value, given by (9.30), exceeds fixed value K_0. The initial price of the investment is $F_0 = 100$, and the rates are taken to be independent normal random variables with yearly mean and variance given by .06 and .0004. The guaranteed rate is assumed to be .04. Based on monthly time periods, $\gamma = .06/12 = .005$, $r_{1/12} = .04/12 = .003333$, and $\beta^2 = .0004/12 = .0000333$. An *SS* consisting of 10,000 repetitions is conducted for various present value prices K_0, and the statistic of interest is the present value (9.32) with $K = K_0 \exp(r_{1/12} j)$. In this way an optimum or efficient value of K_0 can be determined by comparing the simulated means. These quantities are given in Table 9.18.

Based on the *SS* results the maximum value of the present value of the investment (9.32) occurs when the sales price K_0 is 102. This gives the investor an efficient financial investment strategy.

9.4 Stochastic Surplus Application

There are additional simulation applications involving the assessment of financial stability of investment and actuarial models. In this section, a simulation analysis of the discrete surplus model introduced in Section 4.5.1 is demonstrated. The surplus is defined by (4.72), $U_j = u + jc - S_j$, where u is the initial fund, c is the period payment, U_j is the surplus, and S_j is the aggregate loss, withdrawals, or claims for periods $j = 1, 2, \dots$. This setting is interpreted as a stochastic status model where the status is defined by the condition of a positive surplus. The future lifetime random variable is the future time of ruin T_r. Simulation techniques allow for a broader application and analysis of the surplus model, especially in terms of the underlying distribution of losses that is assumed.

Simulation techniques have an advantage over the asymptotic analysis in that models can be examined for any number of periods. Theoretical

TABLE 9.19

SS Period Ruin Probability for Exponential Claims

n	1	2	3	5	10	15	20	25	50
$R(u, n)_s$	.136	.165	.185	.193	.200	.206	.208	.203	.204

analysis in the finite number of period setting is complex and is dependent on the statistical distribution of prospective losses. Data samples are drawn sequentially until either ruin is observed or the maximum number of periods, denoted n, is reached. The simulated probability of ruin within n periods is denoted by $R(u, n)_s$. An SS analysis of a discrete stochastic surplus model is given in Example 9.23.

Example 9.23

Simulation analysis is applied to the discrete surplus model of Example 4.21 where the claim amounts are exponential random variables with mean 1, the period payment is $c = 2$, and the initial fund is $u = 2$. For fixed positive integer n, simulations are drawn sequentially until $U_j < 0$ or n periods is reached, whichever is first. An SS based on 10,000 repetitions is utilized, and the simulated ruin probability, $R(u, n)_s$, is found for various values of n and listed in Table 9.19.

The observed simulation ruin probability using SS is close to the theoretical ruin probability of .203 calculated in Example 4.21. In practical settings the number of samples in an SS can be increased until convergence of the observed simulation ruin probability is achieved.

In the analysis of stochastic surplus models, other variables are of interest to the modeler. One such variable is the magnitude of the loss at the time of ruin, while another is the mean surplus amount over a fixed or random time period. The analysis of general surplus models is theoretically complex, and simulation techniques give an avenue for economic and statistical evaluation.

9.5 Future Directions in Simulation Analysis

The invention of powerful sampling and analysis algorithms allows for the analysis of large data sets that lend themselves to financial and actuarial modeling. Unlike many scientific areas, there is no lack of data. Financial data as well as mortality and survival statistics are observed and published regularly. There exist many possibilities for the applications of computer-intensive modeling procedures in connection with the estimation and statistical analysis of these models. In the framework of simulation techniques, Markov chain and Baysian modeling provide fertile ground for new investigations.

One problem connected with simulation methods is the realization of unstable simulation statistics. This can happen in conjunction with the bootstrapping sampling techniques introduced in Section 9.2.1 where the underlying distribution of the data is heavy tailed or skewed. This type of data is often encountered in actuarial analysis, with one example being in terms of insurance claims. In general terms, a statistic is unstable if small changes in the observed sample can produce large changes in statistical values. Unstable statistics will produce wide prediction intervals using either the SSPI or BCPI methods. Further, convergence rates in connection with aggregate models will be adversely affected by unstable statistics. This situation is encountered in the modeling of future values in Example 9.12 with the Pareto, lognormal, and Weibull distributions. One solution to the unstable statistics problem is a technique introduced for jackknife resampling referred to as bagging; it is based on bootstrap aggregation of the data. This procedure was first introduced by Breiman (1996), and for a general review we refer to Buhlmann and Yu (2002). Bagging, along with associated variations, can be extended with additional theory to the *BS* setting of Section 9.2.1. Many other possibilities, such as classical trimmed or Winsorized mean techniques, exist, and much work is left to be done in the area of stable simulation estimation in the context of financial and actuarial modeling.

Connected with general simulation techniques are associated inference methods. The theory and application of bootstrap confidence intervals continues to be a source of much exploration in scientific literature. New simulation confidence interval methods that improve the accuracy of interval estimates of the form presented in Section 9.2.4 have been proposed. Some of these methods reduce the bias presented by typical bootstrap estimators and have been included in this chapter. These simulation methods include the ABC, BC_a, bootstrap-t, and nonparametric intervals. For a review of these intervals we refer the reader to DiCiccio and Efron (1996). These confidence intervals can be adapted to many settings in financial and actuarial modeling, thereby producing efficient prediction intervals.

The general bootstrap procedure can be extended to the dependent modeling structure. In the general dependent structured case, the construction of *BS* samples is more complicated than in the independent setting. In Example 9.11, *BS* sampling in connection with the dependent modeling situation consisting of an *AR*(1) model with normal variables was presented. Other more general and advanced dependent bootstrapping methods exist. A partial list of these methods includes the block, sieve, and local bootstraps. For a review and comparison of these we refer to Buhlmann (2002). Much work can be done in the dependent simulation and resampling setting.

One popular stochastically dependent modeling technique is that of Markov chain modeling. Chapter 8 gives basic discrete Markov chain theory and applications in connection with actuarial modeling. Markov chain simulation methods have been used extensively in various areas of scientific investigation with areas of application including statistical physics, spatial statistics, and Bayesian modeling. These dependent modeling methods have yet to be widely applied to financial and actuarial models. For example, the statistical variability of life table measurements, such as yearly mortality or survival rates and expected future lifetimes, has yet to be efficiently addressed. Simulation techniques in conjunction with Markov chain approaches provide a wide source of modeling possibilities.

A last avenue for further developments in simulation is in terms of Bayesian applications. The utilization of Markov chain models results in greater opportunities in connection with the Baysian modeling of financial and actuarial systems. With methods such as the Metropolis or Gibbs–Hastings algorithm, a general method for the simulation of stochastic processes has been demonstrated. For review articles we refer to Besag et al. (1995) and Geyer (1992).

Problems

9.1. An investment of $1,000 is made for 5 years where the yearly interest rate is variable. Three different scenarios are considered where the interest rates i_j for $1 \leq j \leq 3$ are listed below:

Year	1	2	3	4	5
$j = 1$	.01	.02	.03	.04	.05
$j = 2$	.05	.04	.03	.02	.01
$j = 3$	.03	.03	.03	.03	.03

Using (3.2) compute the future values for the different scenarios and make comments.

9.2. Consider a stochastic rate δ and function $(d^2/d\delta^2)g(\delta) > 0$ over the support of δ. Jensen's inequality implies

$$E\{g(\delta)\} \geq g(E\{\delta\}) \tag{9.33}$$

We consider a deterministic status over fixed time t.

a. For an investment of P apply (9.33) to the FV defined by (3.4).

b. For a fund of P in future time t apply (9.33) to the PV defined by (3.9).

9.3. In Problem 9.2 let the rate over time t be $\delta_t \sim N(t\mu, t\sigma^2)$. Using inequality (9.33), show (a) $E\{FV(t)\} \geq F_0 \exp(t\mu)$ and (b) $E\{FV(t)\} \geq \exp(-t\mu)$.

9.4. A deterministic status has stochastic rate δ with gamma pdf $f(\delta) = (1/\Gamma(\alpha)) \beta^{-\alpha} \delta^{\alpha-1} \exp(-\delta/\beta)$, with support $S = \{\delta > 0\}$ and parameters $\alpha > 0$ and $\beta > 0$. For fixed time t using the moment generating function (mgf) give formulas for (a) $FV(t)$ and (b) $PV(t)$.

9.5. Scenario testing is used to investigate the effect of interest rate on a short-term annuity. Over a 3-year period 1 unit is deposited in an account at the end of each month. The yearly interest rates investigated are $i_j = .035 + j(.005)$ with future value ${}^jFV(3)$ given by (3.18) for $1 \leq j \leq 5$. Compute the percentage increase of the value of the annuity over the face value as given by $PI_j = {}^jFV(3)/36 - 1$ for $1 \leq j \leq 5$.

9.6. In the scenario setting of Example 9.3 for i_j compute the discrete annuity expectations and EP-based premiums:

$$ {}^j\ddot{a}_x \quad \text{and} \quad {}^jP_x = {}^jA_x/{}^j\ddot{a}_x \quad \text{for} \quad 1 \leq j \leq 5 $$

9.7. Consider scenario testing for unit benefit discrete whole life insurance for a status age x where the rate $\delta_j \sim N(\gamma_j, \beta_j^2)$ for $1 \leq j \leq m$. Based on the life table data, (a) apply (9.5) to show expectation $\sum_{k=0}^{\infty} \exp$,

$$ NSV(j) = E\{{}^jA_x\} = \sum_{k=0}^{\infty} \exp\left(-(k+1)(\gamma_j + \beta_j^2/2)\right) d_{x+k}/l_x \tag{9.34} $$

and (b) apply (9.6) and (9.7) to obtain $(1 - \alpha)100\%$ lower and upper bounds,

$$ LB({}^jA_x) = \sum_{k=0}^{\infty} \exp\left(-(k+1)\gamma_j - z_{1-\alpha/2}\, (k+1)^{1/2}\, \beta_j\right) d_{x+k}/l_x \tag{9.35} $$

and

$$ UB({}^jA_x) \sum_{k=0}^{\infty} \exp\left(-(k+1)\gamma_j + z_{1-\alpha/2}\, (k+1)^{1/2}\, \beta_j\right) d_{x+k}/l_x \tag{9.36} $$

9.8. Consider scenario testing for unit benefit discrete whole life insurance for a status age x where the interest rate is uniformly distributed between positive constants $a_j < b_j$ for $1 \le j \le m$. Based on the life table data (9.8) modify to show expectation:

$$\text{ENSV}(j) = E\{{}^j A_x\}$$
$$= \Sigma \, [\exp(-(k+1)a_j) - \exp(-(k+1)b_j)] \, (k+1)^{-1} \, (b_j - a_j)^{-1} \, d_{x+k}/l_x \quad (9.37)$$

9.9. Consider the *SS* of Example 9.12 where the sample statistics are given in Table 9.15. Construct 50% *PI* for the future value using (9.11) for the (a) Pareto *SS*, (b) lognormal *SS*, and (c) Weibull *SS*.

9.10. Consider the setting of Example 9.12.

   a. Using the inverse of the distribution function and parameters given in Table 9.13 construct exact 50% *PI* for the future time and future value for the (i) Pareto distribution, (ii) lognormal Distribution, and (iii) Weibull distribution.

   b. Comment on the differences between the results of (a) and the SSPI and BCPI computed in Table 9.16.

9.11. Claims amounts X_i for $1 \le i \le 10$ presented in Example 2.24 are examined.

   a. Find a 50% *PI* for individual claim amount X using the BCPI of Section 2.3.

   b. For the *BS* in Table 9.10 find a 50% SBCPI for the aggregate sum of 10 claim amounts.

   c. Compare the *PI* in (b) with the linear extension of the *PI* in (a) for the aggregate sum of 10 claims.

9.12. Consider the sample of 10 lifetimes 11, 15, 21, 25, 27, 30, 32, 36, 39, and 42 presented in Example 3.1.

   a. Compute the normal-based 50% *PI* for individual lifetimes using (2.16).

   b. Compute a 50% SPI presented in Section 2.3.

Excel Problems

9.13. Scenario testing is applied to fully discrete unit benefit whole life insurance for a status age 2 following the life table in Problem 5.17. The interest rates take on values $i_j = .02j - .01$ for $1 \le j \le 5$.

   a. Compute ${}^j A_2$ for $1 \le j \le 5$.

   b. Compute ${}_j a_2$ for $1 \le j \le 5$.

   c. Compute the premiums ${}_j P$ for $1 \le j \le 5$.

   Excel: Utilize basic operations.

9.14. Consider the whole life insurance policy of Problem 9.7 and life table of Problem 5.17. Assume the interest rate $\delta_j \sim N(\mu_j, .01^2)$, where $\mu_j = .02 + .01j$ for $1 \leq j \leq 5$. The expectation (9.34) and the associate annuity and *EP*-based premium are explored.

a. Compute $E\{^j A_2\}$ for $1 \leq j \leq 5$.

b. Compute $E\{^j a_2\}$ for $1 \leq j \leq 5$.

c. Compute the premiums $E\{^j P\}$ for $1 \leq j \leq 5$.

Excel: Utilize basic operations.

9.15. Consider a discrete whole life insurance policy and life table of Problem 9.14. We assume the interest rate $\delta_j \sim N(\mu_j, .01^2)$, where $\mu_j = .02 + .01j$ for $1 \leq j \leq 5$. We explore 90% lower and upper bounds for the present value of life insurance.

a. Compute the lower bounds (9.35) for $1 \leq j \leq 5$.

b. Compute the upper bounds (9.36) for $1 \leq j \leq 5$.

Excel : Utilize basic operations.

9.16. Consider a discrete whole life insurance policy on life table of Problem 5.17, where the interest rate δ_j is uniform over (a_j, b_j), where $a_j = .01j$ and $b_j = .1 - .01j$ for $1 \leq j \leq 4$. Compute the life insurance expectations given by (9.37) for $1 \leq j \leq 4$.

Excel: Utilize basic operations.

9.17. Based on the 10 lifetimes in Problem 9.12, we generate a *BS* for a sample consisting of 20 repetitions of samples of size 20. For each *BS* sample we analyze the aggregate sum S_i for $1 \leq i \leq 20$.

a. Give the order statistics for S_i for $1 \leq i \leq 20$.

b. Construct an 80% SSPI for S.

c. Construct an 80% SBCPI for S.

Excel: Utilize, Tools, Data Analysis, Random Number Generator ok; Number of Variables: 20; Number of Random Numbers: 20; Distribution: Discrete, Value and Probability Input Range and Output Range and Sum.

9.18. For a stochastic status lifetime, *T*, have survival function $S(t) = 1/(1 + t)$ for support $S = (0, \infty)$.

a. Generate an *SS* lifetime of size 20.

b. Find an 80% *PI* for lifetimes applying (9.11).

Excel: Tools, Data Analysis, Random Number Generator; Distribution: Uniform, Output Range.

9.19. For the *SS* of lifetimes generated in Problem 9.1 let the statistic of interest be the future value $S = \exp(.04t)$.

a. Find the *SS* of future values.

b. Find an 80% SBCPI for the future value.

c. Find an exact 80% *PI* for the future value.

Solutions

9.1. NSV(1) = NSV(2) = 1,158.73 and NSV(3) = 1,159.27. Increasing and decreasing interest rates give the same *FV*, while using the mean interest rate yields the largest *FV*.

9.2. a. $g(\delta) = \exp(t\delta)$ gives $E\{FV(t)\} \geq \exp(t\,E\{\delta\}$.

b. $g(\delta) = \exp(-t\delta)$ gives $E\{FV(t)\} \geq \exp(-t\,E\{\delta\})$.

9.3. a. From (3.48), $E\{FV(t)\} = F_0 \exp(t(\mu + \sigma^2/2)) \geq F_0 \exp(E\{\delta_t\})$.

b. From (3.48), $E\{PV(t)\} = F_0 \exp(t(-\mu + \sigma^2/2)) \geq F_0 \exp(E\{-\delta_t\})$.

9.4. a. Using (3.4), $FV(t) = F_0\, M_\delta(t) = F_0\, (1 - \beta\, t)^{-\alpha}$.

b. Using (2.9), $PV(t) = F_0\, M_\delta(-t) = F_0\, (1 + \beta\, t)^{-\alpha}$.

9.5. $PI_1 = .0606$, $PI_2 = .0685$, $PI_3 = .0765$, $PI_4 = .0845$, $PI_5 = .0927$.

9.6. Using $d = i_j/(1 + i_j)$, we find $_1P_x = .0428$, $_2P_x = .0345$, $_3P_x = .0277$, $_4P_x = .0221$, and $_5P_x = .0176$.

9.9. a. $1.147 = 7.046 - .675(8.745) \leq S \leq 7.046 + .675(8.745) = 12.945$.

b. $-1.695 = 4.204 - .675(8.745) \leq S \leq 4.204 + .675(8.745) = 10.102$.

c. $-2.560 = 3.338 - .675(8.745) \leq S \leq 3.338 + .675(8.745) = 9.236$.

9.10.

a. i. $16.330 \leq T \leq 28.284$ and $2.263 \leq FV(11) \leq 4.113$.

ii. $14.279 \leq T \leq 28.029$ and $2.042 \leq FV(11) \leq 4.061$.

ii. $12.884 \leq T \leq 28.284$ and $1.905 \leq FV(11) \leq 4.113$.

9.11. a. From (2.54) and (2.55), $z_o = \Phi^{-1}(.6) = .25335$, $\alpha_l = .4334$, and $\alpha_u = .8812$, and from (2.56), $a = 4.6771$ and $b = 9.6936$. Using (2.51)–(2.53), $LB(X) = 1.8 + .677(1.9 - 1.8) = 1.868$ and $UB(X) = 2.5 + .694(2.8 - 2.5) = 2.708$.

b. From (9.22) and (9.23), $z_o = \Phi^{-1}(.3) = - .5244$, $\alpha_l = .0424$, and $\alpha_u = .3541$, and from (9.24), $a = .4666$ and $b = 3.8949$. Using (2.51)–(2.53), $LB(X) = 18.1$ and $UB(X) = 19.8 + .895(20.4 - 19.8) = 20.334$.

9.12. a. $20.991 = 27.8 - .6745(10.095) \leq X \leq 27.8 + .6745(10.095) = 34.609$.

b. $a = .25(11) = 2.75$, $b = .75(11) = 8.25$, $LB(X) = 15 + .75(21 - 15) = 19.6$, and $UB(X) = 36 + .25(39 - 36) = 37.75$.

9.13. a. .95422, .87130, .79843, .73414, .67720
     b. 4.6234, 4.41876, 4.2330, 4.0638, 3.9094
     c. .20639, .19718, .18862, .18065, .17322

9.14. a. .86934, .83057, .79392, .75926, .72646
     b. 4.4138, 4.3157, 4.2213, 4.1306, 4.0433
     c. 19696, .19246, .18808, .18381, .17967

9.15. a. 84041, .80315, .76786, .73448, .70289
     b. .89968, .85939, .82131, .78529, .75122

9.16. .63853, .63926, .64047, .64217

10

Further Statistical Considerations

In the previous chapters concepts and techniques, both theoretic and applied, concerning the modeling of financial and actuarial systems were introduced and discussed from a statistical point of view. In financial statistics topics presented included investment analysis, asset pricing, and options pricing. Basic topics in actuarial science, such as insurance, annuities, reserves, and pension plans, were explored. Applied statistic inference techniques were introduced and utilized not only to analyze possible future economic actions, but also to come to a better understanding of these economic and financial systems. The statistic inference techniques explored included estimation, percentile formulations, and prediction intervals, and their application shed much light on financial and actuarial models. There is room for many exciting new statistical techniques, as well as some standard ones, in these models. In this chapter we consider a few statistical considerations not addressed in the previous chapters.

Actuarial analysis, whether it is insurance pricing, pension planning, or reserves analysis, depends on associated life and mortality tables. In many analyses of stochastic status models individuals are classified into various risk categories. For insurance purposes, demographics, such as gender, smoking-nonsmoking, and geographic region, influence mortality rates. An introduction to multirisk strata modeling using Markov chains was presented in Section 8.2.4. In the past it was the job of the insurance underwriter to assess these added risks and measure their impact on survivability and insurance rates. This was often done without the aid of modern statistical techniques. In Section 10.1 techniques to adjust the yearly mortality rates to model high-risk strata are presented. These mortality acceleration models have a history in engineering reliability modeling. Two variants, one based on reduced mean future lifetimes and the second based on survival rates over a fixed number of years, are discussed in, respectively, Sections 10.1.2 and 10.1.3.

The mortality tables and resulting statistical measurements for individual statuses assume a steady-state environment. This is not the situation in practice, where as time advances, the mortality rates change for individuals in all risk categories. Modern advances in medicine and other areas produce a decrease in mortality rates as the future time variables increase. In Section 10.2 a method to adjust mortality tables for mortality trends in time based on a simple linear regression model is introduced. There are many methods present in the statistical literature to assess and make adjustments for time trends in data.

Based on a collection of independent and identically distributed (iid) stochastic statuses, the distribution of actuarial statistics, such as life insurance or life annuity present value expectations, can be examined. The concept of actuarial statistics is introduced in Section 10.3 and used in the statistical inference of the collection of statuses. Topics such as statistical estimation, prediction interval construction, and simulation prediction intervals are explored through theory and application in Sections 10.3.1, 10.3.2, and Section 10.3.3, respectively.

In financial and insurance modeling there is no lack of available data. The stock prices as well as various consumer indexes are published daily. The data on human survivability that comprise the basic information in life and mortality table construction are surveyed continuously. It is the goal of many modern statistical methods to make sense of the vast supply of data available to the financial and actuarial advisor. In Section 10.4 we give a brief introduction to statistical methods used in the construction of data classification techniques. Details for these statistical techniques are left to the interested reader to locate, and relevant references are listed.

10.1 Mortality Adjustment Models

The analysis of stochastic status models is dependent on the mortality structure utilized. Individual statuses are classified into separate strata based on survival characteristics. Demographics, such as gender, smoking-nonsmoking, region of the country, and lifestyle analysis, are used to affect the mortality measurements used in actuarial and associated statistical analysis. For some demographics, such as gender, separate mortality tables are currently being constructed and applied in the insurance industry. In a statistical framework these demographics, along with others, partition the population into separate risk strata. This idea was introduced in Section 8.2.4, where the separate risk strata had distinct mortality rates and the entire system was modeled using Markov chains. It is the task of the insurance underwriter to assess these survival risks and adjust the relevant computations. In the future the underwriting and mortality assessment of individual statuses will be aided by the construction and analysis of efficient risk strata models.

In this section we present an introduction to the topic of mortality adjustment for individual status models. The mortality adjustments are made on standard mortality table measurements and are typically associated with higher-risk or increased mortality statuses. In this way, mortality rates for various degrees of risk can be obtained. The resulting various risk strata actuarial models can be analyzed using modern techniques.

10.1.1 Linear Mortality Acceleration Models

The simplest form of mortality adjustment, referred to as a linear acceleration factor, associates the survival rates of two strata, one an increased risk
stratum and the other a standard or control stratum. The concept of linear
acceleration factor modeling is taken from applied engineering reliability
modeling, and as a reference, we refer to Tobias and Trindade (1995, Chapter
7). Lifetime random variables associated with individuals are taken from the
two strata, one a standard stratum and the other a high-risk stratum, and
they are assumed to be linearly associated.

For individuals in the control stratum the future lifetime random variable
is denoted by T, while for the high-risk stratum individuals the future lifetime random variable is ^sT. In a linear mortality acceleration model we relate
the two future lifetime random variables by the linear relation

$$T = AF \, ^sT \tag{10.1}$$

where AF is the linear acceleration factor. If $AF > 1$, then added stress associated with the high-risk stratum produces a shorter future lifetime with $T >$
^sT. Relationship (10.1) can be used to relate statistical functions, such as the
cumulative distribution function (cdf), probability density function (pdf), or
reliability, associated with the two groups.

Linear mortality acceleration modeling can be directly applied to standard
survival techniques. From (10.1) the survival functions corresponding to $T =$
t and $^sT = t$ are related as

$$P(^sT > t) = P(T > t \, AF) \tag{10.2}$$

Relation (10.2) can be reformulated in terms of the force of mortality discussed in Section 5.3. If the force of mortality for the control stratum is μ_{x+t},
then using the form of the survival function in (5.18), the survival probability
beyond future time t for the high-risk stratum is

$$P(^sT > t) = \exp\left(-\int_0^{tAF} \mu_{x+s} \, ds \right) \tag{10.3}$$

To explore the decrease in the survivor function represented by (10.3), the
exponent can be rewritten. Letting $w = s/AF$, we find $s = wAF$, so that $ds =$
$dwAF$ and (10.3) becomes

$$P(^sT > t) = \exp\left(-\int_0^t \mu_{x+w}^s \, dw \right), \quad \text{where} \quad \mu_{x+w}^s = AF\mu_{x+wAF} \tag{10.4}$$

From (10.4) we see that the linear mortality acceleration factor affects survival rates in two ways. The first is a direct multiplication effect, and the second is an adjustment to the force of mortality scale. Two specific examples now follow.

Example 10.1

The future lifetime random variable for an individual from the control stratum follows an exponential distribution with pdf $f(t) = (1/\theta)\exp(-t/\theta)$ for $t > 0$. The survival probability beyond future time $t > 0$ is

$$P(T > t) = \exp(-t/\theta) \quad \text{for} \quad t > 0 \tag{10.5}$$

From Example 5.5 we note that the force of mortality is a constant given by $\mu_{x+t} = \mu = 1/\theta$. For individuals from the high-risk stratum, the survival probabilities from (10.4) become

$$P({}^{s}T > t) = \exp(-t/{}^{s}\theta), \quad \text{where} \quad {}^{s}\theta = \theta/AF \tag{10.6}$$

for $t > 0$. From (10.6) the distribution of the future lifetime random variable associated with high-risk stratum individuals is exponential, with mean given by ${}^{s}\theta$. As stated earlier, decreased survival probabilities of the high-risk stratum, leading to increased mortality rates, are associated with $AF > 1$ and lower mean future lifetimes.

Example 10.2

For a stochastic status model age x let the control stratum be associated with a future lifetime random following a Weibull distribution, as presented in Section 1.6.3, with pdf $f(t) = \alpha\,\beta^{-\alpha}\,t^{\alpha-1}\exp(-(t/\beta)^{\alpha})$ for $t > 0$. The survival probability associated with future time t is

$$P(T > t) = {}_{t}p_{x} = \exp(-(t/\beta)^{\alpha}) \quad \text{for} \quad t > 0 \tag{10.7}$$

for shape parameter $\alpha > 0$ and scale parameter $\beta > 0$. For a status from the high-risk stratum, applying (10.2), the survival probability becomes

$$P({}^{s}T > t) = \exp(-(t/(\beta/AF))^{\alpha}) \tag{10.8}$$

From (10.8) the survival probability associated with the high-risk stratum is modeled through a change in the scale parameter. Clearly, as AF increases, (10.8) decreases and the mortality probabilities increase.

In the modeling of high-risk strata using a linear mortality acceleration approach there are many open statistical questions. One topic that needs to be addressed is the estimation of the acceleration factor constant AF. In a practical sense, we desire robust estimates for increased risk mortalities that can be used in conjunction with constructed mortality tables. Associated with a stochastic status model age x from the high-risk stratum the yearly survival and mortality rates are denoted by

$$^sp_x = P(^sT \geq 1) \text{ and } ^sq_x = 1 - ^sp_x \qquad (10.9)$$

In the next two sections, general techniques for computing or approximating the high-risk stratum survival probabilities of the type given in (10.9) are explored. These techniques are chosen due to their reliance on a minimal set of assumed information.

10.1.2 Mean Mortality Acceleration Models

In this section a general strategy to adjust the mortality rates for high-risk strata status models based on the mean of the future lifetime random variable is introduced. The AF quantity is estimated by computing the ratio of the mean future lifetime of the control stratum to the high-risk stratum. Associated with a stochastic status model age x, the future lifetime random variables for the control stratum and high-risk stratum are denoted by T and ^sT, where their associated expectations are $E\{T\}$ and $E\{^sT\}$. The ratio used in the modeling of the mortality adjustment factor is

$$r = E\{T\}/E\{^sT\} \qquad (10.10)$$

In practice, the means of the future lifetimes for the two strata, or representative approximations, may be readily available. For example, a standard life table and the uniform distribution of death (UDD) assumption may be used to estimate $E\{T\}$, as in (5.85). The high-risk stratum mean $E\{^sT\}$ may be estimated by expert testimony or based on other, smaller data sets.

The choice of the distribution used to model the survival probabilities affects the way (10.10) is applied. In the first case, the future lifetime random variable associated with any age follows an exponential distribution. Using (10.6), the mean future lifetime associated with the increased mortality group is $^s\theta = \theta/AF$, and (10.10) implies

$$r = AF \qquad (10.11)$$

For computed ratios, r, mortality adjustments to the standard mortality table quantities can be made. Combining (10.11) and (10.6), the adjusted yearly survival and mortality rates for the high-risk stratum (10.9) take the form

$$^sp_{x+k} = (p_{x+k})^r \text{ and } ^sq_{x+k} = 1 - ^sp_{x+k} \qquad (10.12)$$

for $k \geq 0$. The above technique suggests an easy method to adjust mortalities where only the ratio (10.11) is required. However, the exponential model assumption requires that the force of mortality associated with the future lifetime random variable be constant. This model may not always fit the data in a given mortality table. This approach is demonstrated in Example 10.3 and Problems 10.1 and 10.2, with Excel instructions outlined in Problem 10.5.

Example 10.3

For a discrete stochastic status model age x yearly survival totals and rates associated with the future lifetime random variable are given in Table 10.1, where the limiting duration is 10. For curtate future lifetime K the mean is $E\{K\} = 4.9315$, and assuming UDD, the mean future lifetime is $E\{T\} = 4.9315 + .5 = 5.4315$. For a high-risk stratum the mean lifetime drops to $E\{{}^sT\} = 2.0$, and for the corresponding curtate future lifetime ${}^sK$, we have $E\{{}^sK\} = 1.5$. From (10.10), $r = 5.4315/2 = 2.7158$, and applying (10.12), the survival and mortality rates for the high-risk stratum are

$$ {}^sp_{x+k} = p_{x+k}{}^{2.7158} \quad \text{and} \quad {}^sq_{x+k} = 1 - p_{x+k}{}^{2.7158} $$

for $k \geq 0$. The survival rates associated with the various ages are also listed in Table 10.1.

The mean lifetime for high-risk stratum individuals overestimates the curtate mean of 1.5, where we compute $E\{{}^sK\} = 2.7990$. The source of this discrepancy includes the inadequacies of modeling the life table mortalities with the exponential distribution where the UDD assumption does not hold.

There are a multitude of choices for the stochastic status distribution utilized in the construction of mortality adjustment functions based on the ratio of mean future lifetimes. We now present a method where the future lifetime random variable for each age is modeled by the Weibull survival function given by (10.7). The theoretical mean for the Weibull distribution is given by

$$ E\{T\} = \beta \, \Gamma(1 + 1/\alpha) \tag{10.13} $$

Utilizing the results of Example 10.2, the control stratum and the high-risk stratum have identical shape parameters α but differing scale parameters, β and ${}^s\beta$, respectively. For a given ratio (10.10), applying (10.13) yields

$$ {}^s\beta = \beta/r \tag{10.14} $$

TABLE 10.1

Mortality and Exponential Adjusted Mortalities

k	0	1	2	3	4	5	6	7	8	9	10
l_{x+k}	100	95	85	68	56	46	38	29	20	12	4
p_{x+k}	.950	.995	.800	.824	.821	.826	.763	.690	.600	.333	0.0
${}^sp_{x+k}$	.870	.986	.546	.591	.585	.595	.480	.365	.250	.050	0.0

Utilizing (10.7), the yearly survival rates (10.31) for the high-risk stratum can be written as

$$^sp_{x+k} = (p_{x+k})^{r(\alpha)} \tag{10.15}$$

where $r(\alpha) = r^\alpha$ for $k \geq 0$. To use (10.15), we need an estimate for the shape parameter α. To estimate α using a particular data set or mortality table we utilize the method of least squares in conjunction with a Weibull probability plot as discussed in Section 2.8.1. For $k \geq 1$, let the survival probabilities corresponding to each year be given by

$$_kp_x = \prod_{j=0}^{k-1} p_{x+j} \tag{10.16}$$

Defining the transformations $a_i = \ln(-\ln(_ip_x))$, $b_i = \ln(i)\ln(-\ln(_ip_x))$, and $c_i = \ln(i)$ for $i \geq 1$, the least squares estimator (2.105) is given by

$$\hat{\alpha} = \left[\sum_{i=1}^n a_i^2 - \left(\sum_{i=1}^n a_i \right) \Big/ n \right]^2 \Big/ \left[\sum_{i=1}^n b_i^2 - \left(\sum_{i=1}^n c_i \right) \left(\sum_{i=1}^n a_i \right) \Big/ n \right] \tag{10.17}$$

Adjustments to mortality tables can be made for individuals in the high-risk stratum using (10.15) and (10.17). This procedure is demonstrated in Example 10.4, and Excel instructions are outlined in Problem 10.6.

Example 10.4

Increased mortality adjustments for a stochastic status model are demonstrated where the mortality structure follows a Weibull distribution. For the high-risk stratum group the mean future lifetime is taken to be $E\{^sT\} = 2.0$, so as before, $E\{^sK\} = 1.5$. Applying the survival rates in Table 10.2, we compute $r = 2.7158$, and formula (10.17) gives the least squares estimate and function

$$\hat{\alpha} = 1.71 \quad \text{and} \quad \hat{r}^\alpha = 5.5203$$

TABLE 10.2

Mortality and Weibull Adjusted Mortalities

k	0	1	2	3	4	5	6	7	8	9	10
l_{x+k}	100	95	85	68	56	46	38	29	20	12	4
p_{x+k}	.950	.995	.800	.824	.821	.826	.763	.690	.600	.333	0.0
$^sp_{x+k}$	.753	.973	.292	.343	.337	.348	.225	.129	.060	.002	0.0

Hence, the increased mortality group survival rates are computed as

$$^sp_{x+k} = (p_{x+k})^{5.5203} \tag{10.18}$$

for $k \geq 0$. The adjusted survival rates following (10.18) are listed in Table 10.2.

Assuming UDD, the mean curtate future lifetime associated with rates (10.18) is computed to be $E\{^sK\} = 1.809$. The Weibull adjusted survival rates (10.18) are more efficient than their exponential counterparts (10.13), producing a closer expected mean curtate value to the target 1.5 value.

In this section, we have presented a general mortality adjustment approach that can be applied on any given mortality table. There are many other statistical reliability models that an experimenter could consider in new investigations. In the next section, we introduce an additional method of constructing adjusted mortalities.

10.1.3 Survival-Based Mortality Acceleration Models

In many survival analysis applications, data are recorded on the survival rates based on a fixed number of future whole years lived. This is common in medical studies where survival rates for various diseases, such as specific forms of cancer, are recorded and analyzed. For the two strata settings introduced in Section 10.1.2, adjustments to a standard mortality table based on survival probabilities are presented. For individuals in the high-risk stratum the m future-year survival rate is fixed at $_m^sp_x$. The adjustment procedure we discuss is based on the exponential distribution, but as with the last section, there are many other directions an investigator could take to construct this type of mortality adjustment.

To construct an adjustment to the mortality rates presented in a mortality table, we use the exponential survival function of the form (10.5). The t-year survival probability associated with the increased mortality status group is a function of the initial age x and is modeled by the relation

$$^s_tp_x = {}_tp_x \exp(-t\, AF) \tag{10.19}$$

for fixed $t > 0$ and acceleration factor AF. The survival rate after m future years is fixed at $_m^sp_x$, and we solve (10.19) for AF as

$$AF = \ln [(_mp_x / {}_m^sp_x)^{1/m}] \tag{10.20}$$

In Problem 10.3 we show the adjusted yearly mortality rates for high-risk strata individuals age $x + k$ can be written as

$$^sp_{x+k} = r\, p_{x+k} \tag{10.21}$$

TABLE 10.3

Mortalities and Survival-Adjusted Mortalities

k	0	1	2	3	4	5	6	7	8	9	10
l_{x+k}	100	95	85	68	56	46	38	29	20	12	4
p_{x+k}	.950	.995	.800	.824	.821	.826	.763	.690	.600	.333	0.0
$^sp_{x+k}$	.828	.867	.697	.718	.715	.720	.665	.601	.523	.290	0.0

where ratio r is

$$r = (^s_m p_x / _m p_x)^{1/m} \tag{10.22}$$

for $k \geq 0$. In this construction, the experiment needs only the ratio (10.22), or a suitable estimate, to project the mortality adjustments over the entire mortality table. This is demonstrated in Example 10.5.

Example 10.5

For expository purposes the mortality data given in Example 10.3 are revisited. The survival mortality acceleration approach is applied to this data set. From the original mortality data the 3-year survival rate is computed to be

$$_3p_x = p_x \, p_{x+1} \, p_{x+2} = .95(.995)(.8) = .7562$$

For the high-risk stratum, the 3-year survival rate is assumed to be only .5, so that from (10.22) the ratio r is

$$r = (.5/.7562)^{1/3} = .8712$$

From (10.21) the adjusted mortality rates are found and listed in Table 10.3. The 3-year survival rate for the high-risk status is, as desired, .500.

The mortality adjustment procedures presented in the last two sections represent possible modeling techniques for statuses associated with increased risk or stress resulting in lower yearly mortalities. In the next section, the topic of incorporating a time trend effect to mortality table construction is considered.

10.2 Mortality Trend Modeling

In the life tables introduced in Section 5.7 it was assumed that the mortality rates for individual ages are constant over time. This is not the case, as reflected in modern survival studies of humans that indicate that individuals are generally living longer. To construct accurate lifetime and mortality tables for many types of statuses, the possibility of modeling for a mortality time trend is required. In this section, we present an introduction to time trend modeling for mortality data by applying a regression approach as discussed in Section 2.6. Similar to the previous section, an adjustment procedure to be applied to a standard mortality table is proposed.

To model for trends of mortality rates over time, survival rates based on differing construction times are required. For this reason, we assume that separate mortality tables are constructed at times indicated by the future lifetime random variable T. Thus, if past tables are constructed at whole years, the set of possible times for T, based on the current time $T = 0$, is given by $S_t = \{-1, -2, \ldots\}$. For any past time t, the mortality tables give the set of mortality values $q_x(t)$ for $x \geq 0$. To model the time trend in the mortality rates, we present a method based on simple linear regression, but various techniques are possible.

To model the time trend in mortality rates, we apply a linear regression model to a collection of mortality tables taken at times in S_t. We modify the simple linear regression model given in (2.103), where the independent or predictor variable is t. The dependent or response variable and the resulting linear regression model are

$$y_t(x) = q_x(t)/q_x(0) \quad \text{and} \quad y_t(x) = 1 + \beta_x t + e_t \tag{10.23}$$

for $t \geq 0$. The distribution of the error term and its effect on statistical inference are an open question. Using the method of least squares introduced in Section 2.6.1, the least squares estimator is given by

$$\hat{\beta}_x = \left(\sum_{S_t} tY_t - \sum_{S_t} t \right) \Big/ \left(\sum_{S_t} t^2 \right) \tag{10.24}$$

Utilizing the least squares estimator (10.24), the predicted mortality rate at future time $T = t$ takes the form

$$q_x(t) = (1 + t\,\hat{\beta}_x)q_x(0) \tag{10.25}$$

In applying formula (10.25) to a given life table, an adjustment for a time trend in mortality rates is spread throughout the entire set of age-based mortalities.

The procedure we have presented is just one choice out of the collection of many possible time trend modeling techniques. Standard methods as well as recent developments in statistical modeling have widened the scope of modeling choices for time trend data. Possible techniques for time trend modeling include nonlinear modeling and generalized linear modeling techniques. For a review we refer to works by Ratkowsky (1983) and McCullagh and

TABLE 10.4

Mortality Trend Data and Response

	Mortality Rates			Response Variable		
x	$q_x(-3)$	$q_x(-1)$	$q_x(0)$	$y_{-3}(x)$	$y_{-1}(x)$	$\hat{\beta}_x$
64	.016	.015	.014	1.1429	1.0714	−.0500
65	.016	.016	.015	1.0667	1.0667	−.0267
66	.017	.016	.015	1.1333	1.0667	−.0467
67	.018	.017	.016	1.1250	1.0625	−.0438
68	.019	.017	.017	1.1176	1.0000	−.0353

Nelder (1983). The choice of selection among competing models was investigated by Borowiak (1989).

Example 10.6

In this example, yearly based mortality rates are collected for base years represented by $T = -3$, $T = -1$, and $T = 0$ and are listed in Table 10.4. The least squares estimators, given by (10.24), are computed for each age and also listed in Table 10.4.

These estimators for each year are all negative, indicating a decreasing mortality rate time trend. For an individual status age x, efficient mortality tables can be constructed in the presence of a time trend. This is accomplished using the linear mortality time trend model given by (10.25). For status age x, the true mortality rates are given by $q_{x+k}(k)$ for $k = 0, 1, \ldots$, and the estimated mortalities are given by

$$\hat{q}_{x+k} = (1 + k\hat{\beta}_x)\, q_x(0) \quad \text{for} \quad k \geq 0 \tag{10.26}$$

Based on (10.26), an adjusted mortality table based on the mortality time trend model can be constructed, and the approach is demonstrated in Example 10.7. Excel work dealing with mortality modeling is presented in Problem 10.8.

Example 10.7

The mortality data and least squares estimators given in Example 10.6 are now utilized to form a time trend-adjusted mortality table. Using (10.26), the adjusted yearly mortality rates are computed and listed in Table 10.5. Further, the expected present value for a unit benefit discrete 5-year term life insurance policy is computed for both the original and adjusted mortality values and is also given.

The lower mortality values in the time trend-adjusted model result in a lower expected value for the insurance policy. Hence, these adjustments are important for accurate actuarial measurements and computations such as premium computations.

TABLE 10.5

Original and Projected Mortality Tables

$K = k$	1	2	3	4	5	$A^1_{x:\overline{5}|}$
Original	.014	.015	.015	.016	.017	.0593
Projected	.014	.0146	.0136	.0139	.0146	.0550

This section represents only an introduction to the modeling of time trends with mortality data. For the linear regression model approach presented in this section, many open questions exist. The theory for the efficient modeling of the error terms and the resulting prediction and confidence intervals are some of the many questions open to investigation.

10.3 Actuarial Statistics

Generally, statistics are functions of observed sample data, and a general discussion of sampling distributions and estimation was presented in Section 2.1. For a discrete stochastic status model, the relevant actuarial computations, such as present value expectations, are functions of yearly mortality and survival rates. For an individual status age x with curtate future lifetime K, the yearly mortality rates are given by

$$P(K = k) = {}_{k|}q_x \quad \text{for} \quad 0 \le k \le \omega - x \tag{10.27}$$

Yearly mortality rates (10.27) are computed using appropriate life tables where the limiting age is denoted ω. For a collection of l_x independent and identically distributed stochastic statuses age x, the number of observed decrements corresponding to curtate future lifetime k is denoted by the statistic $\mathcal{D}_{x+k}$ for $1 \le k \le \omega - x$. The point estimator for (10.27) is the observed decrement rate

$$_{k|}\mathcal{Q}_x = \mathcal{D}_{x+k}/l_x \quad \text{for} \quad 0 \le k \le m - x \tag{10.28}$$

For fixed age x, applying the weak law of large numbers (see Laha and Rohatgi, 1979), the observed decrement rates (10.28) converge in probability to the true decrement rates (10.27).

For a collection of independent and identically distributed stochastic status models, actuarial statistics are defined to be functions of the observed mortality rates (10.28). In particular, for a collection of unit benefit discrete whole life insurances for statuses age x, the observed mean present value is defined

$$\hat{A}_x = \sum_{k=0}^{\omega-x-1} v^{k+1} \, {}_{k|}\mathcal{Q}_x \tag{10.29}$$

As the collection size increases, statistic (10.29) converges in probability to the actuarial computation A_x, as defined in (6.13). Similarly, the observed aggregate present value is a statistic and is defined as

$$l_x \hat{A}_x = \sum_{k=0}^{\omega-x} v^{k+1} \, \mathcal{D}_{x+1} \tag{10.30}$$

In the following sections, the techniques presented in Chapter 8 are utilized to construct prediction intervals for the observed aggregate total (10.30). In particular, the techniques include normality-based prediction interval (*PI*) as well as symmetric simulation prediction interval (SSPI) and the bias correcting prediction interval (BCPI), as described in Section 9.2.4.

10.3.1 Normality-Based Prediction Intervals

The central limit theorem as described in Section 2.4.1 can be applied to a collection of actuarial statistics based on independent and identically distributed samples. For large samples, a general approximate *PI* follows the form of (2.16), and in particular we consider a *PI* for the observed aggregate total defined (10.30). Using the survivorship group theory for a status age x, $E\{_k|Q_x\} = _k|q_x$ for $0 \le k \le \omega - x - 1$ and the expectation is

$$E\{\hat{A}_x\} = \sum_{k=0}^{\omega-x-1} v^{k+1} \, _k|q_x = A_x \tag{10.31}$$

In a similar manner the expected second moment is

$$E\{^2\hat{A}_x\} = \sum_{k=0}^{\omega-x-1} v^{2(k+1)} \, _k|q_x = {}^2A_x \tag{10.32}$$

and the variance is computed as

$$\text{Var}\{\hat{A}_x\} = {}^2A_x - (A_x)^2 \tag{10.33}$$

In a similar manner for the collection of l_x statuses,

$$E\{l_x \hat{A}_x\} = l_x \, A_x \quad \text{and} \quad \text{Var}\{l_x \hat{A}_x\} = l_x^2 \, \text{Var}\{\hat{A}_x\} \tag{10.34}$$

The approximate $(1 - \alpha)$ 100% *PI* for the observed aggregate total as denoted by (10.30) is

$$l_x \, A_x - z_{(1-\alpha/2)} \, l_x \, [{}^2A_x - (A_x)^2]^{1/2} \le l_x \hat{A}_x \le l_x \, A_x + z_{(1-\alpha/2)} \, l_x \, [{}^2A_x - (A_x)^2]^{1/2}$$

$$\tag{10.35}$$

TABLE 10.6

Curtate Future Lifetime Probabilities

k	0	1	2	3	4	5	6	7	8	9
$P(K = k)$	.05	.05	.07	.1	.13	.1	.1	.15	.15	.1

A demonstrational example now follows with Excel work discussed in Problem 10.10.

Example 10.8

For fixed age x, a discrete stochastic status model is associated with curtate future lifetime K where the pdf is given in Problem 6.7 and the probabilities are relisted in Table 10.6.

A collection of $l_x = 10$ independent and identical whole life insurance policies is examined, and a prediction interval for the observed aggregate total present values (10.30) is constructed. The yearly effective interest rate is taken to be $i = .04$. Applying (10.31) and (10.32), the first two moments of the actuarial statistic (10.29) are computed as $E\{\hat{A}_x\} = A_x = .78644$ and $E\{\hat{A}_x^2\} = {}^2A_x = .625159$, and following (10.33), $\text{Var}\{\hat{A}_x\} = .006676$. An 80% PI for the observed aggregate total (10.30) is computed following (10.35) as

$$7.75981 = 7.8644 - 1.28 \,(.08171) \le l_x \hat{A}_x \le 7.8644 + 1.28 \,(.08171) = 7.96899$$

$$(10.36)$$

The present value cost for the collection of 10 whole life insurance policies is estimated to be between 7.760 and 7.969 per unit benefit. The nonlinear nature of the present value function and small sample size makes the normal-based PI (10.36) suspect.

10.3.2 Prediction Set-Based Prediction Intervals

Due to the nonnormality of actuarial statistics, central limit theorem-based prediction intervals based on moderate data sizes, such as (10.36), may be suspect. Other methods for the construction of prediction intervals, such as those based on order statistics discussed in Section 2.3, exist. In this section, the prediction interval construction based on prediction sets introduced in Section 5.8 is applied to an actuarial statistic. The actuarial statistic is the observed aggregate statistic (10.30), and an approximate PI for the associated aggregate mean is the goal.

Consider a collection of independent and identically distributed stochastic status models all age x where the probabilities are associated with the curtate future lifetime K following (10.27). The statistical vector of yearly observed decrements (10.28) is $\mathbf{Q}_x^t = (_0|Q_{x'}\,_1|Q_{x'}\,\ldots,\,_{\omega-x}|Q_x)$. Following the development in Section 5.8, the approximate chi square random variable with $\omega - x$ degrees of freedom given in (5.60) is

$$W_x = l_x \sum_{k=0}^{\omega-x-1} [(_k|Q_x -_k|q_x)^2 /_k|q_x]$$ (10.37)

Modifying prediction set (5.60), an approximate $(1 - \alpha)100\%$ prediction set, for $0 < \alpha < 1$, in terms of the observed decrement rate, is

$$PS_{1-\alpha} = \left\{ Q_x : \sum_{k=0}^{\omega-x-1} {}_k|Q_x = 1,\,_k|Q_x \geq 0 \quad \text{and} \quad W_x \leq \chi^2_{1-\alpha} \right\}$$ (10.38)

Software such as Excel is used to find lower and upper bounds for the actuarial statistic (10.29) over prediction set (10.38):

$$LB(\hat{A}_x) = \min_{Q_x \in PS_{1-\alpha}} (\hat{A}_x) \quad \text{and} \quad UB\{\hat{A}_x\} = \max_{Q_x \in PS_{1-\alpha}} \{\hat{A}_x\}$$ (10.39)

The $(1 - \alpha)100\%$ approximate *PI* for the observed aggregate total (10.30) results from putting the bounds in (10.39) into (10.30) and is

$$l_x\,LB(\hat{A}_x) \leq l_x \hat{A}_x \leq l_x\,UB(\hat{A}_x)$$ (10.40)

A computational example follows with Excel instructions outlined in Problem 10.9.

Example 10.9

A prediction set-based *PI* for the observed aggregate total is constructed for the setting of Example 10.8, where the curtate probabilities are given in Table 10.6. For a collection of $l_x = 10$ independent discrete whole life insurance policies, the prediction set (10.38) is utilized to construct an approximate 80% *PI* for the observed aggregate total (10.30). The bounds (10.39) are found producing the *PI* of the form (10.30):

$$7.0736 = 10\,(.70736) \leq l_x \hat{A}_x \leq 10\,(.87497) = 8.7497$$ (10.41)

The *PI* in (10.41) is much wider than the normality-based *PI* given in (10.36). This is due in major part to the conservative nature of prediction set prediction intervals. The observed yearly decrement rates (10.28) corresponding to the lower and upper bounds utilized in (10.41) are listed in Table 10.7.

TABLE 10.7

Observed Probabilities for Lower and Upper Bounds

$LB(\hat{A}_x)$	.0000	.0000	.0000	.0000	.0000	.0114	.0888	.2447	.3518	.3033
$UB(\hat{A}_x)$	.1799	.1505	.1713	.1905	.1798	.0881	.0400	.0000	.0000	.0000

> The relatively small sample size of 10 results in quite a bit of variability in the observed decrement rates. A larger sample size would result in tighter and more efficient estimation.

The prediction set approach to the construction of prediction intervals for actuarial statistics may be quite conservative in nature. A viable research topic is the construction of more efficient prediction intervals based on asymptotic distributions.

10.3.3 Simulation-Based Prediction Intervals

Another approach for the construction of prediction sets for actuarial statistics is to utilize simulation methods, such as those discussed in Section 9.2. As before, consider a collection of independent and identically distributed stochastic status models all age x with curtate future lifetime K with probabilities (10.27). A collection of $l_x = n$ simulated observed curtate future lifetimes, $k_1, k_2, \ldots, k_n$, is generated with corresponding present value functions. The statistic of interest is the aggregate sum of the generated present values, $l_x\hat{A}_x$, taking the form of (10.30). Based on n repetitions the order statistics for the aggregate sums are $W_{(1)} \leq W_{(2)} \leq \ldots \leq W_{(n)}$. These form the basis of mortality table-based statistical inference.

Simulated probabilities discussed in Section 9.2.3 are computed using the empirical distribution function (9.14). Further, the two types of prediction intervals discussed in Section 9.2.4, namely, SSPI and BCPI, are applied to the simulated aggregate sums. The SSPI is outlined by formulas (9.17)–(9.20), while the BCPI utilizes (9.22)–(9.24).

Example 10.10

The conditions of a stochastic status model are based on an individual age x with curtate future lifetime K and pdf given in Problem 6.7. Thirty simulated samples of size 10 each are generated with the aggregate present values listed in increasing order in Table 10.8.

A 90% SSPI for $l_x\hat{A}_x$ is constructed based on the SS. Following (9.17), we find $a = .05(31) = 1.55$ and $b = .950(31) = 29.45$, and from (9.18) and (9.19), the lower and upper bounds for $FV(11)$ are

$$LB(l_x\hat{A}_x) = 7.267 + .55(7.471 - 7.267) = 7.3792$$

TABLE 10.8

Simulated Samples of Aggregate Present Values

7.267	7.471	7.593	7.602	7.624	7.633	7.634	7.71	7.733	7.746	7.757	7.761
7.807	7.842	7.842	7.875	7.888	7.897	7.951	7.955	7.957	7.972	7.978	8.006
		8.044	8.062	8.081	8.107	8.124	9.306				

and

$$UB\ (l_x \hat{A}_x) = 8.124 + .45(9.306 - 8.124) = 8.6559$$

Thus, the 90% SSPI is computed as

$$7.3792 \le l_x \hat{A}_x \le 8.6559 \tag{10.42}$$

To compute an 80% BCPI, the mean of simulation future values is 7.8742, and in (9.22), $F(7.8742)_s = 16/30$. The quantities in (9.23) are computed as

$$z_o = \Phi^{-1}(16/30) = .08365$$

and

$$\alpha_L = \Phi(2(.08365) - 1.2816) = .13258$$

and

$$\alpha_U = \Phi(2(.08365) + 1.2816) = .92631$$

The percentile ranks defined in (9.24) are $a = .13258(31) = 4.1100$ and $b = .92631(31) = 28.7156$. The lower and upper bounds (9.18) and (9.19) are computed as

$$LB(l_x \hat{A}_x) = 7.602 + .1100(6.624 - 7.602) = 7.4944$$

and

$$UB(l_x \hat{A}_x) = 8.107 + .7156(8.124 - 8.107) = 8.1192$$

The 80% BCPI is computed as

$$7.4944 \le l_x \hat{A}_x \le 8.1192 \tag{10.43}$$

The two simulation-based prediction intervals (10.42) and (10.43) are very similar and are comparable to the normal-based *PI* given in (10.36). By contrast, the prediction set *PI* given by (10.41) is much wider. Excel instructions for simulation of these prediction intervals are given in Problem 10.11.

10.4 Data Set Simplifications

In both fields of financial and actuarial modeling new data sets continue to be created. In financial analysis, consumer indexes, stock prices, and other factors that influence economic systems are continuously observed. In connection with actuarial science, mortality data stratified by demographics, and modes of decrement are compiled and refined. It is the goal of many standards, as well as modern statistical techniques, to make sense of this great body of data. Classical methods exist in the statistical literature in the area of data reduction and the analysis of large data sets. In this section, we present a brief introduction to some of the applicable statistical techniques and their associated references.

In both financial and actuarial modeling, the analysis and reduction of large data sets are important. In investment modeling, the researcher must decide among variables, out of an ever-expanding collection, to form a set that is important to the future value of a portfolio under consideration. The field of insurance underwriting consists of analyzing and computing the adjustment of mortality rates, thereby affecting insurance premiums, for added risk factors. These risk factors include individual demographics such as gender, smoking-nonsmoking, general health conditions, and lifestyle. The introduction of modern statistical techniques into this important field of insurance requires a thorough investigation into the nature of survival data and the efficient construction of efficient classifications.

As time advances, mortality and survival data become higher and higher dimensionally with the introduction of variables affecting these rates. In the future, not only will standard demographics, such as gender, individual health, and lifestyle factors, be measured, but other factors, such as family health histories and specific medical measurements, also will be included in actuarial assessments. More survival data will be collected on these survival factors forming high-dimensional sets of data. The statistical area of high-dimension data reduction and analysis includes standard techniques such as classification, cluster, principal components, and discriminant analysis. In classification and cluster analysis, statistical patterns in the high-dimensional data sets are explored. Graphical techniques such as proximity graphics have been introduced to aid in the analysis of these data sets. In some large data sets, the search for data patterns is enhanced by the construction and application of mathematical and statistical functions, such as dissimilarity measures, that are applied to individual high-dimensional data points. For a review of these techniques we refer to Ryzin (1976).

There are many statistical techniques, classical and new, for the analysis of large high-dimensional data sets. Two classical statistical techniques are principal components and discriminant analysis, where the goals are data reduction and interpretation. Linear combinations or discriminant functions within the high-dimensional data sets are utilized to this end. For a further look at these procedures, we refer to Johnson and Wichern (1982) and Morrison (1976).

Problems

10.1. For a stochastic status model age x let the future lifetime random variable T have a lognormal distribution as presented in Section 1.6.2. Thus, $Y = \ln(T) \sim N(\mu, \sigma^2)$.

    a. Apply the linear adjustment (10.1). Using (10.2), what is the distribution of T_s, the future lifetime for a high-risk stratum status?

    b. Using a mean mortality acceleration model, what is the value of r in (10.10)?

10.2. For a stochastic status model age x let the future lifetime random variable T have a Pareto distribution as presented in Section 1.6.1 with nonnegative parameters α and β. Thus, $_tp_x = (\beta/t)^\alpha$ for $t > \beta$.

    a. Apply the linear adjustment (10.1). Using (10.2), what is the distribution of ^sT, the future lifetime for a high-risk stratum status?

    b. Using a mean mortality acceleration model, what is the value of r in (10.10)?

10.3. For a stochastic status model in a high-risk stratum show (10.21) where (10.22) holds.

10.4. Using the linear regression model given by (10.23), use the method of least squares to derive the estimator defined by (10.24).

Excel Problems

10.5. Consider the survival data listed below for a status age x:

k	0	1	2	3	4	5	6	7	8	9
l_{x+k}	1,000	982	931	825	557	421	330	215	87	0

a. Compute p_{x+k} for the ages listed in the table.

b. Assuming UDD, compute $E\{K\}$ and $E\{T\}$.

c. Apply the mean acceleration method of Section 10.2 based on the exponential distribution with $E\{^sT\} = E\{T\}/1.5$ to find adjusted mortalities $q_{x+k,s}$ using (10.11) and (10.12).

d. In part (c), compute $E\{^sK\} = \Sigma\, l_k\,{}^sp_k/l_0$.

Excel: Basic operations.

10.6. Apply the mean acceleration method of Section 10.2 based on the Weibull distribution on the life table of Problem 10.5.

a. Compute (10.17).

b. Using $E\{^sT\} = E\{T\}/1.5$, find adjusted survival rates ${}^sp_{x+k}$ using (10.15).

c. In part (a), find $E\{^sK\} = \Sigma\, l_k\,{}^sp_k/l_0$.

Excel Problem 10.5 extension: Basic operations.

10.7. In Problem 10.5 we apply the survival acceleration factor method of Section 10.1.3, where $m = 3$ and $_3p_x/2 = {}^s_3p_x$.

a. Compute r in (10.22).

b. Use (10.21) to compute adjusted survival rates ${}^sp_x$ for ages listed in the table.

c. Find $E\{^sK\}$.

Excel: Basic operations.

10.8. Mortality data were compiled for three different years corresponding to $-2, -1$, and the present time 0. The mortality data are

x	$q_x(-2)$	$q_x(-1)$	$q_x(0)$
0	.014	.013	.011
1	.018	.017	.015
2	.021	.020	.018
3	.022	.021	.019
4	.024	.022	.020

a. Using (10.24) compute the response variable $Y_t(x)$ for the ages and $t = -2$ and $t = -1$.

b. Use (10.24) to compute the least squares estimator for each age.

c. Find the projected mortality rates (10.26) for $0 \le k \le 4$.

d. Using $i = .05$ compute $A^1_{0:\overline{5}|}$ for the original mortalities.

e. Using $i = .05$ compute $A^1_{0:\overline{5}|}$ for the projected mortalities.

Excel: Basic operations.

10.9. A collection of 20 iid unit benefit whole life insurance polices have mortality rates following the partial life table of Problem 10.5. An approximate 80% *PI* for the observed aggregate total present value (10.30) is constructed using the prediction set techniques of Section 10.3.2 following (10.39) to (10.41), where the interest rate is $i = .5$. Excel: CHIINV and Data, Solver; Set target cell: Equal to Max or Min: Options: Assume Nonnegative.

10.10. Construct an approximate 80% *PI* for the observed aggregate total present value (10.30) using the normality-based approach of Section 10.3.1 for the 20 policies discussed in Problem 10.9. We follow (10.35) where the interest rate is $i = .5$.

Excel: NORMSINV, SUMPRODUCT.

10.11. Using the SSPI and BCPI approach of Section 10.3.3 construct an approximate 80% *PI* for the observed aggregate total present value (10.30) for the setting of Problem 10.9 where the interest rate is $i = .5$. Excel: Data, Data Analysis, Random Number Generator, OK, Value, and Probability Input Range Output Options.

Solutions

10.1 a. $^sY = \ln(^sT) \sim N(^s\mu, \sigma^2)$, where $^s\mu = \mu - \ln(AF)$.

   b. Using (1.52), $r = AF$.

10.2. a. Pareto with parameters $\alpha_s = \alpha$ and $^s\beta = \beta/AF$.

   b. Using (1.47), $r = AF$.

10.5 b. 4.348, 4.845.

   c. .9731, .9231, .8342, .5548, .6571, .6940, .5259, .2574.

   d. 4.00097.

10.6. a. 1.9610.

   b. $r = 1.5$ and $r(\alpha) = 2.214744$ producing .9606, .8886, .7651, .4189, .5379, .5831, .3872, .1348.

   c. 3.5930.

10.7. a. .7973.

   b. .7794, .7525, .7033, .5359, .6000, .6221, .5171, .3212.

   c. 3.45101.

10.8. b. $-.1454, -.10667, -.08889, -.08421, -.1$.

    c.   .011, .0134, .0148, .0142, .012.

    d.   .068739.

    e.   .05512.

10.9.   For decrements 0, 0, .114, 2.533, 2.731, 2.278, 3.711, 5.014, and 3.979, $LB(\hat{A}_x) = 14.336$, and for decrements .991, 2.343, 3.954, 7.860, 2.889, 1.257, .765, 0, and 0, $UB(\hat{A}_x) = 16.647$; thus, $14.336 \le T_x \le 16.647$.

10.10. In (10.34) $E\{\hat{A}_x\} = A_x = .77436$, $\text{Var}\{\hat{A}_x\} = .006175$, leading to $13.472 \le l_x \hat{A}_x \le 17.502$.

Appendix A: Excel Statistical Functions, Basic Mathematical Functions, and Add-Ins

Excel Statistical Functions

CHIINV(1-α,df): Chi square random variable inverse; percentile (2.1.3).

GAMMADIST(x,α,β,a): Gamma random variable; Problem 1.19, a = o pdf, a = 1 cdf.

NORMDIST(x,μ,σ,a): Normal random variable; a = 0 pdf, a = 1 (1.18), cdf (1.19).

NORMSDIST(z): Standard normal cdf (1.19).

NORMSINV(1-α): Standard normal random variable inverse; percentile (2.1.3).

POISSON(x,λ,a): Poisson random variable; a = 0 pdf (1.12), a = 1 cdf (1.13).

TINV(1-α,df): T random variable inverse; percentile (2.1.3).

Excel Basic Mathematical Functions

$A1{*}A2$: Multiply $A1$ and $A2$.

$A1/A2$: Divide $A1$ by $A2$.

$A1{\wedge}A2$: $A1$ raised to the power $A2$.

exp($A1$): e to the power $A1$.

Fill: Formula computations extend over highlighted cells. Aj changes while $A\$j$ does not change.

LN($A1$): Natural log of $A1$.

MINVERSE($A1:B2$): Matrix inverse of $A1:B2$ presented in $D1:E1$. Highlight $D1:E2$. MINVERSE Array $A1:B2$: Hold down Ctrl and Shift and press Enter.

MMULT(*A*1:*B*2,*A*4:*B*5): Matrix multiplication of *A*1:*B*2 and *A*4:*B*5 presented in *D*1:*E*1. Highlight *D*1:*E*2, MMULT array1 *A*1:*B*2, array *A*4:*B*5, hold down Ctrl and Shift and press Enter.

PRODUCT(*A*1:*An*): Multiply *A*1 through *An*.

SUM(*A*1:*An*): Sum *A*1 through *An*.

SUMPRODUCT(*A*1:*An*, *B*1:*Bn*): Sums the product of $Aj*Bj$ over $j = 1, 2, ..., n$.

Excel Add-Ins: Enable Descriptive Statistics, Data Analysis, and Solver

Data graph for data: Chart wizard, scatter, dots.

Statistics for a data set: Data, descriptive statistics; data analysis, descriptive statistics.

Generate uniform random numbers between 0 and 1: Data, data analysis, random number generator, distribution: uniform.

Generate *k* sample of size *n* from a given discrete distribution: Data, data analysis, random number generator, discrete, number of variables *k*, number of random numbers *n*, value and probability input range, output range.

Optimization of a function based on selecting parameters: Data, Solver, set target cell, min or max, by changing cells, subject to constraints.

Percentile C1 from a given cdf: Data, Solver, set target cell, equal to C1, by changing cells, subject to constraint.

Regression of *X* on *Y*: Data, data analysis, regression, input *Y*, input *X*.

Roots of equations: Data Solver, set target cell, equal to 0, by changing cells, subject to constraints.

Appendix B: Acronyms and Principal Sections

AF: Acceleration factor; Sections 10.1.1, 10.1.3.

$AF(t)$: Accumulation function; Sections 3.1.3, 3.2.2.

$AR(1)$: Auto regressive model of order 1; Section 2.7.

AS_{x+h}: Actual salary; Sections 7.3.3, 7.3.4.

BCPI: Bias correcting prediction; Section 2.3.

BS: Bootstrap; Section 9.2.1.

cdf: Cumulative distribution function; Section 1.2.2.

CLT: Central limit theorem; Section 2.4.1.

$DF(t)$: Discount function; Sections 3.1.2, 3.1.3.

$ENSV(j)$: jth scenario expected net single value; Section 9.1.3.

EP: Equivalence principle; Sections 4.1.1, 4.2.1.

ES_{x+h+i}: Expected salary; Section 3.3.3.

$FVE(t)$: Future value of expenditures; Sections 4.1, 4.1.2

$FVR(t)$: Future value of revenue; Sections 4.2, 4.1.2.

$FV(t)$: Future value; Sections 3.1.1, 3.1.2, 3.2.

$FV(t, j)$: jth scenario future value; Section 9.1.1.

G: Gross premium; Section 4.3.1.

GLSM: Generalized least squares measure; Section 2.8.3.

HAA: Haldane type A approximation; Section 2.4.2.

iid: Independent and identically distributed; Section 2.1.1.

JLS: Joint life status; Section 7.1.1.

LD: Loading of premium; Section 4.3.1.

$LF(t)$: Loss function; Section 4.1.

$_iLF(W)$: Reserve-based loss function; Section 6.9.

LSS: Last survivor status; Section 7.1.2.

mgf: Moment generating function; Section 1.4.

mpf: Mixed probability function; Section 1.2.3.

MS: Markov state; Section 8.1.

NSP: Net single premium; Section 6.4.1.

NSV: Net single value; Sections 4.3, 4.3.1.

NSV-B: Net single value of the benefit; Sections 7.3.1, 7.3.3.

NSV-C: Net single value of contributions; Sections 7.3.2, 7.3.3.

NSV(j): jth scenario net single value; Section 9.1.2.

PI: Prediction interval; Section 2.1.3.

PC(.25): Percentile criteria; Section 4.2.2.

pdf: Probability density function; Section 1.2.2.

pmf: Probability mass function; Section 1.2.1.

PS: Prediction set; Section 2.1.4.

PV: Present value; Sections 3.1.1, 3.1.2, 3.1.3.

PVE(t): Present value of expenditures; Section 4.1.

PVR(t): Present value of revenues; Sections 4.1, 4.1.2.

PV(t)$_{1-\alpha}$: Present value percentile; Section 3.3.2.

PV(t, j): jth scenario present value; Section 9.1.1.

RC: Risk criteria; Section 4.2.1.

R(t): Risk; Section 4.2.1.

SBCPI: Simulation bias-correcting prediction interval; Section 9.2.4.

SPA: Saddlepoint approximation; Section 2.4.3.

SPI: Simulation prediction interval; Section 2.3.

SS: Simulation sampling; Section 9.2.2.

SSPI: Symmetric simulation prediction interval; Section 9.2.4.

TFS$_{x+h}$: Total future salaries; Section 7.3.4.

TPS$_{x+h}$: Total past salaries; Section 7.3.4.

UDD: Uniform distribution of deaths; Sections 5.4, 6.2.3.

References

Actuarial Society of America. (1947). International actuarial notation. *Transactions of the Actuarial Society of America*, XLVIII 166–176.

Aichison, J., and Brown, J.A.C. (1957). *The lognormal distribution, with special references to its use in economics*. Cambridge University Press, New York.

Allen, L.J.S. (2003). *An introduction to stochastic processes with applications to biology*, Second Edition. CRC Press, Boca Raton, FL.

Arrow, K.J. (1963). Uncertainty and the welfare of medical care. *American Economic Review*, 53: 305–307.

Bartlett, D.K. (1965). Excess ratio distribution in risk theory. *Transactions of the Society of Actuaries*, XVII: 435–463.

Basu, H., and Ghosh, M. (1980). Identifiability of distributions under competing risks and complementary risks model. *Communications in Statistics*, A9(14): 1515–1525.

Beekman, J.A., and Bowers, N.L. (1972). An approximation to the finite time ruin function. *Skandinavisk Aktuarietidskrift*, 128–137.

Benard, A., and Bos-Levenbach, E.D. (1953). The plotting of observations on probability paper. *Statistica Neerlandica*, 7: 163–173.

Besag, J., Green, P., Higdon, D., and Mengersen, K. (1995). Bayesian computation and stochastic systems. *Statistical Science*, 10(1): 3–66.

Bickel, P.J., and Doksum, K.A. (2001). *Mathematical statistics*. 2nd ed. Prentice Hall, Upper Saddle River, NJ.

Black, F., and Scholes, M. (1973). The pricing of options and corporate liabilities. *Journal of Political Economy*, 81: 637–659.

Bohman, H., and Esscher, F. (1964). Studies in risk theory with numerical illustrations concerning distribution functions and stop-loss premiums. *Skandinavisk Aktuarietidskrift*, 1–40.

Borowiak, D.S. (1989). *Model discrimination for nonlinear regression models*. Marcel Dekker, New York.

Borowiak, D.S. (1998). Upper and lower bounds in the estimation of single decrement rates. *Expanding Horizons*, 18: 9–10.

Borowiak, D.S. (1999a). A saddlepoint approximation for tail probabilities in collective risk models. *Journal of Actuarial Practice*, 7(1): 239–249.

Borowiak, D.S. (1999b). Insurance and annuity calculations in the presence of stochastic interest rates. *ARCH*, 1: 445–450.

Borowiak, D.S. (2001). Life table saddlepoint approximations for collective risk models. Fifth International Congress on Insurance: Mathematics and Economics, Penn State University, July 23–25.

Borowiak, D.S., and Das, A. (2007). Sensitivity analysis of the T-distribution under truncated normal populations. Indian Statistical Institute.

Bowers, N.L. (1969). An upper bound for the net stop-loss premium. *Transactions of the Society of Actuaries*, XXI: 211–218.

Bowers, N.L., Gerber, H.S., Hickman, J.C., Jones, D.A., and Nesbitt, C.J. (1997). *Actuarial mathematics*. 2nd ed. Society of Actuaries, Schaumburg, IL.

Box, G.E.P., and Jenkins, G.M. (1976). *Time series analysis*. 2nd ed. Holden-Day, San Francisco, CA.

Breiman, L. (1996). Bagging predictors. *Machine Learning*, 24: 123–140.

Breiman, L. (2001a). Random forests. *Machine Learning*, 45: 5–32.

Breiman, L. (2001b). Statistical modeling: Two cultures. *Statistical Science*, 16(3): 199–215.

Buhlmann, P. (2002). Bootstraps for time series. *Statistical Science*, 17(1): 52–72.

Butler, R.J., Gardner, H., and Gardner, H. (1998). Workers compensation costs when maximum benefits change. *Journal of Risk and Uncertainty*, 15: 259–269.

Butler, R.J., and Sutton, R. (1998). Saddlepoint approximations for multivariate cumulative distribution functions and probability computations in sampling theory and outlier testing. *Journal of the American Statistical Association*, 19(442): 596–604.

Butler, R.J., and Worall, J.D. (1991). Claim reporting and risk bearing moral hazard in workers compensation. *Journal of Risk and Insurance*, 53: 191–204.

Chiang, C.L. (1968). *Introduction to stochastic processes in biostatistics*. John Wiley & Sons, New York.

Christensen, R. (1984). *Data distributions, a statistical handbook*. Entropy Limited, Lincoln, MA.

Christianini, N., and Shawe-Taylor, J. (2000). *An introduction to support vector machines*. Cambridge University Press, Cambridge.

Cinlar, E. (1975). *Introduction to stochastic processes*. Dover Publications, Englewood Cliffs, NJ.

Cox, D.R., and Oakes, D. (1990). *Analysis of survival data*. Chapman & Hall, New York.

Cummins, J.D., and Tennyson, S. (1996). Moral hazard in insurance claiming: Evidence from automobile insurance. *Journal of Risk and Uncertainty*, 12: 29–50.

Daniels, H.E. (1954). Saddlepoint approximations in statistics. *Annals of Mathematical Statistics*, 25: 631–650.

Daniels, H.E. (2008). A saddlepoint approximation for tail probabilities in collective risk models. *Journal of Actuarial Practice*, 7: 1–11.

Diaconnis, P., and Efron, B. (1983). Computer intensive methods in statistics. *Scientific American*, 248: 116–131.

DiCiccio, T.J., and Efron, B. (1996). Bootstrap confidence intervals. *Statistical Science*, 11(3), 189–228.

Dickson, D.C.M., Hardy, M.R., and Waters, H.R. (2009). *Actuarial mathematics for life contingent risks*. Cambridge University Press, Cambridge.

Draper, N.R., and Smith, H. (1998). *Applied regression analysis*. 2nd ed. Wiley & Sons, New York.

Dropkin, L.B. (1959). Some actuarial considerations in automobile rating systems utilizing individual driving records. *Proceedings of the Casual Actuarial Society*, XLVI: 165–176.

Dufresne, D. (2009). *Sums of lognormals*. Centre for Actuarial Studies, University of Melbourne.

Efron, B. (1979). Bootstrap methods: Another look at the jackknife. *Annals of Statistics*, 7: 1–26.

Efron, B., and Tibshirani, R. (1986). Bootstrap methods for standard errors, confidence intervals, and other measures of statistical accuracy. *Statistical Science*, 1(1): 54–77.

Elandt-Johnson, R.C., and Johnson, N.L. (1980). *Survival models and data analysis*. John Wiley, New York.

Esscher, F. (1932). On the probability function in collective risk theory. *Scandinavian Actuarial Journal*, 15: 175–195.

Feller, W. (1971). *An introduction to probability and its application*. Vol. 2, 2nd ed. Wiley & Sons, New York.

Field, C., and Ronchetti, E. (1990). *Small sample asymptotics*. IMS Lecture Notes Monograph Series 13. IMS, Hayward, CA.

Frees, E.W. (1990). Stochastic life contingencies with solvency considerations. *Transactions of the Society of Actuaries*, XLLI: 91–129.

Genest, C. (1987). Franks Family of bivariate distributions. *Biometrika*, CXXIV: 549–555.

Genest, C., and McKay, J. (1986). The joy of copulas: Bivariate distributions with uniform marginals. *American Statistician*, 40: 280–283.

Gerber, H.U. (1976). A probabilistic model for life contingencies and a delta-free approach to contingency reserves. *Transactions of the Society of Actuaries*, XXVIII: 12–41.

Gerber, H.U. (1979). *An introduction to mathematical risk theory*. Hubener Foundation Monograph 8. Distributed by Richard D. Irwin, Homewood, IL.

Gerber, H.U., and Shiu, E.S. (1997). The time value of ruin. *Actuarial Research Clearing House*, 1: 25–36.

Geyer, C.J. (1992). Practical Markov chain Monte Carlo. *Statistical Science*, 7(4): 473–511.

Gollier, C. (1996). Optimum insurance of approximate losses. *Journal of Risk and Insurance*, 63(3): 369–380.

Gompertz, B. (1825). On the nature of the function expressive of the law of human mortality, and on a new mode of determining the value of life contingencies. *Philosophical Transactions of the Royal Society of London*, 115: 513–583.

Goovaerts, M.J., and De Pril, N. (1980). Survival probabilities based on the Pareto claim distribution. *ASTIN Bulletin*, 11: 154–157.

Goovaerts, M.J., and DeVylder, F. (1980). Upper bounds on stop-loss premiums under constraints on claim size distributions as derived from representative theorems for distribution functions. *Scandinavian Actuarial Journal*, 141–148.

Goutis, C., and Casella, G. (1999). Explaining the saddlepoint approximation. *American Statistician*, 53(3): 216–224.

Graybill, F.A. (1976). *Theory and application of the linear model*. Wadsworth Publishing, Belmont, CA.

Grosen, A., and Jorgensen, P.L. (1997). Valuation of early exercisable interest rate guarantees. *Journal of Risk and Insurance*, 63(3): 481–503.

Guthrie, L.G., and Lemon, L.D. (2004). *Mathematics of interest rates and finance*. Prentice Hall, Upper Saddle River, NJ.

Guiahi, F. (2007). Pricing multiple property cover based on a bivariate lognormal distribution. *Variance*, 1(2): 273–291.

Hahn, G.J., and Shapiro, S.S. (1967). *Statistical models in engineering*. Wiley & Sons, New York.

Hand, D.J., Blunt, G., Kelly, M.G., and Adams, N.M. (2000). Data mining for fun and profit. *Statistical Science*, 15(2): 111–126.

Helms, L.L. (1997). *Introduction to probability theory*. W.H. Freeman and Company, New York.

Hogg, R.V., McKean, J.W., and Craig, A.T. (2005). *Introduction to mathematical statistics*. 6th ed. Prentice Hall, Englewood Cliffs, NJ.

Hogg, R.V., and Tanis, E.A. (2010). *Probability and statistical inference*. 8th ed. Prentice Hall, Englewood Cliffs, NJ.

Huzurbazar, S. (1999). Practical saddlepoint approximations. *American Statistician*, 53(3): 225–232.

Jennrich, R.J. (1969). Asymptotic properties of nonlinear least squares estimators. *Annals of Mathematical Statistics*, 40(2): 633–643.

Johnson, L.G. (1951). The median ranks of sample values in their population with an application to certain fatigue studies, *Industrial Mathematics*, 2: 1–9.

Johnson, R.A., and Wichern, D.W. (2007). *Applied multivariate statistical analysis*. 6th ed. Prentice Hall, Englewood Cliffs, NJ.

Jordan, C.W. (1967). *Life contingencies*. 2nd ed. Society of Actuaries, Itasca, IL.

Kellison, S.G. (2009). *The theory of interest*, 3rd ed. McGraw-Hill, New York.

Klugman S.A., Panjer, H.H., and Willmot, G.E. (2008). *Loss models: From data to decisions*. 3rd ed. John Wiley & Sons, Hoboken, NJ.

Laha, R.G., and Rohatgi, V.K. (1979). *Probability theory*. John Wiley & Sons, New York.

Langberg, N., Proshan, F., and Quinzi, A.J. (1978). Converting dependent models into independent ones. *Annals of Probability*, 6: 174–181.

Larson, H.J. (1995). *Introduction to probability*. Advanced Series in Statistics. Addison-Wesley, Reading, MA.

Levy, P. (1954). *Theorie de l'addition des variables aleatoires*. 2nd ed. Gauthier-Villars, Paris.

London, D. (1997). *Survival models and their estimation*. 3rd ed. ACTEX Publications, Winsted, CT.

Lukacs, E. (1948). On the mathematical theory of risk. *Journal of the Institute of Actuaries Students' Society*, 8: 20–37.

Makeham, W.M. (1860). On the law of mortality and the construction of annuity tables. *Journal of the Institute of Actuaries*, 8(6): 301–310.

Marshall, A.W., and Olkin, I. (1967). A multivariate exponential distribution. *JASA*, 62: 30–44.

Marshall, A.W., and Olkin, I. (1988). Families of multivariate distributions. *JASA*, 83: 834–841.

McBean, E.A., and Rovers, F.A. (1998). *Statistical procedures for analysis of environmental monitoring data and risk assessment*. Prentice Hall, Englewood Cliffs, NJ.

McCullagh, P., and Nelder, J.A. (1983). *Generalized linear models*. Chapman & Hall, London.

Miller, R.G. (1974). The jackknife: A review. *Biometrika*, 61: 1–15.

Mood, A.M., Graybill, F.A., and Boes, D.C. (1974). *Introduction to the theory of statistics*. McGraw-Hill, New York.

Morrison, D.F. (1976). *Multivariate statistical methods*. 2nd ed. McGraw-Hill, New York.

Myers, G.G (2008). Estimating predictive distributions for loss reserve models. *Variance*, 2: 248–272.

Myers, R.H. (1990). *Classical and modern regression with applications*. 3rd ed. Duxbury Press, Boston, MA.

Myers, R.H., and Milton, J.S. (1998). *A first course in the theory of linear statistical models*. McGraw-Hill, New York.

Neill, A. (1977). *Life contingencies*. Heinemann, London.

Nelson, W. (2005). *Applied life data analysis*. John Wiley & Sons, New York.

Page, A.N., and Kelley, A.M. (1971). *Translation of Manuale di economia politica*. A.M. Kelley, New York.

Panjer, H.H. (1981). Recursive evaluation of a family of compound distributions. *ASTIN Bulletin*, 12: 2–26.

Panjer, H.H., and Bellhouse, J. (1980). Stochastic modelling of interest rates with applications to life contingencies. *Journal of Risk and Insurance*, 27: 91–110.

Panjer, H.H., and Wang, S. (1993). On the stability of recursive formulas. *Astin Bulletin*, 23(2): 227–258.

Panjer, H.H., and Willmot, G.E. (1992). *Insurance risk models*. Society of Actuaries, Schaumburg, IL.

Pentikainen, T. (1987). Approximate evaluation of the distribution function of aggregate claims. *ASTIN Bulletin*, 17: 15–40.

Quenouille, M.H. (1949). Approximate tests of correlation in time series. *Journal of the Royal Statistical Society, B*, 11: 68–84.

Ratkowsky, D.A. (1983). *Nonlinear regression modeling, a unified approach*. Marcel Dekker, New York.

Rohatgi, V.K. (1976). *An introduction to probability theory and mathematical statistics*. Wiley & Sons, New York.

Ross, S. (2002). *A first course in probability*. Prentice Hall, Englewood Cliffs, NJ.

Scheerer, A.E. (1969). *Probability on discrete sample spaces with applications*. International Textbook Co., Scranton, PA.

Seal, H.L. (1969). *Stochastic theory of a risk business*. John Wiley & Sons, New York.

Seal, H.L. (1977). Studies in the history of probability and statistics, multiple decrement or competing risks. *Biometrika*, LXIV: 429–439.

Seal, H.L. (1978). From aggregate claims distribution to probability of ruin. *ASTIN Bulletin*, X: 47–53.

Searle, S.R. (1971). *Linear models*. Wiley, New York.

Simon, L.J. (1960). The negative binomial and the Poisson distributions compared. *Proceedings of the Casualty Society*, XLVII: 20–24.

SOA. (2000). *The RP—2000 mortality tables*. www.soa.org.

SOA. (2001). *Report of the Individual Life Insurance Valuation Mortality Task Force*. www.soa.org.

Sundt, B., and Jewell, W.S. (1981). Further results of recursive evaluation of compound distributions. *ASTIN Bulletin*, 12: 27–39.

Taylor, G.C. (1977). Upper bounds on stop-loss premiums under constraints on claim size distributions. *Scandinavian Actuarial Journal*, 94–105.

Tobias, P.A., and Trindade, D.C. (1995). *Applied reliability*. 2nd ed. Chapman & Hall, New York.

Tsuchiya, T., and Konishi, S. (1997). General saddlepoint approximations and normalizing transformations for multivariate statistics. *Communications in Statistics, Theory and Methods*, 26(11): 2541–2563.

Tukey, J.W. (1958). Bias and confidence in not-quite large samples, *Annals of Mathematical Statistics*, 29: 614.

Walpole, R.E., Myers, R.H., and Myers, S.L. (1998). *Probability and statistics for engineers and scientists*. 6th ed. Prentice Hall, Englewood Cliffs, NJ.

Willemse, W.S., and Koppelaar, H. (2000). Knowledge elicitation of Gompertz law of mortality. *Scandinavian Actuarial Journal*, 2: 168–179.

Willmot, G. (1998). Statistical independence and fractional age assumption. *North American Actuarial Journal*, 1(1): 84–100.

Wood, A.T., Booth, J.G., and Butler, R.W. (1993). Saddlepoint approximations to the CDF of some statistics with nonnormal limit distributions. *JASA*, 88(442): 680–686.

Symbol Index

Symbol	Page	Symbol	Page
$\ddot{a}_{\overline{k}\rceil}$	104	$_n\mathrm{d}_x^{(j)}$	275
$\overline{a}_x$	218	$DF(t_n)$	99
$\ddot{a}_x$	218	D_{kG}	132
$\overline{a}_{x:\overline{n}\rceil}$	218	D_x^t	186
$\ddot{a}_{x:\overline{n}\rceil}$	219	$_nD_x^{(j)}$	275
$_{n\rceil}\ddot{a}_x$	219	$_nD_x^{(\tau)}$	275
$\ddot{a}_x^{\{1\}}$	222	$_nD_x$	180
$\ddot{a}_x^{(m)}$	246	e_x	309
$\ddot{a}_{x:\overline{n}\rceil}^{(m)}$	247	$\overset{o}{e}_x$	188
AS_{x+h}	287	$ENSV(j)$	329
AF	355	EP	124
$\overline{A}_x$	212	ES_{x+h+t}	287
A_x	212	$^jE\{FV(t_n)\}$	328
$\hat{A}_x$	365	$E[g(X)]$	14
^jA_x	325	$E\{g(X,Y)\}$	34
$^j\overline{A}_x$	326	$E\{S_N\}$	65
$\overline{A}_{x:\overline{n}\rceil}^1$	213	fv	207
$A_{x:\overline{n}\rceil}^1$	213	$f(x)$	8
$\overline{A}_{x:\overline{n}\rceil}$	213	$f(x,y)$	31
$A_x^{(j)}$	280	$f(x\,|\,Y{=}y)$	32
$A_x^{(m)}$	247	$f(x\,|\,X{>}c)$	29
$_{n\rceil}\overline{A}_x$	213	$f_X(x)$	164
$^2\overline{A}_x$	236	$f_S(s)$	50
^2A_x	236	$f_{XY}(t)$	258
$AF(t)$	100	$f_{\overline{xy}}(t)$	260
$Cov(x,y)$	34	F_0	94
$CS_{1-\alpha}$	47	F_i^*	79
$c(t)$	152	$FV(1)$	94
d	216	$FV(t_n)$	95
d_x	180	$FV(t)$	96
$d^{(m)}$	245	$FV(t,j)$	324
$d_x^{(j)}$	275	$F(x)$	10
$_nd_x$	180	$F(w)_S$	335

Symbol	Page	Symbol	Page
$FVE(t)$	126	$L_{[x]}$	194
$F_X(x)$	164	M_n	296
$FVR(t_n)$	126	$m(x)$	188
$F_S(s)$	50	$MX(t)$	21
$F_{XY}(t)$	258	$M_{Sm}(t)$	52
$F_{\overline{xy}}(t)$	260	NSP	303
$g(t)$	265	$NSP^{(w)}$	284
$g(x,h,t)$	288	NSV	131
G	132	$NSV^{(j)}$	326
$G(t)$	265	$NSV^{(j)}$	283
$h(j)$	265	$NSV\text{-}C$	285
$h(y_i)$	79	$NSV\text{-}B$	283
$h(j|t)$	267	$NSV\text{-}B_{x+h}$	288
i	97	$^s p_x$	357
i_j	95	$_t p_{xy}$	258
$i^{(m)}$	244	$_t p_{\overline{xy}}$	260
$I_j(x)$	179	$_t p_x^{s(j)}$	270
L_x	192	p_x	165
LD	132	$_t p[x]+u$	178
$LB(x_p)$	48	$_t p_x$	164
$LF(t)$	124	$_t p_x^{(\tau)}$	266
$_t LF(W)$	238	$^{(i)} p_x^{(j)}$	303
$_k LF(U)$	240	$_t^s p_x$	360
K	163	$\hat{p}$	43
$K(x)$	167	$\overline{P}(\overline{A}_x)$	234
$K_{(1)}$	259	P_x	234
K_u	16	$_h P(\overline{A}_{x:\overline{n}|}^1)$	234
$l(\theta)$	42	$_h P(\overline{A}_{x:\overline{n}|})$	234
$L(\theta)$	42	$_h P(_n E_x)$	234
l_x	179	$_h P(_{n|} \ddot{a}_x)$	234
$l_{\underset{x}{}}$	313	$_4 P(A_{50:\overline{4}|}^1)$	234
$^{(i)} l_x$	315	$PV(t)$	98
$l_x^{(\tau)}$	275	$PV(t_n)$	97
$l_{[x]}$	194	$PV(t,j)$	324
L_x	179	$P(A)$	2
$L_x^{(\tau)}$	275	$PC(\alpha)$	130

Symbol	Page	Symbol	Page
$PV(T)$	93	$S(t)$	152
$PVE(t)$	124	$S(x)$	22
$PVR(t)$	124	S_d	145
$P(A|B)$	4	S_c	13
$PS_{1-\alpha}$	47	S_d	13
q_x	165	S_N	65
$_t q_x$	164	S_m	50
$_k|q_x$	167	S_K	16
$_{t|u}q_x$	165	S^2	43
$_t q_x^{s(j)}$	270	$S_{W,BS}^2$	331
$^{(i)}q_x$	303	$\ddot{s}_{\overline{k}|}$	104
$^s q_x$	357	T_n	95
$_n q_{xy}^1$	263	$^s T$	355
$_n q_{xy}^2$	263	$t_n(i,j)$	296
$_t q_x^{(j)}$	265	$t_x(i,j)^{(u)}$	304
$_t q_x^{(\tau)}$	265	$t_n^{(u)}(i,j)$	296
$q_x^{(w)}$	281	T	163
$q_x^{(d)}$	281	$\mathbf{T}_n$	296
$q_x^{(i)}$	281	$\mathbf{T}_n^{(u)}$	296
$q_x^{(r)}$	281	TPS_{x+h}	289
$q_x(t)$	362	$T(x)$	164
$_k|Q_x$	186	$T_{(1)}$	55
r	72	$T_{(2)}$	260
r_a	149	$T_{(m)}$	55
$r_{i,r+1}^{(u)}$	311	$T_{(x,1)}$	192
$r_{j/m}$	97	T_x	193
$r_x(1,1)$	299	T_p	46
$r_x(1,1)^{(u)}$	300	U_j	148
RC	129	$UB(x_p)$	48
$R(u,n)$	344	$U(t)$	152
$\mathbf{R}_x$	299	v	204
$\mathbf{R}_x^{(u)}$	299	v_m	245
$R(t)$	129	v_j	325
$R(x,h,t)$	288	$Var(X)$	15
R_d	145	$Var\{S_N\}$	65
S	177	$_t\bar{V}(\bar{A}_x)$	241

Symbol	Page	Symbol	Page
$_kV_x$	241	λ	144
$_t^h\bar{V}(\bar{A}_{x:\overline{n}|}^1)$	241	μ	15
$_k^hV(\bar{A}_{x:\overline{n}|})$	241	μ_t	112
$_k^hV(_nE_x)$	241	μ_x	169
$_k^hV(_{n|}\ddot{a}_x)$	241	μ_{xy+t}	258
$\bar{W}_{BS}$	331	$\mu_{\overline{xy}+t}$	261
$w(F_i^*)$	79	$\mu_{x+t}^{(\tau)}$	266
$W(\hat{\boldsymbol{\theta}}, \boldsymbol{\theta})$	47	$\mu_{x+t}^{(j)}$	266
$X_{(j)}$	54	$\mu_{x+t}^{(w)}$	281
x_p	28	$\mu_{x+t}^{(d)}$	281
$\bar{X}$	43	$\mu_{x+t}^{(i)}$	281
(x)	6	$\mu_{x+t}^{(r)}$	281
$[x]$	178	π	102
Y_j	54	π_{x+u}	304
$y_t(x)$	362	π_j	101
$\hat{y}$	72	$\pi^{(m)}$	245
Z	12	$^{(i)}\pi_x$	304
$1-\alpha$	44	ρ	140
β_0	71	ρ	140
β_1	71	σ	15
$\hat{\beta}_0$	72	σ^2	15
$\hat{\beta}_1$	72	σ_t^2	112
δ	97	σ_{XY}	34
δ_n	99	ν	140
δ_t	100	ϕ_j	75
Γ	139	$\hat{\phi}$	77
Δ	139	$\Phi(x)$	12
θ	140	$\Phi^{-1}(x)$	56
$\hat{\boldsymbol{\theta}}$	47	$\psi(u)$	149

Index

A

Acronyms and principal sections, 377–378
Actuarial life tables, 227–230
  actuarial life table for selected ages, 228
  annuity payments, 229
  benefit paid, 229
  example, 228
  group survivorship tables, 227
  net single premiums, 230
Actuarial statistics, 364–369
  aggregate present value, 365
  bias correcting prediction interval, 365
  definition of, 181
  discrete stochastic status model, 364
  normality-based prediction intervals, 365–366
    curtate future lifetime probabilities, 366
    example, 366
    expectation, 365
    survivorship group theory, 365
  prediction set-based prediction intervals, 366–368
    actuarial statistic, 366
    central limit theorem, 366
    example, 376
    observed probabilities for lower and upper bounds, 368
    prediction interval construction, 366
    software, 367
    statistical vector, 366–367
  simulation-based prediction intervals, 368–369
    aggregate sum of present values, 368
    bias correcting prediction interval, 368

    example, 368
    mortality table-based statistical inference, 368
    percentile ranks, 369
    simulated samples of aggregate present values, 369
    symmetric simulation prediction interval, 368
  symmetric simulation prediction interval, 365
  weak law of large numbers, 364
  yearly mortality rates, 364
Advanced stochastic status models, 257–293
  Excel problems, 291–292
  future salary-based benefits and contributions, 287–288
    approximate expectation formulas, 288
    benefit retirement rate, 287
    example, 287, 288
  general contingent status, 263
    numerical integration methods, 263
    stochastic status model conditions, 263
  joint life status, 258–260
    actuarial science notation, 258
    discrete probability formulas, 259
    example, 259
    force of mortality, 258
    probability density function, 258
  last survivor status, 260–263
    actuarial science notation, 260
    economic action, 260
    example, 261
    force of mortality, 262
    fractional ages, 261
    fractional lifetime probabilities, 262
    multiple survival probabilities, 262

probability construction, 261
survival function, 260
multiple-decrement benefits,
    281–285
  aggregate benefit expectation,
      285
  death benefit, 284
  differing levels of benefits, 281
  example, 281, 283
  group survivorship theory, 281
  pension actuarial life table, 282
  unit benefit expectation, 284
  yearly effective interest rate, 283
multiple-decrement models, 264–280
  competing risk models, 264
  continuous multiple decrements,
      264–266
  discrete multiple decrements,
      268–269
  example, 264
  forces of mortality, 266–268
  multiple-decrement
      computations, 279–280
  multiple-decrement life tables,
      275–278
  single-decrement life tables,
      278–279
  single-decrement probabilities,
      269–271
  single-decrement probability
      bounds, 273–274
  status decrements, 264
  uniformly distributed single-
      decrement rates, 271–273
multiple future lifetimes, 257–263
  decomposition of future lifetime
      random variable, 257
  general contingent status, 263
  joint life status, 258–260
  last survivor status, 260–263
  stochastic status model, 257
pension contributions, 285–287
  basic plans for contributions, 285
  contribution payments, 285
  example, 285, 286
  flat contributions, 286
pension plans, 280–290
  contributions, 280, 283
  financing, 280

future salary-based benefits and
    contributions, 287–288
  multiple-decrement benefits,
      281–285
  other benefits, 280
  pension contributions, 285–287
  retirement benefit, 280
  survivorship group, 280
  yearly based retirement benefits,
      288–290
  problems, 290
  solutions, 292–293
  yearly based retirement benefits,
      288–290
    age-based retirement annuity, 288
    benefit rate based on final year's
        salary, 289
    benefit rate function, 288
    example, 289
    net single premium formula, 289
Aggregate distributions,
        approximating, 57–64
  central limit theorem, 57–61
  Haldane type A approximation, 57,
      61–62
  Panjer recursion, 57
  saddlepoint approximation, 62–64
  Wilson–Hilferty approximation, 57
Aggregate models (collective), 140–148
  aggregate stop-loss reinsurance and
      dividends, 145–148
    aggregate claim, decomposed,
        145
    ceding company, 145
    claim obligations of insurer, 145
    dividend, 147
    example, 146
    excess of loss coverage, 145
    gross premium, 147
    stop-loss coverage, 145
    stop-loss insurance, 145, 147
  fixed number of variables, 141–143
    benefit, 142
    central limit theorem, 141
    example, 142, 144
    fire-related house insurance
        claims, 142
    Haldane type A approximation,
        141

net single value, 141
saddlepoint approximation, 141
types of insurance policies, 143
stochastic number of variables,
143–145
compound Poisson aggregate
variable, 143
differentiation techniques, 143
fixed benefits policy types, 144
prediction intervals for aggregate
variable, 145
Aggregate variables, compound, 65–70
applications, 65
collective risk modeling, 65
expectations, 65–66
compound geometric random
variable, 66
compound Poisson process, 66
confidence and prediction
intervals, 65
construction of statistical
inference, 65
example, 66
moment generating function, 65
limiting distributions, 66–70
compound Poisson
approximations, 70
compound Poisson distribution,
67, 68
estimators of true probability, 70
example, 67
moments, 67
saddlepoint approximation, 68, 69
standardized variable, 67
three-moment approximation,
68, 69
statistical properties, 65
stochastic structure, 65
American call option, 137, 342
Annuities, fixed-rate, 101–106
amount of the annuity, 101
continuous annuity models, 104–106
accumulation function, 104, 105
cumulative rate, 104
discrete time period model, 104
example, 105
level annuity, 105
payment pattern, 104
potential flexibility, 105

definition of annuity, 101
discrete annuity models, 101–104
actuarial settings, 102
aggregate value of annuity, 103
applications, 102
bank interest, 103
disjoint time periods, 101
example, 102, 103
future value, 101, 103
present value of annuity, 104
principal investments, 103
summation formula, 103
time line, 101
unit payments, 104
period, 101
Annuities, life, 215–222
apportionable annuities, 220–222
analysis, 221
annuity overpayment, 220
example, 222
expected value of adjusted
payment, 221
future value quantity, 221
mathematical modeling example,
222
NSP expectation, 220
unit payment, 222
continuous whole life annuity, 217
deterministic status annuities, 215
discrete life annuity, 215
example, 216
expectation formula, 216
force of interest, 215
stochastic status life annuity, 215
types of unit payment life annuities,
217–220
continuous n-year temporary life
annuity, 218
continuous whole life annuity,
218
discrete n-year deferred whole
life annuity-due, 219
discrete n-year temporary life
annuity-due, 219
discrete whole life annuity-due,
218
economic complication, 220
example, 219, 220
interest rate, 218

Autoregressive systems, 75–78
  ad hoc approximate prediction
      interval, 77
  application, 78
  ARMA model, 76
  asymptotic normality of least
      squares estimators, 77
  dependent variable techniques, 75
  error terms, 76
  example, 78
  least squares point estimate, 77
  relation between variables, 75
  unknown parameters, 76, 77

B

Bagging, 345
Bayesian modeling, 346
Bayesian statistics, 41
BCPI, *see* Bias correcting prediction
      interval
Bernoulli random variables, 9
Bias correcting prediction interval
      (BCPI), 56, 337, 365, 368
Biostatistics, 185
Black–Scholes option pricing formula,
      138, 139
Bootstrap sampling, 331–332
  claim values, 331
  example, 331, 332
  proxy sampling distribution, 331
  validity problem, 332
Box–Muller transformation, 334

C

Call option, 133
cdf, *see* Cumulative distribution
      function
Ceding company, 145
Central limit theorem (CLT), 57–61
  example, 58, 59, 60
  fixed number of variables, 141
  large sample size, 60
  Levy central limit theorem, 58
  Lindeberg–Feller central limit
      theorem, 57
  prediction set-based prediction
      intervals, 366

relaxed conditions, 57
  sample standard deviation,
      statistical consistency of, 58
  small sample setting, 59
Chapman–Kolmogorov equations, 297
Chi squared goodness-of-fit test
      statistic, 86
CLT, *see* Central limit theorem
Collective aggregate models, 14
Competing risk models, 264
Complete expectation of life, 188
Compound aggregate variables,
      see Aggregate variables,
      compound
Compounding interest, 95
Conditional distributions, 29–31
  example, 29, 30
  geometric random variable, 30
  lack of memory, 30
  lifetime variable, 30
  magnitude of truncation, 31
  random variables, 29
Confidence intervals, 44–45
  confidence coefficient, 44
  pivot, 44, 45
  random set, 44
Confidence and prediction sets, 46–49
  approximate chi square random
      variable, 47
  confidence set, 46, 47
  construction of confidence intervals,
      46
  diagnostic measurements, 46
  example, 47
  observed parameter estimates, 48
  prediction set, 46
  sample coefficient of variation, 48
  statistical pivot, 47
Continuous random variables, 10–12
  cumulative distribution function,
      10
  example, 11, 12
  graphs, 11
  probability density function, 10
  status condition, 12
Continuous-status whole life
      insurance, 212
Cumulative distribution function (cdf),
      10

aggregate stop-loss reinsurance and
dividends, 146
conditional distributions, 29
continuous future lifetime, 164
discrete future lifetime, 168
expectations, 17
force of mortality, 169
future lifetime random variables
and life tables, 163
generalized least squares diagnostic,
83
Gompertz distribution, 27
Haldane type A approximation, 61
investment pricing, 136
joint life status, 258
last survivor status, 260
linear mortality acceleration models,
355
lognormal distribution, 26
multiple-decrement models, 265
order statistics, 54
Pareto distribution, 25
percentiles and prediction intervals,
46
Poisson compound sums, 64
probability plotting, 79
random variables, 14
saddlepoint approximation, 62, 69
simulation prediction intervals, 338
simulation probabilities, 335
simulation sampling, 333
sums of independent variables, 50
survival functions, 22
Curtate expectation of life, 191

D

Data mining techniques, 330
Data set simplifications, 370–371
analysis of high-dimensional data
sets, 370–371
financial and actuarial modeling,
370
goal of standards, 370
graphical techniques, 370
investment modeling, 370
Deterministic status models, 123–161
aggregate stop-loss reinsurance and
dividends, 145–148

aggregate claim, decomposed,
145
ceding company, 145
claim obligations of insurer, 145
dividend, 147
example, 146
excess of loss coverage, 145
gross premium, 147
stop-loss coverage, 145
stop-loss insurance, 145, 147
basic loss model, 123–128
applications, 124
choice of expenditures, 124
deterministic loss models,
124–126
future time, 123
future values of revenues, 124
reference point, 124
stochastic rate models, 126–128
collective aggregate models, 140–148
aggregate stop-loss reinsurance
and dividends, 145–148
fixed number of variables,
141–143
stochastic number of variables,
143–145
continuous surplus model
(stochastic surplus model),
152–155
adjustment coefficient, 154
compound Poisson aggregate
claims, 154
example, 153
initial fund, 152
moment generating function, 153
probability of future ruin, 152
reinsurance setting, 154
deterministic loss models, 124–126
equivalence principal, 124
example, 124, 125, 126
expense term, 125
house purchase, 125
premium payment, 125
reference point, 124
unknown constants, 124, 126
discrete surplus model (stochastic
surplus model), 148–152
adjustment coefficient, 149
aggregate claim variable, 148

example, 150, 151
exponential surplus model, 152
financial actions, 148
payment of claims, 148
probability of ruin bounds for
   normal claims, 151
stop-loss model, 149
Excel problems, 158–159
fixed number of variables (collective
   aggregate models), 141–143
benefit, 142
central limit theorem, 141
example, 142, 144
fire-related house insurance
   claims, 142
Haldane type A approximation,
   141
net single value, 141
saddlepoint approximation, 141
types of insurance policies, 143
insurance pricing, 131–134
claim amount, 132
claim variable, 133
deductible insurance, 132, 133
disability insurance, 134
distribution of loss amounts, 133
example, 132, 133
gross premium, 132
loading, 132
moment formulas, 134
overall cost, 131
investment pricing, 135–136
conservative approach, 136
example, 136
investment growth, 135
normality assumption, 135
return rates, 135
risk-free rate, 135
option pricing, 136–139
American call option, 137
Black–Scholes formula, 137, 138
call option, 133, 137
European call option, 137
example, 138
expiration date, 136
indicator function, 137
investment pricing option
   formula, 137
options pricing formula, 138

single-risk models, 137, 138
strike price, 136
option pricing diagnostics,
   139–140
call option, 139
European call option, 140
example, 140
Greeks, 139
rho values, 140
risk-free interest rate, 140
sensitivity of delta, 139
vega, 140
percentile criteria (stochastic loss
   criterion), 130–131
associated stochastic loss models,
   130
example, 130
unknown constants, 130
problems, 155–158
risk criteria (stochastic loss
   criterion), 129–130
application, 129
equilibrium principle, 129
example, 129
investment example, 129
statistical point estimators,
   130
single-risk models, 123, 131–140
insurance pricing, 131–134
insurance setting, 131
investment pricing, 135–136
net single premium, 131
net single value, 131
option pricing, 136–139
option pricing diagnostics,
   139–140
solutions, 159–161
stochastic loss criterion, 128–131
main approaches, 128–129
percentile calculations, 129
percentile criteria, 130–131
risk analysis, 129
risk criteria, 129–130
stochastic number of variables
   (collective aggregate models),
   143–145
compound Poisson aggregate
   variable, 143
differentiation techniques, 143

fixed benefits policy types, 144
prediction intervals for aggregate
variable, 145
stochastic rate models, 126–128
accepted risk tolerance, 127
example, 127
fixed-rate annuity, 127
house mortgage computation,
126
investment types, 126
probability, 127, 128
stochastic status model, 123
stochastic surplus model, 148–155
continuous surplus model,
152–155
discrete surplus model, 148–152
joint distribution, 148
simulation techniques, 148
ultimate ruin, 148
Discrete monthly period model, 244
Discrete random variables, 8–10
Bernoulli random variables, 9
binomial random variable, 9
Poisson postulates, 9
Discrete-status whole life insurance,
212

E

Empirical prediction intervals, *see*
Order statistics and empirical
prediction intervals
Equilibrium principle, 129
European call option, 137, 140, 342
Excel
add-ins, 375–376
application, probability plotting, 81
basic mathematical functions, 37376
problems
advanced stochastic status
models, 291–292
deterministic status models,
158–159
financial computational models,
119–120
further statistical considerations,
371–373
future lifetime random variables
and life tables, 200

Markov chain methods, 317–319
scenario and simulation testing,
348–350
statistical concepts, 38
statistical techniques, 88–90
stochastic status models, 254
Solver, 63, 187
statistical functions, 375
survivorship chain computation, 315
Expiration date, 136

F

Financial computational models,
93–121
continuous annuity models, 104–106
accumulation function, 104, 105
cumulative rate, 104
discrete time period model, 104
example, 105
level annuity, 105
payment pattern, 104
potential flexibility, 105
continuous-rate models, 100–101
example, 100
formulas, 100
simplest case, 100
time effect on monetary values,
100
continuous stochastic annuity
models, 116–117
constant payment, 116
continuous annuity, 117
example, 117
prediction intervals, 116
yearly parameters, 117
continuous stochastic rate models,
112–114
cumulative rate, 112
example, 112, 113
expectation formulas, 113
future obligation, 114
normal random variables, 113
present value function, 112
yearly investment return rate, 114
deterministic status models, 93
discrete annuity models, 101–104
actuarial settings, 102
aggregate value of annuity, 103

applications, 102
bank interest, 103
disjoint time periods, 101
example, 102, 103
future value, 101, 103
present value of annuity, 104
principal investments, 103
summation formula, 103
time line, 101
unit payments, 104
discrete stochastic annuity models,
    114–116
    accumulation function, 115
    annuities with varying
        payments, 114
    cumulative interest rate, 115
    example, 115
    statistical approximation
        techniques, 115
    stochastic nature of financial
        rates, 116
discrete stochastic rate model,
    106–112
    distribution parameters, 109
    empirical return rate, 110
    equal-width time periods, 106
    example, 107, 108, 109
    Excel example, 111
    expectations and variances for
        future and present value
        functions, 108
    future value, 107
    independent rate model, 110
    moment generating function, 108
    prediction intervals, 107, 109
    sample correlation coefficient, 111
    statistical model fitting, 111
    substituted estimator procedures,
        110
economic actions, collection of, 93
Excel problems, 119–120
financial rate-based calculations,
    94–99
    annual interest rate, 97
    compounding interest, 95, 96
    continuous future time structure,
        96
    equal-length discrete time period
        model, 95

example, 95, 96, 98
force of interest, 96
future value of annuity-
    immediate, 95
general discrete model, 97
modeling of monetary growth, 94
present value of annuity-
    immediate, 98
principal, 94, 95
time line, 99
fixed financial rate models, 94–101
    additional monetary worth of
        investment, 94
    continuous-rate models, 100–101
    financial rate-based calculations,
        94–99
    general period discrete rate
        models, 99–100
    interest or investment return, 94
    principal, 94
fixed-rate annuities, 101–106
    amount of the annuity, 101
    continuous annuity models,
        104–106
    definition of annuity, 101
    discrete annuity models, 101–104
    period, 101
future value, 93
general period discrete rate models,
    99–100
    continuous-time structures, 99
    cumulative interest rate, 99
    discrete time period settings, 99
    return rates, 100
monthly mortgage payments, 93
problems, 117–119
reference time, 93
series of monetary payments, 93
solutions, 120–121
status model, 93
stochastic rate models, 106–117
    continuous stochastic annuity
        models, 116–117
    continuous stochastic rate
        models, 112–114
    discrete stochastic annuity
        models, 114–116
    discrete stochastic rate model,
        106–112

graphical assessments, 106
varying interest and return rates, 106
types of financial actions, 93
value of investment, 93
Fixed financial rate models, 94–101
additional monetary worth of investment, 94
continuous-rate models, 100–101
example, 100
formulas, 100
simplest case, 100
time effect on monetary values, 100
financial rate-based calculations, 94–99
annual interest rate, 97
compounding interest, 95, 96
continuous future time structure, 96
equal-length discrete time period model, 95
example, 95, 96, 98
force of interest, 96
future value of annuity-immediate, 95
general discrete model, 97
modeling of monetary growth, 94
present value of annuity-immediate, 98
principal, 94, 95
time line, 99
general period discrete rate models, 99–100
continuous-time structures, 99
cumulative interest rate, 99
discrete time period settings, 99
return rates, 100
interest or investment return, 94
principal, 94
Fixed-rate annuities, 101–106
amount of the annuity, 101
continuous annuity models, 104–106
accumulation function, 104, 105
cumulative rate, 104
discrete time period model, 104
example, 105
level annuity, 105

payment pattern, 104
potential flexibility, 105
definition of annuity, 101
discrete annuity models, 101–104
actuarial settings, 102
aggregate value of annuity, 103
applications, 102
bank interest, 103
disjoint time periods, 101
example, 102, 103
future value, 101, 103
present value of annuity, 104
principal investments, 103
summation formula, 103
time line, 101
unit payments, 104
period, 101
Force of mortality, 169–175
applications, 172
classical distribution, 173
conditional survival probability, 170
continuous type probability density function, 171
definition, 169
example, 171, 172, 174
function uniqueness, 170
Gompertz law of mortality, 173
hazard function, 169
infant mortality, 169, 174
instantaneous failure, 169
joint life status, 258
last survivor status, 262
law of geometrical progression, 173
linear mortality acceleration models, 356
Makeham law of mortality, 173
mortality probabilities, 174
multiple-decrement models, 266–268
employment system, 266
example, 267
overall force of decrement, 266
single-decrement setting, 267
summation, 266
two-decrement example, 267
probability of failure, 175
survival probability, 170
wearout mortality, 174
Weibull distribution, 27

Fractional ages
    adjustments, 193–194
        assumptions, 194
        expectations, 194
        survival and failure probabilities,
            193
        UDD assumption, 193
    aggregate parameters, 192–193
    assumption, 176
    conditional mortality, 176
    example, 176, 177
    interpolation techniques, 175
    last survivor status, 261
    natural decomposition, 175
    uniform distribution of death, 176
Frequentist statistics, 41
Further statistical considerations,
        353–374
    actuarial statistics, 364–369
        aggregate present value, 365
        bias correcting prediction
            interval, 365
        discrete stochastic status model,
            364
        normality-based prediction
            intervals, 365–366
        prediction set-based prediction
            intervals, 366–368
        simulation-based prediction
            intervals, 368–369
        symmetric simulation prediction
            interval, 365
        weak law of large numbers,
            364
        yearly mortality rates, 364
    classification of individuals, 353
    data set simplifications, 370–371
        analysis of high-dimensional
            data sets, 370–371
        financial and actuarial modeling,
            370
        goal of standards, 370
        graphical techniques, 370
        investment modeling, 370
    Excel problems, 371–373
    linear mortality acceleration models,
        355–337
        example, 356
        force of mortality, 356

        high-risk stratum individuals,
            356
        lifetime random variables, 355
        linear acceleration factor, 355
        standard survival techniques,
            355
        stochastic status model, 356
        survival functions, 355
    mean mortality acceleration models,
        357–360
        choice of distribution, 357
        example, 358, 359
        lifetime random variables, 357
        method to adjust mortalities, 357
        mortality and exponential
            adjusted mortalities, 358
        mortality and Weibull adjusted
            mortalities, 359
        UDD assumption, 357
        yearly survival rates, 359
    mortality adjustment models,
        354–361
        demographics, 354
        individual status models, 354
        linear mortality acceleration
            models, 355–337
        mean mortality acceleration
            models, 357–360
        survival-based mortality
            acceleration models, 360–361
        system modeling, 354
    mortality tables, 353
    mortality trend modeling, 361–364
        distribution of error term, 362
        example, 363
        generalized linear modeling,
            362
        least squares estimator, 362, 363
        model, 362
        mortality trend data and
            response, 363
        nonlinear modeling, 362
        original and projected mortality
            tables, 364
        time trend, 362
        yearly based mortality rates, 363
    multirisk strata modeling, 353
    normality-based prediction
        intervals, 365–366

curtate future lifetime
  probabilities, 366
  example, 366
  expectation, 365
  survivorship group theory, 365
prediction set-based prediction
    intervals, 366–368
  actuarial statistic, 366
  central limit theorem, 366
  example, 376
  observed probabilities for lower
    and upper bounds, 368
  prediction interval construction,
    366
  software, 367
  statistical vector, 366–367
problems, 371
simulation-based prediction
    intervals, 368–369
  aggregate sum of present values,
    368
  bias correcting prediction
    interval, 368
  example, 368
  mortality table-based statistical
    inference, 368
  percentile ranks, 369
  simulated samples of aggregate
    present values, 369
  symmetric simulation prediction
    interval, 368
solutions, 373–374
statistic inference techniques, 353
survival-based mortality
    acceleration models, 360–361
  adjustment procedure, 360
  example, 361
  medical studies, 360
  mortalities and survival-adjusted
    mortalities, 361
  strata settings, 360
Future lifetime random variables and
    life tables, 163–202
  aggregate parameters (life models
    and life table parameters),
    191–193
  discrete stochastic status models,
    191
  example, 193

fractional age techniques, 192–193
  group survivorship concepts, 191
  mortality structure of
    survivorship group, 192
concepts and formulas, 163
continuous future lifetime, 164–167
  conditioning argument, 164
  example, 165, 166
  expected sale price, 167
  future lifetime random variable,
    164
  mortality and survival rates, 165
  standard notations, 164
  survival function, 165
discrete future lifetime, 167–169
  curtate future lifetime, 167
  discrete setting, 168
  example, 168
  failure probability, 167
  lack of memory property, 168
Excel problems, 200
force of mortality, 169–175
  applications, 172
  classical distribution, 173
  conditional survival probability,
    170
  continuous type probability
    density function, 171
  example, 171, 172, 174
  function uniqueness, 170
  Gompertz law of mortality, 173
  hazard function, 169
  infant mortality, 169, 174
  instantaneous failure, 169
  law of geometrical progression,
    173
  Makeham law of mortality, 173
  mortality probabilities, 174
  probability of failure, 175
  survival probability, 170
  wearout mortality, 174
fractional age adjustments (life
    models and life table
    parameters), 193–194
  assumptions, 194
  expectations, 194
  survival and failure probabilities,
    193
  UDD assumption, 193

fractional ages, 175–177
  assumption, 176
  conditional mortality, 176
  example, 176, 177
  interpolation techniques, 175
  natural decomposition, 175
  uniform distribution of death, 176
life models and life table
     parameters, 187–194
  aggregate parameters, 191–193
  fractional age adjustments,
    193–194
  population parameters, 188–191
life models and life tables, 182–185
  adjusted survival rate, 185
  alternate form of life table, 183
  application, 182
  discrete stochastic status model,
    182
  example, 182, 184
  force of mortality, 185
  high-risk strata, 185
  life table listing mortality rates,
    184
  UDD assumption, 182
life table confidence sets and
     prediction intervals, 185–187
  approximate chi square random
    variable, 186
  asymptotic statistical theory, 186
  biostatistics, 185
  discrete stochastic status, 186
  example, 186
  generalized least squares
    measure, 186
  life table prediction interval, 187
  prediction sets, 185
population parameters (life models
     and life table parameters),
     188–191
  complete expectation of life, 188
  continuous future lifetime
    random variable, 188
  curtate expectation of life, 191
  curtate future lifetime, 190
  discrete-status age, 191
  discrete stochastic status model,
    190
  example, 189, 190

  expectation, 188
  median, 188
  moments computed, 188
  probability density function, 187
  underlying mortality structure,
    189
  variance, 188
problems, 198–199
select future lifetimes, 177–179
  application, 179
  example, 178
  fixed age, 178
  hypothetical mathematical
    modeling, 178
  life insurance policy, 178
  mathematical modeling example,
    178
  newborns, 177
  select lifetime, 178
  select probabilities, 178
  time of related events, 177
select and ultimate life tables, 194–198
  constraints, 197
  example, 195, 196, 197
  future lifetime random variables,
    194
  graduated select and ultimate life
    table, 197
  group survivorship theory, 194
  mortality probabilities, 194
  select life table, 196
  select period, 196
  select-ultimate life table, 196
solutions, 200–202
statistical concepts, 163
survivorship groups, 179–182
  actuarial statistics, definition of,
    181
  closed group, 179
  expectations, 181
  force of mortality, 180
  graduation of the table, 181
  group survivorship expectations,
    181
  independent lifetimes, 179
  insurance pricing theory, 181
  mortality notations, 180
  number of survivors, 179
  probability laws, 180

G

Gamma distribution, sums of
        independent variables, 53
Gamma function, 27
Generalized least squares measure
        (GLSM)
    model diagnostics, 84
    observed decrement totals, 186
Gibbs–Hastings algorithm, 346
GLSM, *see* Generalized least squares
        measure
Gompertz mortality law, 27, 173
Greeks, 139
Gross premiums, 249
Group survivorship
    aggregate parameters, 191
    pension plan, 281
    select and ultimate life tables, 194
    theory, 281

H

HAA, *see* Haldane type A
        approximation
Haldane type A approximation (HAA),
        57, 61–62
    claim values, 62
    example, 62
    fixed number of variables, 141
    nonnegative random variables, 61
    skewness of aggregate, 61
Hazard function, 169
Homogeneous stochastic status chains,
        307–312
    actuarial chains, 310–312
        discrete life insurance present
            value expectations, 312
        example, 311
        multistep transition matrix, 311
        nonzero components, 311
        stochastic status model age, 310
    construction, 307
    example, 308
    expected curtate future lifetime,
        309–310
        discrete stochastic status age,
            310
        example, 309

homogeneous chains, advantage
        of, 309
    mean curtate future lifetimes, 310
    nonzero entries, 307
    states in stochastic chain, 307
    transition matrices, 308

I

iid random variables, *see* Independent
        and identically distributed
        random variables
Independent and identically
        distributed (iid) random
        variables, 42
Independent variables, sums of, 49–54
    aggregate claims, 52
    applications, 49
    calculation of relevant probabilities,
        53
    collective risk modeling, 49
    convolution process, 50, 51
    example, 51, 52
    gamma distribution, 53
    moment generating function, 52
    sums resulting from random
        samples, 53
    two-variable case, 50
Infant mortality, 174
Insurance pricing, 131–134
    claim amount, 132
    claim variable, 133
    deductible insurance, 132, 133
    disability insurance, 134
    distribution of loss amounts, 133
    example, 132, 133
    gross premium, 132
    loading, 132
    moment formulas, 134
    overall cost, 131
Interval data diagnostic, 84–87
    chi squared goodness-of-fit test
        statistic, 86
    discrete random variables, 85
    disjoint sets, 85
    dispersion matrix for random
        vector, 85
    empirical and discrete distributions,
        86

example, 86
interval data, 84
observed data, 84
Poisson distribution, 87
Investment pricing applications
    (scenario and simulation
    testing), 340–343
    advantage, 342
    American call option, 342
    asset pricing, 343
    European call option, 342
    example, 340, 342
    financial investment strategy, 341
    future time of sale, 341
    present value of investment, 341
    stock purchasing strategy, 343

J

Jackknife resampling, 345
Jensen's inequality, 324
Joint distributions, 31–35
    computation of statistical
        measurements, 34
    covariance, 34
    example, 32, 33
    functions of multiple random
        variables, 35
    independent variables, 32
    linear relationship, 34
    relationships between variables, 32
    risk categories, frequencies of, 33
    statistical concepts, 31–32
    two-variable stochastic modeling, 21
Joint life conditions, 6
Joint life status, 258–260
    actuarial science notation, 258
    discrete probability formulas, 259
    example, 259
    force of mortality, 258
    probability density function, 258

L

Last survivor status (LSS), 260–263
    actuarial science notation, 260
    economic action, 260
    example, 261
    force of mortality, 262

fractional ages, 261
fractional lifetime probabilities,
    262
multiple survival probabilities, 262
probability construction, 261
survival function, 260
Least squares estimation, 71–73
    application, 72
    coefficient of determination, 72
    example, 73
    measure of linear fit, 72
    regression of stock prices, 73
    sample correlation coefficient, 72
    scatter plot, 73
Levy central limit theorem, 58
Life annuities, 215–222
    apportionable annuities, 220–222
        analysis, 221
        annuity overpayment, 220
        example, 222
        expected value of adjusted
            payment, 221
        future value quantity, 221
        mathematical modeling example,
            222
        NSP expectation, 220
        unit payment, 222
    continuous whole life annuity, 217
    deterministic status annuities, 215
    discrete life annuity, 215
    example, 216
    expectation formula, 216
    force of interest, 215
    stochastic status life annuity, 215
    types of unit payment life annuities,
        217–220
        continuous n-year temporary life
            annuity, 218
        continuous whole life annuity,
            218
        discrete n-year deferred whole
            life annuity-due, 219
        discrete n-year temporary life
            annuity-due, 219
        discrete whole life annuity-due,
            218
        economic complication, 220
        example, 219, 220
        interest rate, 218

Life insurance (stochastic status models), 210–215
  continuous-status whole life insurance, 212
  discrete-status whole life insurance, 212
  endowment policy, 214
  example, 211, 214
  financing, 210
  general nomenclature, 210
  single-status models, 215
  stochastic structures, 211
  types of unit benefit life insurance, 212–215
  unit benefit, 213
  variance of present value function, 211
Life models and life table parameters, 187–194
  aggregate parameters, 191–193
    discrete stochastic status models, 191
    example, 193
    fractional age techniques, 192–193
    group survivorship concepts, 191
    mortality structure of survivorship group, 192
  fractional age adjustments, 193–194
    assumptions, 194
    expectations, 194
    survival and failure probabilities, 193
    UDD assumption, 193
  population parameters, 188–191
    complete expectation of life, 188
    continuous future lifetime random variable, 188
    curtate expectation of life, 191
    curtate future lifetime, 190
    discrete-status age, 191
    discrete stochastic status model, 190
    example, 189, 190
    expectation, 18
    median, 188
    moments computed, 188
    probability density function, 187
    underlying mortality structure, 189
    variance, 188

Life models and life tables, 182–185
  adjusted survival rate, 185
  alternate form of life table, 183
  application, 182
  discrete stochastic status model, 182
  example, 182, 184
  force of mortality, 185
  high-risk strata, 185
  life table listing mortality rates, 184
  UDD assumption, 182
Life table confidence sets and prediction intervals, 185–187
  approximate chi square random variable, 186
  asymptotic statistical theory, 186
  biostatistics, 185
  discrete stochastic status, 186
  example, 186
  generalized least squares measure, 186
  life table prediction interval, 187
  prediction sets, 185
Lindeberg–Feller central limit theorem, 57
Loss model, basic, 123–128
  applications, 124
  choice of expenditures, 124
  deterministic loss models, 124–126
    equivalence principal, 124
    example, 124, 125, 126
    expense term, 125
    house purchase, 125
    premium payment, 125
    reference point, 124
    unknown constants, 124, 126
  future time, 123
  future values of revenues, 124
  reference point, 124
  stochastic rate models, 126–128
    accepted risk tolerance, 127
    example, 127
    fixed-rate annuity, 127
    house mortgage computation, 126
    investment types, 126
    probability, 127, 128
Loss models, insurance premiums and, 230–237
  curtate future lifetime, 232
  example, 230, 231

precedent of notations, 232
premiums, 230
standard insurance loss function
    constructions, 233
stochastic loss model, 230, 232
typical insurance policy, 230
unit benefit premium notation,
    232–235
    constant interest rate, 232
    example, 234, 235
    standard unit benefit level
        premium notation, 234
    unit benefit yearly premium, 235
variance of the loss function,
    235–237
    discrete-status whole life
        insurance policy, 236
    example, 235, 236
    standard life insurance loss
        functions, 235
    variance, 236
LSS, *see* Last survivor status

M

Makeham mortality law, 28, 173
Markov chain methods, 295–321
    actuarial chains, 299–300
        actuarial transition matrix, 299
        example, 300
        force of interest, 299
        nonhomogeneous Markov
            chains, 300
        unit benefit pure endowment, 300
    Excel problems, 317–319
    homogeneous stochastic status
        chains, 307–312
        actuarial chains, 310–312
        construction, 307
        discrete life insurance present
            value expectations, 312
        discrete stochastic status age, 310
        example, 308, 309, 311
        expected curtate future lifetime,
            309–310
        homogeneous chains, advantage
            of, 309
        mean curtate future lifetimes, 310
        multistep transition matrix, 311

nonzero components, 311
nonzero entries, 307
states in stochastic chain, 307
stochastic status model age, 310
transition matrices, 308
introduction to Markov chains,
    296–297
    Chapman–Kolmogorov
        equations, 297
    homogeneous Markov chain, 297
    initial state, 296
    stochastic matrix, 296
    transition probabilities, 296
multiple-decrement chains, 300–303
    discrete stochastic status model,
        300
    example, 301, 302
    multiple-decrement setting, 302
    nonnegative components, 301
    transition matrix, 301
multirisk strata chains, 303–307
    example, 304, 306
    group survivorship structure,
        partitioning of, 303
    multirisk strata mortality and
        survival rates, 305
    nonhomogeneous Markov chain
        theory, 303
    nonnegative components, 304
    risk strata, 303
    survival and mortality rates, 306
    u-step transition matrices, 305
nonhomogeneous stochastic status
    chains, 297–307
    actuarial chains, 299–300
    life tables, 297
    multiple-decrement chains,
        300–303
    multirisk strata chains, 303–307
    single-decrement chains, 298–299
    status age, 297
problems, 316–317
single-decrement chains, 298–299
    discrete stochastic status age, 298
    example, 298
    matrix components, 298
    nonhomogeneous stochastic
        chain theory, 298
    transition matrix, 298

solutions, 320–321
survivorship chains, 312–316
  closed-group survivorship
    structure, 312
  example, 313, 314, 315
  expected number of survivors,
    314
  initial vector, 314
  multiple-decrement models,
    314–315
  multirisk strata models, 315–316
  number of newborns, 313
  single-decrement models,
    313–314
  stochastic status age, 315
  survivorship chain, 314
  transition probabilities, 315
Metropolis algorithm, 346
mgf, *see* Moment generating function
Mixed random variables, 7, 13–14
  applications, 13
  collective aggregate models, 14
  example, 13
  expectations, 15
  moment generating function, 21
  probabilities, computation of, 13
  support parts, 13
Model diagnostics, 78–87
  assumed continuous distribution,
    efficiency of, 78
  generalized least squares diagnostic,
    83–84
    assumed continuous statistical
      distribution, 83
    dispersion matrix, 83
    empirical cdf vector, 83
    example, 84
    generalized least squares
      measure, 84
  graphical technique, 78
  interval data diagnostic, 84–87
    chi squared goodness-of-fit test
      statistic, 86
    discrete random variables, 85
    disjoint sets, 85
    dispersion matrix for random
      vector, 85
    empirical and discrete
      distributions, 86

    example, 86
    interval data, 84
    observed data, 84
    Poisson distribution, 87
  probability plotting, 79–82
    claim data, 80
    comparison for Pareto,
      lognormal, and Weibull, 82
    error term, 79
    Excel application, 81
    linear regression model, 80
    main concept, 79
    normal probability plot, 81
    parameter components, 79
    probability plot, 80
    scatter plot of transformed data,
      80
    transformation quantities, 82
Moment generating function, 20–22, 52
  complex-valued function, 22
  compound aggregate variables, 65
  continuous-risk calculations, 206
  continuous surplus model, 153
  discrete stochastic rate model, 108
  example, 21
  mixed random variable, 21
  random variable, 22
  saddlepoint approximation, 63
  sums of independent variables, 52
Mortality adjustment models, 354–361
  demographics, 354
  individual status models, 354
  linear mortality acceleration models,
    355–337
    example, 356
    force of mortality, 356
    high-risk stratum individuals,
      356
    lifetime random variables, 355
    linear acceleration factor, 355
    standard survival techniques,
      355
    stochastic status model, 356
    survival functions, 355
  mean mortality acceleration models,
    357–360
    choice of distribution, 357
    example, 358, 359
    lifetime random variables, 357

method to adjust mortalities, 357
mortality and exponential
adjusted mortalities, 358
mortality and Weibull adjusted
mortalities, 359
UDD assumption, 357
yearly survival rates, 359
survival-based mortality
acceleration models, 360–361
adjustment procedure, 360
example, 361
medical studies, 360
mortalities and survival-adjusted
mortalities, 361
strata settings, 360
system modeling, 354
Mortality trend modeling, 361–364
distribution of error term, 362
example, 363
generalized linear modeling, 362
least squares estimator, 362, 363
model, 362
mortality trend data and response,
363
nonlinear modeling, 362
original and projected mortality
tables, 364
time trend, 362
yearly based mortality rates, 363
Multiple-decrement models, 264–280
competing risk models, 264
continuous multiple decrements,
264–266
cumulative probability, 265
decrement probabilities, 266
future lifetime random variable,
264
joint distributions, 265
discrete multiple decrements,
268–269
integer-valued future lifetime,
268
nonnegative integers, 269
example, 264
forces of mortality, 266–268
employment system, 266
example, 267
overall force of decrement, 266
single-decrement setting, 267

summation, 266
two-decrement example, 267
multiple-decrement computations,
279–280
continuous stochastic status
model, 279
discrete-status setting, 279
example, 280
net single premium, 280
multiple-decrement life tables, 275–278
alternate form, 277
decrement probabilities, 276
decrement rates by year, 277
example, 276
expected number of decrements,
275
group survivorship notations,
275
single-decrement life tables, 278–279
single-decrement probabilities,
269–271
probability system, 270
single-decrement rate, 270
unique probability distribution,
270
single-decrement probability
bounds, 273–274
example, 274
lower bound, 273
multiple-decrement probabilities,
274
Taylor series expansion, 273
UDD multiple-decrement
probabilities and upper
bounds, 274
upper bound, 274
status decrements, 264
uniformly distributed single-
decrement rates, 271–273
example, 272
force of decrement, 271
multiple-decrement probabilities,
272
single-decrement rate, 271

N

Net single value (NSV), 205
definition, 131

discrete risk calculations, 207
fixed number of variables 141
risk evaluations, 205
single-risk models, 131
stochastic status scenarios, 325
Nonhomogeneous stochastic status
    chains, 297–307
    actuarial chains, 299–300
    life tables, 297
    multiple-decrement chains, 300–303
    multirisk strata chains, 303–307
    single-decrement chains, 298–299
    status age, 297
Nonnegative random variables, 25–29
    force of mortality, 27
    gamma function, 27
    Gompertz distribution, 27–28
    lognormal distribution, 26
    Makeham distribution, 28–29
    modeling of insurance claim
        amounts, 26
    Pareto distribution, 25
    Weibull distribution, 26–27
NSV, *see* Net single value

O

Option pricing, 136–139
    American call option, 137
    Black–Scholes formula, 137, 138
    call option, 133, 137
    European call option, 137
    example, 138
    expiration date, 136
    indicator function, 137
    investment pricing option formula,
        137
    options pricing formula, 138
    single-risk models, 137, 138
    strike price, 136
Order statistics and empirical
        prediction intervals, 54–57
    bias-correcting percentile method,
        56
    example, 54, 56
    independence property, 54
    indicator function, 56
    linear interpolation, 55
    observed variables, 54

specialized cases, 54
stochastic actions, 55

P

Panjer recursion, 57
pdf, *see* Probability density function
Pension plans, 280–290
    contributions, 280, 283
    financing, 280
    future salary-based benefits and
        contributions, 287–288
        approximate expectation
            formulas, 288
        benefit retirement rate, 287
        example, 287, 288
    multiple-decrement benefits,
        281–285
        aggregate benefit expectation, 285
        death benefit, 284
        differing levels of benefits, 281
        example, 281, 283
        group survivorship theory, 281
        pension actuarial life table, 282
        unit benefit expectation, 284
        yearly effective interest rate, 283
    other benefits, 280
    pension contributions, 285–287
        basic plans for contributions, 285
        contribution payments, 285
        example, 285, 286
        flat contributions, 286
    retirement benefit, 280
    survivorship group, 280
    yearly based retirement benefits,
        288–290
        age-based retirement annuity,
            288
        benefit rate based on final year's
            salary, 289
        benefit rate function, 288
        example, 289
        net single premium formula, 289
Percentiles and prediction intervals,
        45–46
    one-sided prediction intervals, 45
    outcomes of random variables, 45
    population measurements, 45
    prediction interval, 45

Point estimation, 42–44
    consistency, 44
    example, 43
    independent and identically
        distributed random variables,
        42
    minimum variance unbiased
        estimators, 44
    random sample, 42
    sample variance, 43
    statistics, 42
    weak law of large numbers, 44
Poisson postulates, 9
Prediction sets, *see* Confidence and
        prediction sets
Probability, 1–7
    axioms of probability, 2, 5
    basic set operations, 2
    collection of disjoint events, 2
    conditional probabilities, 4
    construction of probability
        measures, 3
    events, 2
    example, 2, 3
    integer future lifetime, 3
    joint life conditions, 6
    likelihood of events occurring, 2
    probability measure, 2
    probability system, 2
    random process, 2
    risk categories, 4
    Venn diagrams, 3, 5
    weak law of large numbers, 2
Probability density function (pdf)
    aggregate stop-loss reinsurance and
        dividends, 146
    apportionable annuities, 222
    compound aggregate variables, 65
    conditional distributions, 29
    confidence and prediction sets, 46
    continuous future lifetime, 164
    continuous random variables, 10
    continuous-risk calculations, 205
    continuous type, 171
    discrete future lifetime, 167
    discrete multiple decrements, 269
    discrete risk calculations, 206
    expectations, 16
    force of mortality, 169, 266

fractional ages, 176
future lifetime random variables
        and life tables, 163
general contingent status, 263
Gompertz distribution, 27
insurance pricing, 132
joint distributions, 31, 34
joint life status, 258
last survivor status, 260
life insurance, 214
life models and life table
        parameters, 187, 194
linear mortality acceleration models,
        355
lognormal distribution, 26
Makeham distribution, 29
mixed random variables, 13
moment generating function, 21
multiple-decrement life tables, 275
nonnegative random variables, 25
normality-based prediction
        intervals, 366
point estimation, 42
population parameters, 188
reserves, 239
saddlepoint approximation, 64
simulation-based prediction
        intervals, 368
simulation sampling, 333
stochastic present value functions,
        204
stochastic status scenarios, 326
survival functions, 23
Probability plotting, 79–82
    claim data, 80
    comparison for Pareto, lognormal,
        and Weibull, 82
    error term, 79
    Excel application, 81
    linear regression model, 80
    main concept, 79
    normal probability plot, 81
    parameter components, 79
    probability plot, 80
    scatter plot of transformed data, 80
    transformation quantities, 82
Problems
    advanced stochastic status models,
        290

deterministic status models, 155–158
financial computational models,
117–119
further statistical considerations, 371
future lifetime random variables
and life tables, 198–199
Markov chain methods, 316–317
scenario and simulation testing,
346–348
statistical concepts, 36–37
statistical techniques, 87–88
stochastic status models, 252–254

Q

Quenouille–Tukey jackknife, 330

R

Random variables, 7–14
continuous random variables, 10–12
discrete random variables, 8–10
mixed random variables, 7, 13–14
probability mass function, 8
status, 7
types, 7
Regression modeling, 70–75
error term, 71
fitted regression line, 72
least squares estimation, 71–73
application, 72
coefficient of determination, 72
example, 73
measure of linear fit, 72
regression of stock prices, 73
sample correlation coefficient, 72
scatter plot, 73
observed data sets, 71
predictor variable, 71
regression model-based inference,
74–75
confidence intervals, 74, 75
example, 75
minimum variance, 74
normality of the error term, 74
prediction interval, 74
statistical distribution theory, 74
unbiasedness, 74
scatter plot, 71

simple linear regression model, 71
variables of interest, 71
Rho values, 140
Risk evaluations, 205–208
continuous-risk calculations,
205–206
components, 205
example, 206
moment generating function, 206
single-status model, 205
discrete risk calculations, 206–207
example, 206
financial action, 206
future time structure, 206
model versatility, 207
insurance premium payments, 205
mixed risk calculations, 207–208
common actuarial usage, 207
example, 208
expectation, 207
life table data mortality rates, 207
UDD assumption, 207
net single value, 205

S

Saddlepoint approximation (SPA),
62–64
accuracy, 62
applications, 63
example, 64
Excel Solver routine, 63
fixed number of variables, 141
information drawn, 62
moment-based, 68
moment generating function, 63
Poisson compound sums, 64
prediction intervals, 63
Sample variance, 43
Sampling distributions and estimation,
41–49
confidence intervals, 44–45
confidence coefficient, 44
pivot, 44, 45
random set, 44
confidence and prediction sets,
46–49
approximate chi square random
variable, 47

confidence set, 46, 47
construction of confidence
    intervals, 46
diagnostic measurements, 46
example, 47
observed parameter estimates, 48
prediction set, 46
sample coefficient of variation, 48
statistical pivot, 47
examples of consistent estimators,
    44
observed samples, 41
percentiles and prediction intervals,
    45–46
    coverage probabilities, 46
    one-sided prediction intervals, 45
    outcomes of random variables, 45
    population measurements, 45
    prediction interval, 45
point estimation, 42–44
    consistency, 44
    example, 43
    independent and identically
        distributed random variables,
        42
    minimum variance unbiased
        estimators, 44
    random sample, 42
    sample variance, 43
    statistics, 42
    weak law of large numbers, 44
sampling distribution of the
    statistic, 42
Scatter plot, 71
Scenario and simulation testing,
    323–351
applications, 323
bootstrap sampling, 331–332
    claim values, 331
    example, 331, 332
    proxy sampling distribution, 331
    validity problem, 332
deterministic status scenarios,
    324–325
    application, 325
    example, 324, 325
    goal of investment, 324
    scenario rates, 324
    stochastic rates, 324

Excel problems, 348–350
future directions in simulation
    analysis, 344–346
    bagging, 345
    Bayesian applications, 346
    Bayesian modeling, 346
    confidence intervals, 345
    data encountered in actuarial
        analysis, 345
    Gibbs–Hastings algorithm, 346
    inference methods, 345
    jackknife resampling, 345
    Markov chain modeling, 346
    Metropolis algorithm, 346
    unstable simulation statistics,
        345
    unstable statistics problem, 345
investment pricing applications,
    340–343
    advantage, 342
    American call option, 342
    asset pricing, 343
    European call option, 342
    example, 340, 342
    financial investment strategy, 341
    future time of sale, 341
    present value of investment, 341
    stock purchasing strategy, 343
problems, 346–348
proxy data sets, 323
scenario testing, 323–330
    criteria, 324
    deterministic status scenarios,
        324–325
    Jensen's inequality, 324
    present value of future quantity,
        323
    stochastic rate scenarios, 328–330
    stochastic status scenarios,
        325–327
simulation prediction intervals,
    337–340
    application, 340
    bias-correcting percentile
        method, 338
    bias-correcting prediction
        interval, 337
    construction of complex
        scenarios, 339

cumulative distribution function, 335

empirical cumulative probability, 338

example, 337, 338, 339

future lifetime random variable, 339

lower bound, 337

order statistics, 337

prediction interval method, 338

symmetric prediction interval, 337

symmetric simulation prediction interval, 337

upper bound, 337

simulation probabilities, 335–336

cumulative distribution function, 335

empirical-based statistic measurements, 336

example, 335, 336

interpolation methods, 335

large samples, 336

simulated median, 335

simulation sampling, 331, 332–335

algorithm, 332–333

Box–Muller transformation, 334

continuous random variable, 333

distributions, 332

example, 333, 334

future values, 333, 335

nonnegative random variables, 334

normal random variables, 333

random data generation methods, 332

sample production, 332

simulation techniques, 330–340

algorithmic models, 330

bootstrap sampling, 331–332

data mining techniques, 330

hypothetical distribution, 331

models based on stochastic statuses, 330

Quenouille–Tukey jackknife, 330

simulation prediction intervals, 337–340

simulation probabilities, 335–336

simulation sampling, 331, 332–335

solutions, 350–351

statistical simulation techniques, 323

stochastic rate scenarios, 328–330

distributional parameters, 328

example, 328, 329

expected future value calculations, 328

prediction interval for normal rate random variable, 329

uniform interest rate whole life insurance, 330

stochastic status scenarios, 325–327

equivalence principal, 325

example, 325, 326

fully discrete life insurance premium scenarios, 327

life table-based term life insurance scenario testing, 327

net single value, 325

Poisson curtate future lifetime whole life insurance, 326

probability density function, 326

risk criteria, 325

whole life annuity, 327

stochastic surplus application, 343–344

advantage of simulation techniques, 343

discrete stochastic surplus model, 344

example, 344

future lifetime random variable, 343

period ruin probability for exponential claims, 344

setting interpretation, 343

Simulation testing, *see* Scenario and simulation testing

Single-risk models, 131–140

insurance pricing, 131–134

claim amount, 132

claim variable, 133

deductible insurance, 132, 133

disability insurance, 134

distribution of loss amounts, 133

example, 132, 133

gross premium, 132
  loading, 132
  moment formulas, 134
  overall cost, 131
insurance setting, 131
investment pricing, 135–136
  conservative approach, 136
  example, 136
  investment growth, 135
  normality assumption, 135
  return rates, 135
  risk-free rate, 135
net single premium, 131
net single value, 131
option pricing, 136–139
  American call option, 137
  Black–Scholes formula, 137, 138
  call option, 133, 137
  European call option, 137
  example, 138
  expiration date, 136
  indicator function, 137
  investment pricing option
    formula, 137
  options pricing formula, 138
  single-risk models, 137, 138
  strike price, 136
option pricing diagnostics, 139–140
  call option, 139
  European call option, 140
  example, 140
  Greeks, 139
  rho values, 140
  risk-free interest rate, 140
  sensitivity of delta, 139
  vega, 140
SPA, *see* Saddlepoint approximation
SPI, *see* Symmetric prediction interval
SSPI, *see* Symmetric simulation
    prediction interval
Statistical concepts, 1–39, *see*
    also Further statistical
    considerations
  conditional distributions, 29–31
    example, 29, 30
    geometric random variable, 30
    lack of memory, 30
    lifetime variable, 30
    magnitude of truncation, 31
    random variables, 29
  continuous random variables, 10–12
    cumulative distribution function,
      10
    example, 11, 12
    graphs, 11
    probability density function, 10
    status condition, 12
  discrete random variables, 8–10
    Bernoulli random variables, 9
    binomial random variable, 9
    Poisson postulates, 9
  Excel problems, 38
  expectations, 14–20
    claim variable, 20
    conditional probabilities,
      computation of, 18
    derivations, 17
    distribution of random variable,
      dispersion, 16
    example, 16, 17, 18
    identity function, 15
    insurance claims, 19
    mixed random variable, 15
    moment computations, 16
    Poisson distribution, 15
    possible types of random
      variables, 14
    weighted average, 15
  joint distributions, 31–35
    computation of statistical
      measurements, 34
    covariance, 34
    example, 32, 33
    functions of multiple random
      variables, 35
    independent variables, 32
    linear relationship, 34
    relationships between variables,
      32
    risk categories, frequencies of, 33
    statistical concepts, 31–32
    two-variable stochastic
      modeling, 21
  mixed random variables, 7, 13–14
    applications, 13
    collective aggregate models, 14

example, 13
probabilities, computation of, 13
support parts, 13
moment generating function, 20–22
complex-valued function, 22
example, 21
mixed random variable, 21
random variable, 22
nonnegative random variables, 25–29
force of mortality, 27
gamma function, 27
Gompertz distribution, 27–28
lognormal distribution, 26
Makeham distribution, 28–29
modeling of insurance claim amounts, 26
Pareto distribution, 25
Weibull distribution, 26–27
probability, 1–7
axioms of probability, 2, 5
basic set operations, 2
collection of disjoint events, 2
conditional probabilities, 4
construction of probability measures, 3
events, 2
example, 2, 3
integer future lifetime, 3
joint life conditions, 6
likelihood of events occurring, 2
probability measure, 2
probability system, 2
random process, 2
risk categories, 4
Venn diagrams, 3, 5
weak law of large numbers, 2
problems, 36–37
random variables, 7–14
continuous random variables, 10–12
discrete random variables, 8–10
mixed random variables, 7, 13–14
probability mass function, 8
status, 7
types, 7
solutions, 38–39
stochastic actions, examples of, 1

survival functions, 22–24
example, 23, 24
lifetime random variable, 24
moment computation, 23
normality-based prediction intervals, 24
random variable, 22
types of actions, 1
Statistical techniques, 41–92
aggregate variable, 41
approximating aggregate distributions, 57–64
central limit theorem, 57–61
Haldane type A approximation, 57, 61–62
Panjer recursion, 57
saddlepoint approximation, 62–64
Wilson–Hilferty approximation, 57
autoregressive systems, 75–78
ad hoc approximate prediction interval, 77
application, 78
ARMA model, 76
asymptotic normality of least squares estimators, 77
dependent variable techniques, 75
error terms, 76
example, 78
least squares point estimate, 77
relation between variables, 75
unknown parameters, 76, 77
Bayesian statistics, 41
central limit theorem, 57–61
example, 58, 59, 60
large sample size, 60
Levy central limit theorem, 58
Lindeberg–Feller central limit theorem, 57
relaxed conditions, 57
sample standard deviation, statistical consistency of, 58
small sample setting, 59
compound aggregate variables, 65–70
applications, 65

collective risk modeling, 65
  expectations, 65–66
  limiting distributions, 66–70
  statistical properties, 65
  stochastic structure, 65
confidence intervals, 44–45
  confidence coefficient, 44
  pivot, 44, 45
  random set, 44
confidence and prediction sets, 46–49
  approximate chi square random
    variable, 47
  confidence set, 46, 47
  construction of confidence
    intervals, 46
  diagnostic measurements, 46
  example, 47
  observed parameter estimates, 48
  prediction set, 46
  sample coefficient of variation, 48
  statistical pivot, 47
Excel problems, 88–90
expectations of compound
    aggregate variables, 65–66
  compound geometric random
    variable, 66
  compound Poisson process, 66
  confidence and prediction
    intervals, 65
  construction of statistical
    inference, 65
  example, 66
  moment generating function, 65
frequentist statistics, 41
generalized least squares diagnostic,
    83–84
  assumed continuous statistical
    distribution, 83
  dispersion matrix, 83
  empirical cdf vector, 83
  example, 84
  generalized least squares
    measure, 84
Haldane type A approximation, 57,
    61–62
  claim values, 62
  example, 62
  nonnegative random variables, 61
  skewness of aggregate, 61

interval data diagnostic, 84–87
  chi squared goodness-of-fit test
    statistic, 86
  discrete random variables, 85
  disjoint sets, 85
  dispersion matrix for random
    vector, 85
  empirical and discrete
    distributions, 86
  example, 86
  interval data, 84
  observed data, 84
  Poisson distribution, 87
least squares estimation, 71–73
  application, 72
  coefficient of determination, 72
  example, 73
  measure of linear fit, 72
  regression of stock prices, 73
  sample correlation coefficient, 72
  scatter plot, 73
limiting distributions for compound
    aggregate variables, 66–70
  compound Poisson
    approximations, 70
  compound Poisson distribution,
    67, 68
  estimators of true probability, 70
  example, 67
  moments, 67
  saddlepoint approximation, 68,
    69
  standardized variable, 67
  three-moment approximation,
    68, 69
model diagnostics, 78–87
  assumed continuous distribution,
    efficiency of, 78
  generalized least squares
    diagnostic, 83–84
  graphical technique, 78
  interval data diagnostic, 84–87
  probability plotting, 78, 79–82
order statistics and empirical
    prediction intervals, 54–57
  bias-correcting percentile
    method, 56
  example, 54, 56
  independence property, 54

indicator function, 56
linear interpolation, 55
observed variables, 54
specialized cases, 54
stochastic actions, 55
parameters, 41
percentiles and prediction intervals,
45–46
coverage probabilities, 46
one-sided prediction intervals, 45
outcomes of random variables, 45
population measurements, 45
prediction interval, 45
point estimation, 42–44
consistency, 44
example, 43
independent and identically
distributed random variables,
42
minimum variance unbiased
estimators, 44
random sample, 42
sample variance, 43
statistics, 42
weak law of large numbers, 44
probability plotting, 79–82
claim data, 80
comparison for Pareto,
lognormal, and Weibull, 82
error term, 79
Excel application, 81
linear regression model, 80
main concept, 79
normal probability plot, 81
parameter components, 79
probability plot, 80
scatter plot of transformed data,
80
transformation quantities, 82
problems, 87–88
regression model-based inference,
74–75
confidence intervals, 74, 75
example, 75
minimum variance, 74
normality of the error term, 74
prediction interval, 74
statistical distribution theory, 74
unbiasedness, 74

regression modeling, 70–75
error term, 71
fitted regression line, 72
least squares estimation, 71–73
observed data sets, 71
predictor variable, 71
regression model-based
inference, 74–75
scatter plot, 71
simple linear regression model,
71
variables of interest, 71
saddlepoint approximation, 62–64
accuracy, 62
applications, 63
example, 64
Excel Solver routine, 63
information drawn, 62
moment generating function, 63
Poisson compound sums, 64
prediction intervals, 63
sampling distributions and
estimation, 41–49
confidence intervals, 44–45
confidence and prediction sets,
46–49
examples of consistent
estimators, 44
observed samples, 41
percentiles and prediction
intervals, 45–46
point estimation, 42–44
sampling distribution of the
statistic, 42
solutions, 90–92
sums of independent variables,
49–54
aggregate claims, 52
applications, 49
calculation of relevant
probabilities, 53
collective risk modeling, 49
convolution process, 50, 51
example, 51, 52
gamma distribution, 53
moment generating function, 52
sums resulting from random
samples, 53
two-variable case, 50

Stochastic loss criterion, 128–131
  main approaches, 128–129
  percentile calculations, 129
  percentile criteria, 130–131
    associated stochastic loss models,
      130
    example, 130
    unknown constants, 130
  risk analysis, 129
  risk criteria, 129–130
    application, 129
    equilibrium principle, 129
    example, 129
    investment example, 129
    statistical point estimators, 130
Stochastic rate models, 106–117
  continuous stochastic annuity
    models, 116–117
    constant payment, 116
    continuous annuity, 117
    example, 117
    prediction intervals, 116
    yearly parameters, 117
  continuous stochastic rate models,
    112–114
    cumulative rate, 112
    example, 112, 113
    expectation formulas, 113
    future obligation, 114
    normal random variables, 113
    present value function, 112
    yearly investment return rate, 114
  discrete stochastic annuity models,
    114–116
    accumulation function, 115
    annuities with varying
      payments, 114
    cumulative interest rate, 115
    example, 115
    statistical approximation
      techniques, 115
    stochastic nature of financial
      rates, 116
  discrete stochastic rate model,
    106–112
    distribution parameters, 109
    empirical return rate, 110
    equal-width time periods, 106
    example, 107, 108, 109

Excel example, 111
    expectations and variances for
      future and present value
      functions, 108
    future value, 107
    independent rate model, 110
    moment generating function, 108
    prediction intervals, 107, 109
    sample correlation coefficient, 111
    statistical model fitting, 111
    substituted estimator procedures,
      110
  graphical assessments, 106
  varying interest and return rates,
    106
Stochastic status models, 203–256,
    see also Advanced stochastic
    status models
  actuarial life tables, 227–230
    actuarial life table for selected
      ages, 228
    annuity payments, 229
    benefit paid, 229
    example, 228
    group survivorship tables, 227
    net single premiums, 230
  apportionable annuities, 220–222
    analysis, 221
    annuity overpayment, 220
    example, 222
    expected value of adjusted
      payment, 221
    future value quantity, 221
    mathematical modeling example,
      222
    NSP expectation, 220
    unit payment, 222
  continuous-risk calculations,
    205–206
    components, 205
    example, 206
    moment generating function, 206
    single-status model, 205
  discrete risk calculations, 206–207
    example, 206
    financial action, 206
    future time structure, 206
    model versatility, 207
  Excel problems, 254

expense models and computations, 249–252
  claim administration investigation, 249
  classification of costs, 250
  example, 250, 252
  general costs, 249
  gross premiums, 249, 251
  loading, 249
  maintenance costs, 249
  reserve computations, 251
  stochastic loss function, 250
general period expectation, 245–246
  discrete stochastic status model, 245
  monthly period life insurance, 245
general period expectation, relations among, 246–249
  computational relationship, 247
  continuous whole life insurance, 246
  example, 248
  relational formulas, 247
  UDD assumption, 246
general time period models, 244–249
  discrete monthly period model, 244
  general period expectation, 245–246
  investment present values, 244
  life insurance settings, stochastic benefit in, 245
  monthly period model, 245
  relations among general period expectations, 246–249
life annuities, 215–222
  apportionable annuities, 220–222
  continuous whole life annuity, 217
  deterministic status annuities, 215
  discrete life annuity, 215
  example, 216
  expectation formula, 216
  force of interest, 215
  stochastic status life annuity, 215
  types of unit payment life annuities, 217–220

life insurance, 210–215
  continuous-status whole life insurance, 212
  discrete-status whole life insurance, 212
  endowment policy, 214
  example, 211, 214
  financing, 210
  general nomenclature, 210
  single-status models, 215
  stochastic structures, 211
  types of unit benefit life insurance, 212–215
  unit benefit, 213
  variance of present value function, 211
loss models and insurance premiums, 230–237
  curtate future lifetime, 232
  example, 230, 231
  precedent of notations, 232
  premiums, 230
  standard insurance loss function constructions, 233
  stochastic loss model, 230, 232
  typical insurance policy, 230
  unit benefit premium notation, 232–235
  variance of the loss function, 235–237
mixed risk calculations, 207–208
  common actuarial usage, 207
  example, 208
  expectation, 207
  life table data mortality rates, 207
  UDD assumption, 207
percentile evaluations, 208–210
  assumptions, 209
  example, 209, 210
  future lifetime random variable, 209
  net single premium, 209
  percentile measurement, 210
  probability of positive loss, 209
  stochastic status, 209
premium evaluation for life insurance, 203
problems, 252–254
relating risk calculations, 223–227

decomposition structure, 223
relations among annuity
  expectations, 226–227
relations among insurance and
  annuity expectations, 225–226
relations among insurance
  expectations, 223–225
UDD assumptions, 223
relations among annuity
  expectations, 226–227
apportionable annuity, 227
expectation computation, 227
present value computations, 227
relational formulas, 226
UDD assumption, 226
whole life annuity expectations,
  226
relations among insurance and
  annuity expectations, 225–226
continuous whole life annuity,
  226
example, 226
standard notations, 225
temporary life annuities, 225
relations among insurance
  expectations, 223–225
discrete-status whole life
  insurance policy, 225
endowment insurance, 223
example, 224
expectation for life insurance,
  223
exponential pdf, 224
stochastic status model, 224
reserves, 237–244
cash flow table, 244
curtate future lifetime, 243
endowment insurance policy, 238
example, 238, 239
graph of reserve, 242
group survivorship structure, 243
life insurance status model, 237
relations among reserve
  calculations, 241–243
reserve structure, 237
reserves for standard life
  insurance policies, 241
stochastic lifetime random
  variable, 241

survivorship group approach to
  reserve calculations, 243–244
target population, 240
translated future lifetime
  random variable, 237
unit benefit reserves notations,
  240–241
unit benefit whole life insurance
  policy, 241
year-by-year reserve calculations,
  239
risk evaluations, 205–208
continuous-risk calculations,
  205–206
discrete risk calculations,
  206–207
insurance premium payments,
  205
mixed risk calculations, 207–208
net single value, 205
solutions, 254–256
stochastic present value functions,
  204–205
actuarial stochastic status
  models, 205
future lifetime random variable,
  204
probability density function,
  204
unit benefit premium notation,
  232–235
constant interest rate, 232
example, 234, 235
standard unit benefit level
  premium notation, 234
unit benefit yearly premium, 235
unit payment life annuities, types
  of, 217–220
continuous n-year temporary life
  annuity, 218
continuous whole life annuity,
  218
discrete n-year deferred whole
  life annuity-due, 219
discrete n-year temporary life
  annuity-due, 219
discrete whole life annuity-due,
  218
economic complication, 220

example, 219, 220
interest rate, 218
variance of the loss function,
235–237
discrete-status whole life
insurance policy, 236
example, 235, 236
standard life insurance loss
functions, 235
variance, 236
Stochastic surplus model, 148–155
continuous surplus model, 152–155
adjustment coefficient, 154
compound Poisson aggregate
claims, 154
example, 153
initial fund, 152
moment generating function,
153
probability of future ruin, 152
reinsurance setting, 154
discrete surplus model, 148–152
adjustment coefficient, 149
aggregate claim variable, 148
example, 150, 151
exponential surplus model, 152
financial actions, 148
payment of claims, 148
probability of ruin bounds for
normal claims, 151
stop-loss model, 149
joint distribution, 148
simulation techniques, 148
ultimate ruin, 148
Strike price, 136
Sums of independent variables,
49–54
aggregate claims, 52
applications, 49
calculation of relevant probabilities,
53
collective risk modeling, 49
convolution process, 50, 51
example, 51, 52
gamma distribution, 53
moment generating function, 52
sums resulting from random
samples, 53
two-variable case, 50

Survival functions, 22–24
example, 23, 24
lifetime random variable, 24
moment computation, 23
normality-based prediction
intervals, 24
random variable, 22
Survivorship chains, 312–316
closed-group survivorship
structure, 312
multiple-decrement models,
314–315
example, 314
expected number of survivors, 314
survivorship chain, 314
multirisk strata models, 315–316
example, 315
stochastic status age, 315
transition probabilities, 315
single-decrement models, 313–314
example, 313
initial vector, 314
number of newborns, 313
Survivorship groups, 179–182
actuarial statistics, definition of, 181
closed group, 179
expectations, 181
force of mortality, 180
graduation of the table, 181
group survivorship expectations, 181
independent lifetimes, 179
insurance pricing theory, 181
mortality notations, 180
number of survivors, 179
probability laws, 180
theory, 365
Symmetric prediction interval (SPI), 337
Symmetric simulation prediction
interval (SSPI), 337, 365, 368

T

Three-moment approximation, 68, 69
Time effect on monetary values, 100
Transformation quantities, 82
Transition
matrices, 308
probabilities, 296
Two-variable stochastic modeling, 21

U

UDD, *see* Uniform distribution of death
Uniform distribution of death (UDD),
      176
   fractional ages, 176, 193
   life models and life tables, 182
   mean mortality acceleration models,
      357
   mixed risk calculations, 207
   relating risk calculations, 223
   relations among general period
      expectations, 246
   single-decrement rates, 271
Unit payment life annuities, types of,
      217–220
   continuous *n*-year temporary life
      annuity, 218
   continuous whole life annuity, 218
   discrete *n*-year deferred whole life
      annuity-due, 219
   discrete *n*-year temporary life
      annuity-due, 219
   discrete whole life annuity-due, 218
   economic complication, 220

   example, 219, 220
   interest rate, 218
Unstable statistics problem, 345

V

Vega, 140
Venn diagrams, 3, 5

W

Weak law of large numbers, 2, 44, 364
Wearout mortality, 174
Wilson–Hilferty approximation, 57

Y

Year-by-year reserve calculations, 239
Yearly based mortality rates, 363, 364
Yearly based retirement benefits,
      288–290
Yearly effective interest rate, 283
Yearly investment return rate, 114
Yearly survival rates, 359